Springer Texts in Statistics

Advisors:
George Casella Stephen Fienberg Ingram Olkin

Springer
New York
Berlin
Heidelberg
Barcelona
Budapest
Hong Kong
London
Milan
Paris
Santa Clara
Singapore
Tokyo

Springer Texts in Statistics

Alfred: Elements of Statistics for the Life and Social Sciences
Berger: An Introduction to Probability and Stochastic Processes
Blom: Probability and Statistics: Theory and Applications
Brockwell and Davis: An Introduction to Times Series and Forecasting
Chow and Teicher: Probability Theory: Independence, Interchangeability, Martingales, Second Edition
Christensen: Plane Answers to Complex Questions: The Theory of Linear Models, Second Edition
Christensen: Linear Models for Multivariate, Time Series, and Spatial Data
Christensen: Log-Linear Models
Creighton: A First Course in Probability Models and Statistical Inference
du Toit, Steyn and Stumpf: Graphical Exploratory Data Analysis
Edwards: Introduction to Graphical Modelling
Finkelstein and Levin: Statistics for Lawyers
Jobson: Applied Multivariate Data Analysis, Volume I: Regression and Experimental Design
Jobson: Applied Multivariate Data Analysis, Volume II: Categorical and Multivariate Methods
Kalbfleisch: Probability and Statistical Inference, Volume I: Probability, Second Edition
Kalbfleisch: Probability and Statistical Inference, Volume II: Statistical Inference, Second Edition
Karr: Probability
Keyfitz: Applied Mathematical Demography, Second Edition
Kiefer: Introduction to Statistical Inference
Kokoska and Nevison: Statistical Tables and Formulae
Lehmann: Testing Statistical Hypotheses, Second Edition
Lindman: Analysis of Variance in Experimental Design
Lindsey: Applying Generalized Linear Models
Madansky: Prescriptions for Working Statisticians
McPherson: Statistics in Scientific Investigation: Its Basis, Application, and Interpretation
Mueller: Basic Principles of Structural Equation Modeling
Nguyen and Rogers: Fundamentals of Mathematical Statistics: Volume I: Probability for Statistics
Nguyen and Rogers: Fundamentals of Mathematical Statistics: Volume II: Statistical Inference
Noether: Introduction to Statistics: The Nonparametric Way
Peters: Counting for Something: Statistical Principles and Personalities
Pfeiffer: Probability for Applications
Pitman: Probability
Robert: The Bayesian Choice: A Decision-Theoretic Motivation

Continued at end of book

Ashish Sen Muni Srivastava

Regression Analysis
Theory, Methods, and Applications

With 38 Illustrations and a Diskette

Springer

Ashish Sen
College of Architecture, Art, and Urban Planning
School of Urban Planning and Policy
The University of Illinois
Chicago, IL 60680
USA

Muni Srivastava
Department of Statistics
University of Toronto
Toronto, Ontario
Canada M5S 1A1

Mathematical Subject Classification: 62Jxx, 62-01

Library of Congress Cataloging-in-Publication Data
Sen, Ashish K.
Regression analysis: Theory, methods, and applications/Ashish Sen, Muni Srivastava.
p. cm.—(Springer texts in statistics)
ISBN 0-387-97211-0 (alk. paper)
1. Regression analysis. I. Srivastava, M.S. II. Title.
III. Series.
QA278.2.S46 1990
519.5′36—dc20 89-48506

Printed on acid-free paper.

Photcomposed copy prepared from the authors' LATEX file.
Printed and bound by R.R. Donnelley & Sons, Harrisonburg, Virginia.
Printed in the United States of America.

9 8 7 6 5 4 (Corrected fourth printing, 1997)

ISBN 0-387-97211-0 Springer-Verlag New York Berlin Heidelberg
ISBN 3-540-97211-0 Springer-Verlag Berlin Heidelberg New York SPIN 10569690

To

Ashoka Kumar Sen

and the memory of
Jagdish Bahadur Srivastava

Preface

Any method of fitting equations to data may be called regression. Such equations are valuable for at least two purposes: making predictions and judging the strength of relationships. Because they provide a way of empirically identifying how a variable is affected by other variables, regression methods have become essential in a wide range of fields, including the social sciences, engineering, medical research and business.

Of the various methods of performing regression, least squares is the most widely used. In fact, linear least squares regression is by far the most widely used of *any* statistical technique. Although nonlinear least squares is covered in an appendix, this book is mainly about linear least squares applied to fit a single equation (as opposed to a system of equations).

The writing of this book started in 1982. Since then, various drafts have been used at the University of Toronto for teaching a semester-long course to juniors, seniors and graduate students in a number of fields, including statistics, pharmacology, engineering, economics, forestry and the behavioral sciences. Parts of the book have also been used in a quarter-long course given to Master's and Ph.D. students in public administration, urban planning and engineering at the University of Illinois at Chicago (UIC). This experience and the comments and criticisms from students helped forge the final version.

The book offers an up-to-date account of the theory and methods of regression analysis. We believe our treatment of theory to be the most complete of any book at this level. The methods provide a comprehensive toolbox for the practicing regressionist. The examples, most of them drawn from 'real life', illustrate the difficulties commonly encountered in the practice of regression, while the solutions underscore the subjective judgments the practitioner must make. Each chapter ends with a large number of exercises that supplement and reinforce the discussions in the text and provide valuable practical experience. When the reader has mastered the contents of this book, he or she will have gained both a firm foundation in the theory of regression and the experience necessary to competently practice this valuable craft.

A first course in mathematical statistics, the ability to use statistical computer packages and familiarity with calculus and linear algebra are prerequisites for the study of this book. Additional statistical courses and a good knowledge of matrices would be helpful.

This book has twelve chapters. The Gauss-Markov Conditions are assumed to hold in the discussion of the first four chapters; the next five chapters present methods to alleviate the effects of violations of these conditions. The final three chapters discuss the somewhat related topics of multicollinearity, variable search and biased estimation. Relevant matrix and distribution theory is surveyed in the first two appendices at the end of the book, which are intended as a convenient reference. The last appendix covers nonlinear regression.

Chapters and sections that some readers might find more demanding are identified with an asterisk or are placed in appendices to chapters. A reader can navigate around these without losing much continuity. In fact, a reader who is primarily interested in applications may wish to omit many of the other proofs and derivations. Difficult exercises have also been marked with asterisks.

Since the exercises and examples use over 50 data sets, a disk containing most of them is provided with the book. The READ.ME file in the disk gives further information on its contents.

This book would have been much more difficult, if not impossible, to write without the help of our colleagues and students. We are especially grateful to Professor Siim Sööt, who examined parts of the book and was an all-round friend; George Yanos of the Computer Center at UIC, whose instant E-mail responses to numerous cries for help considerably shortened the time to do the numerical examples (including those that were ultimately not used); Dr. Chris Johnson, who was a research associate of one of the authors during the time he learnt most about the practical art of regression; Professor Michael Dacey, who provided several data sets and whose encouragement was most valuable; and to Professor V. K. Srivastava whose comments on a draft of the book were most useful. We also learnt a lot from earlier books on the subject, particularly the first editions of Draper and Smith (1966) and Daniel and Wood (1971), and we owe a debt of gratitude to their authors.

Numerous present and former students of both authors contributed their time in editing and proof-reading, checking the derivations, inputting data, drawing diagrams and finding data-sets. Soji Abass, Dr. Martin Bilodeau, Robert Drozd, Andrea Fraser, Dr. Sucharita Ghosh, Robert Gray, Neleema Grover, Albert Hoang, M.R. Khavanin, Supin Li, Dr. Claire McKnight, Cæsar Singh, Yanhong Wu, Dr. Y. K. Yau, Seongsun Yun and Zhang Tingwei constitute but a partial list of their names. We would like to single out for particular mention Marguerite Ennis and Piyushimita Thakuriah for their invaluable help in completing the manuscript. Linda Chambers TEXed an earlier draft of the manuscript, Barry Grau was most helpful identifying computer programs, some of which are referred to in the text, Marilyn Engwall did the paste-up on previous drafts, Ray Brod drew one of the figures and Bobbie Albrecht designed the cover. We would like to express our gratitude to all of them. A particular thanks is due to Dr. Colleen Sen who painstakingly edited and proofread draft after draft.

We also appreciate the patience of our colleagues at UIC and the University of Toronto during the writing of this book. The editors at Springer-Verlag, particularly Susan Gordon, were most supportive. We would like to gratefully acknowledge the support of the Natural Sciences and Engineering Research Council of Canada and the National Science Foundation of the U.S. during the time this book was in preparation. The help of the Computer Center at UIC which made computer time freely available was indispensable.

Preface to the Fourth Printing

We have taken advantage of this as well as previous reprintings to correct several typographic errors. In addition, two exercises have been changed. One because it required too much effort and another because we were able to replace it with problems we found more interesting.

In order to keep the price of the book reasonable, the data disk has is no longer included. Its contents have been placed at web sites from which they may be downloaded. The URLs are `http://www.springer-ny.com` and `http://www.uic.edu/~ashish/regression.html`.

Contents

1 Introduction **1**

1.1 Relationships . . . 1
1.2 Determining Relationships: A Specific Problem . . . 2
1.3 The Model . . . 5
1.4 Least Squares . . . 7
1.5 Another Example and a Special Case . . . 10
1.6 When Is Least Squares a Good Method? . . . 11
1.7 A Measure of Fit for Simple Regression . . . 13
1.8 Mean and Variance of b_0 and b_1 . . . 14
1.9 Confidence Intervals and Tests . . . 17
1.10 Predictions . . . 18
Appendix to Chapter 1 . . . 20
Problems . . . 23

2 Multiple Regression **28**

2.1 Introduction . . . 28
2.2 Regression Model in Matrix Notation . . . 28
2.3 Least Squares Estimates . . . 30
2.4 Examples . . . 31
2.5 Gauss-Markov Conditions . . . 35
2.6 Mean and Variance of Estimates Under G-M Conditions . . . 35
2.7 Estimation of σ^2 . . . 37
2.8 Measures of Fit . . . 39
2.9 The Gauss-Markov Theorem . . . 41
2.10 The Centered Model . . . 42
2.11 Centering and Scaling . . . 44
2.12 *Constrained Least Squares . . . 44
Appendix to Chapter 2 . . . 46
Problems . . . 49

3 Tests and Confidence Regions **60**

3.1 Introduction . . . 60
3.2 Linear Hypothesis . . . 60

3.3 *Likelihood Ratio Test 62
3.4 *Distribution of Test Statistic 64
3.5 Two Special Cases 65
3.6 Examples 66
3.7 Comparison of Regression Equations 67
3.8 Confidence Intervals and Regions 71
3.8.1 C.I. for the Expectation of a Predicted Value 71
3.8.2 C.I. for a Future Observation 71
3.8.3 *Confidence Region for Regression Parameters . . . 72
3.8.4 *C.I.'s for Linear Combinations of Coefficients . . . 73
Problems 74

4 Indicator Variables **83**
4.1 Introduction 83
4.2 A Simple Application 83
4.3 Polychotomous Variables 84
4.4 Continuous and Indicator Variables 88
4.5 Broken Line Regression 89
4.6 Indicators as Dependent Variables 92
Problems 95

5 The Normality Assumption **100**
5.1 Introduction 100
5.2 Checking for Normality 101
5.2.1 Probability Plots 101
5.2.2 Tests for Normality 105
5.3 Invoking Large Sample Theory 106
5.4 *Bootstrapping 107
5.5 *Asymptotic Theory 108
Problems 110

6 Unequal Variances **111**
6.1 Introduction 111
6.2 Detecting Heteroscedasticity 111
6.2.1 Formal Tests 114
6.3 Variance Stabilizing Transformations 115
6.4 Weighting 118
Problems 128

7 *Correlated Errors **132**
7.1 Introduction 132
7.2 Generalized Least Squares: Case When Ω Is Known 133
7.3 Estimated Generalized Least Squares 134
7.3.1 Error Variances Unequal and Unknown 134
7.4 Nested Errors 136

7.5 The Growth Curve Model 138
7.6 Serial Correlation 140
7.6.1 The Durbin-Watson Test 142
7.7 Spatial Correlation 143
7.7.1 Testing for Spatial Correlation 143
7.7.2 Estimation of Parameters 144
Problems 146

8 Outliers and Influential Observations 154
8.1 Introduction 154
8.2 The Leverage 155
8.2.1 *Leverage as Description of Remoteness 156
8.3 The Residuals 156
8.4 Detecting Outliers and Points That Do Not Belong to the Model 157
8.5 Influential Observations 158
8.5.1 Other Measures of Influence 160
8.6 Examples 161
Appendix to Chapter 8 173
Problems 176

9 Transformations 180
9.1 Introduction 180
9.1.1 An Important Word of Warning 180
9.2 Some Common Transformations 181
9.2.1 Polynomial Regression 181
9.2.2 Splines 182
9.2.3 Multiplicative Models 182
9.2.4 The Logit Model for Proportions 186
9.3 Deciding on the Need for Transformations 188
9.3.1 Examining Residual Plots 189
9.3.2 Use of Additional Terms 190
9.3.3 Use of Repeat Measurements 192
9.3.4 Daniel and Wood Near-Neighbor Approach 194
9.3.5 Another Method Based on Near Neighbors 195
9.4 Choosing Transformations 197
9.4.1 Graphical Method: One Independent Variable 197
9.4.2 Graphical Method: Many Independent Variables . . 200
9.4.3 Analytic Methods: Transforming the Response . . . 204
9.4.4 Analytic Methods: Transforming the Predictors . . . 209
9.4.5 Simultaneous Power Transformations for Predictors and Response 209
Appendix to Chapter 9 211
Problems 213

10 Multicollinearity **218**
10.1 Introduction 218
10.2 Multicollinearity and Its Effects 218
10.3 Detecting Multicollinearity 222
10.3.1 Tolerances and Variance Inflation Factors 222
10.3.2 Eigenvalues and Condition Numbers 223
10.3.3 Variance Components 224
10.4 Examples 225
Problems 231

11 Variable Selection **233**
11.1 Introduction 233
11.2 Some Effects of Dropping Variables 234
11.2.1 Effects on Estimates of β_j 235
11.2.2 *Effect on Estimation of Error Variance 236
11.2.3 *Effect on Covariance Matrix of Estimates 236
11.2.4 *Effect on Predicted Values: Mallows' C_p 237
11.3 Variable Selection Procedures 238
11.3.1 Search Over All Possible Subsets 239
11.3.2 Stepwise Procedures 240
11.3.3 Stagewise and Modified Stagewise Procedures 243
11.4 Examples 243
Problems 251

12 *Biased Estimation **253**
12.1 Introduction 253
12.2 Principal Component Regression 253
12.2.1 Bias and Variance of Estimates 255
12.3 Ridge Regression 256
12.3.1 Physical Interpretations of Ridge Regression 257
12.3.2 Bias and Variance of Estimates 258
12.4 Shrinkage Estimator 261
Problems 263

A Matrices **267**
A.1 Addition and Multiplication 267
A.2 The Transpose of a Matrix 268
A.3 Null and Identity Matrices 268
A.4 Vectors 269
A.5 Rank of a Matrix 270
A.6 Trace of a Matrix 271
A.7 Partitioned Matrices 271
A.8 Determinants 272
A.9 Inverses 273
A.10 Characteristic Roots and Vectors 276

A.11 Idempotent Matrices 277
A.12 The Generalized Inverse 278
A.13 Quadratic Forms 279
A.14 Vector Spaces 280
Problems 282

B Random Variables and Random Vectors **284**
B.1 Random Variables 284
B.1.1 Independent Random Variables 284
B.1.2 Correlated Random Variables 285
B.1.3 Sample Statistics 285
B.1.4 Linear Combinations of Random Variables 285
B.2 Random Vectors 286
B.3 The Multivariate Normal Distribution 288
B.4 The Chi-Square Distributions 290
B.5 The F and t Distributions 292
B.6 Jacobian of Transformations 293
B.7 Multiple Correlation 295
Problems 297

C Nonlinear Least Squares **298**
C.1 Gauss-Newton Type Algorithms 299
C.1.1 The Gauss-Newton Procedure 299
C.1.2 Step Halving 300
C.1.3 Starting Values and Derivatives 301
C.1.4 Marquardt Procedure 302
C.2 Some Other Algorithms 302
C.2.1 Steepest Descent Method 303
C.2.2 Quasi-Newton Algorithms 303
C.2.3 The Simplex Method 305
C.2.4 Weighting 305
C.3 Pitfalls 306
C.4 Bias, Confidence Regions and Measures of Fit 308
C.5 Examples 310
Problems 316

Tables **319**

References **327**

Index **336**

Author Index **345**

CHAPTER 1

Introduction

1.1 Relationships

Perception of relationships is the cornerstone of civilization. By understanding how certain phenomena depend on others we learn to predict the consequences of our actions and to manipulate our environment. Most relationships we know of are based on empirical observations. Although some relationships are postulated on theoretical grounds, usually the theories themselves were originally obtained from empirical observations. And even these relationships often need to be empirically tested.

Some relationships are relatively easy to discover or verify. This is particularly true when chance plays little or no role in them. But when chance does play a role, the task of discovering relationships often requires fairly careful analysis of data. This book is devoted to the study of the analysis of data aimed at discovering how one or more variables (called *independent* variables, *predictor* variables or *regressors*) affect other variables (called *dependent* variables or *response* variables).

Such analysis is called *regression*. This nomenclature is somewhat unfortunate since it has little to do with going backwards, as the word regression implies. The name comes from an early (1885) application by Sir Francis Galton, which dealt with the relationship of heights of parents and heights of offsprings. He showed that unusually tall parents ('taller than mediocrity', as he put it) had children who were shorter than themselves, and parents who were 'shorter than mediocrity' had children taller than themselves. This led to his theory of 'regression toward mediocrity' and eventually led to its use with other studies involving relationships. This choice of word is doubly unfortunate, since it might tend to date regression from Galton's work. Actually, regression is much older than that. Eighteenth century French mathematicians (particularly Laplace) and others (particularly Boscovich, in 1757) were clearly doing what we would call regression (Stigler, 1975, 1984) and if one is willing to claim two-sample testing as a subcase of regression (as we do in Chapter 4), its history goes back to Biblical times.

1.2 Determining Relationships: A Specific Problem

Example 1.1
We know that the more cars there are on a road the slower the speed of traffic flow becomes. A fairly precise understanding of this is important to the transportation planner since reducing travel time is frequently the main purpose behind increasing transportation facilities. Exhibit 1.1 shows data on density in vehicles per mile and the corresponding speed in miles per hour.

Density	Speed	$(\text{Speed})^{1/2}$	Density	Speed	$(\text{Speed})^{1/2}$
20.4	38.8	6.229	29.5	31.8	5.639
27.4	31.5	5.612	30.8	31.6	5.621
106.2	10.6	3.256	26.5	34.0	5.831
80.4	16.1	4.012	35.7	28.9	5.376
141.3	7.7	2.775	30.0	28.8	5.367
130.9	8.3	2.881	106.2	10.5	3.240
121.7	8.5	2.915	97.0	12.3	3.507
106.5	11.1	3.332	90.1	13.2	3.633
130.5	8.6	2.933	106.7	11.4	3.376
101.1	11.1	3.332	99.3	11.2	3.347
123.9	9.8	3.130	107.2	10.3	3.209
144.2	7.8	2.793	109.1	11.4	3.376

EXHIBIT 1.1: Data on Density of Vehicles and Average Speed
SOURCE: Huber (1957). Reproduced with permission from Transportation Research Board, National Research Council, Washington, D.C.

Since congestion affects speed (and not the other way around) we are interested in determining the effect of density on speed. For reasons that need not concern us at the moment (but will be discussed in Chapter 6), we shall set the dependent variable as the square root of speed. Exhibit 1.2 shows a plot (or a scatter plot as it is sometimes called) with the independent variable (density) on the horizontal axis and the dependent variable (square root of speed) on the vertical axis — as is usual.

Exhibit 1.3 is the same as Exhibit 1.2 except that now a line has been fitted by eye to the data. If desired the equation of the line can be obtained using straightforward analytic geometry methods; e.g., pick any two points $(x^{(1)}, y^{(1)})$ and $(x^{(2)}, y^{(2)})$ on the line and substitute these values in

$$y - y^{(1)} = \frac{y^{(1)} - y^{(2)}}{x^{(1)} - x^{(2)}}(x - x^{(1)}).$$

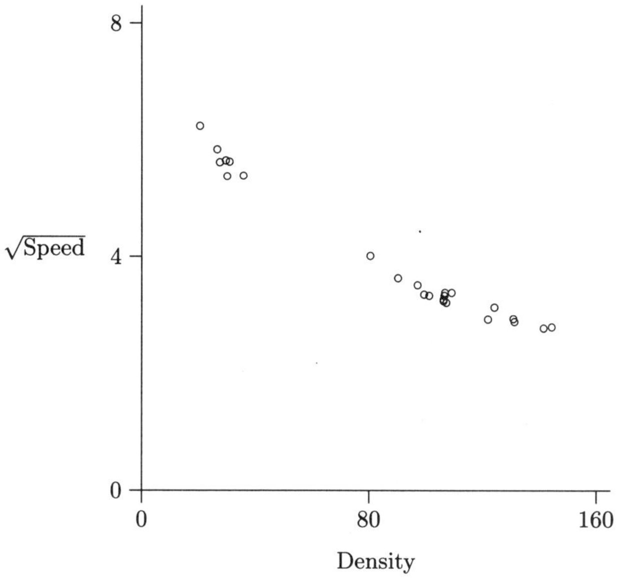

EXHIBIT 1.2: Plot of Square Root of Speed Against Density

In this case we get

$$y - 3.6 = \frac{3.6 - 5.5}{100 - 30}(x - 100) \quad \text{or} \quad y = 6.3 - .027x, \tag{1.1}$$

using the points (100, 3.6) and (30, 5.5). ■

The first question that comes to mind at this stage is: How well does our line fit the observed points? One way to assess this is to compute for each value x_i of our independent variable the value $\hat{y}_i$ of the dependent variable as predicted by our line. Then we can compare this *predicted* value $\hat{y}_i$ with the corresponding *observed* value y_i. This is usually done by computing the *residuals*

$$e_i = y_i - \hat{y}_i. \tag{1.2}$$

Example 1.1 ctd.
For the first value x_1 of x, $\hat{y}_1 = 6.3 - .027(20.4) = 5.7492$ and $e_1 = 6.229 - 5.749 = .48$. Exhibit 1.4 displays a plot of the residuals e_i against the x_i's. While Exhibit 1.3 shows that the residuals are fairly small relative to the original y_i's, indicating a fairly good fit, Exhibit 1.4 shows that we can do better, since there is a slight pattern in the points (the points in the middle are lower than those at either end), which we should be able to account for.

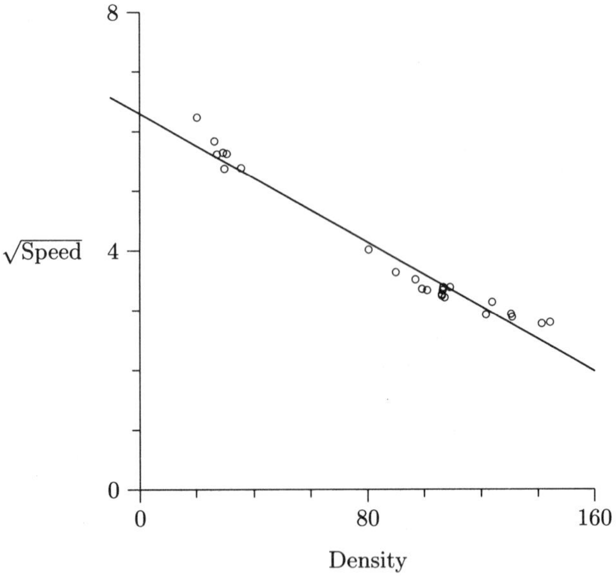

EXHIBIT 1.3: Plot of Square Root of Speed Against Density with Regression Line Drawn

One way to do this is to use, in addition to x_i, a term involving x_i^2. Here we obtained the equation

$$y_i = 7 - 0.05x_i + 0.00015x_i^2, \tag{1.3}$$

using a least squares procedure that we shall describe shortly. We could have tried to fit a parabola to our original data points and found its equation by using three points on the curve, but least squares is simpler. Exhibit 1.5 shows a plot of the residuals against the predicted values (sometimes called 'predicteds') for (1.3), while Exhibit 1.6 gives a plot of the residuals against x_i. Since they do not show any obvious pattern, they indicate that there is perhaps little more we can do and, indeed, we may have done quite well already. Therefore, we choose

$$(\text{speed})^{1/2} = 7 - 0.05 \text{ density} + 0.00015 \text{ density}^2$$

as our final equation.

Traffic flow is defined as speed×density. If we express this flow in terms of density alone, using the regression equation given above, and plot it we will find a curve that is increasing for low values of density and decreasing for higher values. The maximum value of flow is an estimate of the capacity of the road. It is interesting to note that this capacity is reached for a fairly low density. ■

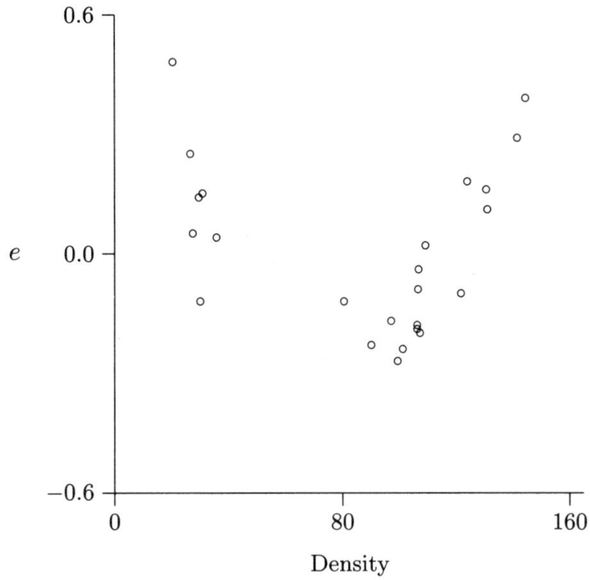

EXHIBIT 1.4: Plot of Residuals Against Density

1.3 The Model

When we set out in the last section to fit a straight line, we had implicitly hypothesized that the data had an underlying linear pattern, i.e., one of the form

$$y = \beta_0 + \beta_1 x. \tag{1.4}$$

However, we also knew that our observations would not fit this pattern exactly (this book is not intended for situations where the fit is exact!). Thus we hypothesized that we had a relationship of the form

$$y_i = \beta_0 + \beta_1 x_i + \epsilon_i, \tag{1.5}$$

where $i = 1, 2, \ldots, n$ and n is the number of data points. Equation (1.5) is called a *regression model* and since we have only one independent variable, it is called a *simple regression* model. Later we found that we could do better with the model

$$y_i = \beta_0 + \beta_1 x_i + \beta_2 x_i^2 + \epsilon_i$$

or, equivalently,

$$y_i = \beta_0 + \beta_1 x_{i1} + \beta_2 x_{i2} + \epsilon_i,$$

where $x_{i1} = x_i$ and $x_{i2} = x_i^2$ and $i = 1, \ldots, n$.

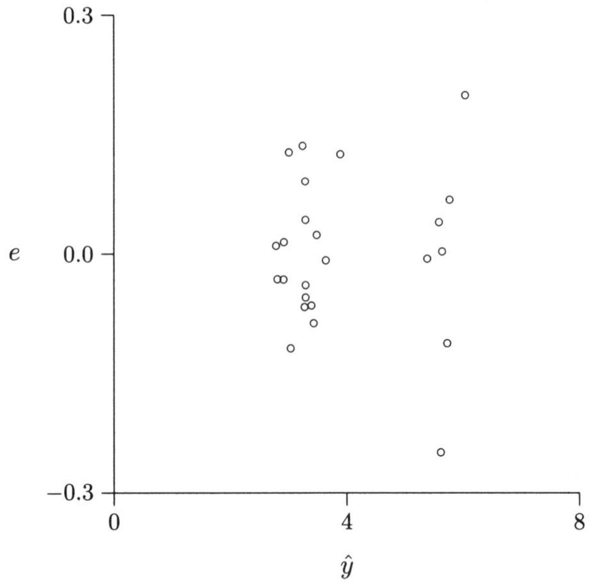

EXHIBIT 1.5: Plot of Residuals Against Predicted Speed (Density[2] Included in Model)

In general, models of the form

$$y_i = \beta_0 + \beta_1 x_{i1} + \beta_2 x_{i2} + \cdots + \beta_k x_{ik} + \epsilon_i \quad \text{where} \quad i = 1, \ldots, n, \qquad (1.6)$$

with $k > 1$ independent variables, are called *multiple* regression models. The β_j's are called *parameters* and the ϵ_i's *errors*. The values of neither the β_j's nor the ϵ_i's can ever be known. However, they can be estimated as we did in the last section (the ϵ_i's by the e_i's).

In a simple regression model the β_0 and β_1 have simple interpretations. When $x = 0$ in the equation (1.4), $y = \beta_0$. The term β_0 is frequently called the *intercept*. For every unit increase in x, y increases by β_1, which is often referred to as the *slope*.

It is important to note that in our regression model the x_{ij}'s are simply numbers — not random variables. Therefore, it is pointless to talk of the distribution of the x_{ij}'s. The ϵ_i's are random variables as are the y_i's, since they depend on the ϵ_i's. The y_i's are called observations, $x_{i1}, \ldots, x_{ik}$ are said to constitute the *design point* corresponding to y_i (or, simply, the ith design point), and together $y_i, x_{i1}, \ldots, x_{ik}$ constitute a *case* or a *data point*.

We often say that y_i's are observations of the dependent variable y and for $i = 1, \ldots, n$, x_{ij}'s are the values of the independent variable x_j. However, we very rarely treat y as a single random variable, and x_j is not a random variable at all. Moreover, more often than not, x_j's are related to

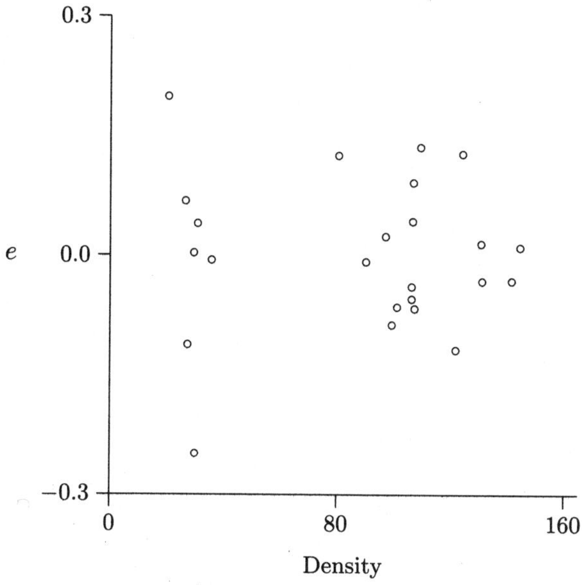

EXHIBIT 1.6: Plot of Residuals Against Density (Density^2 Included in Model)

each other. This makes the terms 'independent variable' and 'dependent variable' rather unfortunate terminological choices, but at this stage they are too entrenched in our vocabulary to change.

In Section 1.2 we fitted an expression involving x_i^2 to the data. This is obviously non-linear in x_i. But we were doing linear regression and for our purposes, linearity means linearity in the parameters $\beta_0, \beta_1, \ldots, \beta_k$; i.e., for linear regression (1.6) needs to be a linear function *of the* β_j*'s*.

1.4 Least Squares

Obviously the smaller the residuals the better the fit. Of all possible values of the β_j's, the least squares (LS) estimates are those that minimize

$$\mathcal{S} = \sum_{i=1}^{n} \epsilon_i^2 = \sum_{i=1}^{n} (y_i - \beta_0 - \beta_1 x_{i1} - \beta_2 x_{i2} - \cdots - \beta_k x_{ik})^2, \tag{1.7}$$

which in the case of simple regression becomes

$$\mathcal{S} = \sum_{i=1}^{n} (y_i - \beta_0 - \beta_1 x_{i1})^2. \tag{1.8}$$

If the reader is at all confused by us acting as if the β_j's in (1.5) and (1.6) are fixed but unknown numbers, and then talking about all possible values of the β_j's, he or she has every reason to be. The fact is that the β_j's have two meanings: one being a generic point in a space of possible values, and the other a specific point in this space, which may be called the 'true value.' It is conventional to be sloppy and mix up this dual role of the β_j's in linear least squares (although in nonlinear least squares the distinction is usually made explicit — see Appendix C, p. 298). But this should cause no confusion, and besides, we only consider β_j's in their 'true value' meaning except when we derive least squares estimates.

Since the partial derivatives of (1.8) with respect to β_0 and β_1 are

$$\partial S/\partial\beta_0 = -2\sum_{i=1}^{n}(y_i - \beta_0 - \beta_1 x_{i1}) \tag{1.9}$$

$$\text{and } \partial S/\partial\beta_1 = -2\sum_{i=1}^{n}(y_i - \beta_0 - \beta_1 x_{i1})\,x_{i1}, \tag{1.10}$$

we obtain the least squares estimates b_0 and b_1 of β_0 and β_1 by setting (1.9) and (1.10) equal to zero and replacing β_0 and β_1 by b_0 and b_1. Thus from (1.9)

$$\sum_{i=1}^{n} y_i - nb_0 - b_1\sum_{i=1}^{n} x_{i1} = 0$$

and setting $\bar{y} = n^{-1}\sum_{i=1}^{n} y_i$ and $\bar{x}_1 = n^{-1}\sum_{i=1}^{n} x_{i1}$, we get

$$b_0 = \bar{y} - b_1\bar{x}_1. \tag{1.11}$$

From (1.10) it follows that

$$\sum_{i=1}^{n} y_i x_{i1} - nb_0\bar{x}_1 - b_1\sum_{i=1}^{n} x_{i1}^2 = 0,$$

which, when we use (1.11) to substitute for b_0, yields

$$\sum_{i=1}^{n} y_i x_{i1} - n\bar{x}_1(\bar{y} - b_1\bar{x}_1) - b_1\sum_{i=1}^{n} x_{i1}^2 = 0. \tag{1.12}$$

Therefore,

$$b_1 = \Big(\sum_{i=1}^{n} y_i x_{i1} - n\bar{x}_1\bar{y}\Big)\Big/\Big(\sum_{i=1}^{n} x_{i1}^2 - n\bar{x}_1^2\Big). \tag{1.13}$$

The alternative expression

$$b_1 = \sum_{i=1}^{n}(y_i - \bar{y})(x_{i1} - \bar{x}_1)\Big/\sum_{i=1}^{n}(x_{i1} - \bar{x}_1)^2 \tag{1.14}$$

is sometimes useful. The equivalence of (1.14) and (1.13) is proved in Appendix 1A.[1]

These derivations would have been shortened somewhat if we rewrote (1.5) as

$$y_i = (\beta_0 + \beta_1 \bar{x}_1) + \beta_1(x_{i1} - \bar{x}_1) + \epsilon_i. \tag{1.15}$$

This model is called a *centered* model or, more specifically, the *centered version of* (1.5). If $\gamma_0 = \beta_0 + \beta_1 \bar{x}_1$, then there is a one-to-one correspondence between (β_0, β_1) and (γ_0, β_1) and the value of either pair determines the other. Minimizing

$$\sum_{i=1}^{n} [y_i - \gamma_0 - \beta_1(x_{i1} - \bar{x}_1)]^2$$

with respect to γ_0 and β_1 and noting that $\sum_{i=1}^{n}(x_{i1} - \bar{x}_1) = 0$ we see that the least squares estimate of γ_0 is $\bar{y}$ and that of β_1 is still (1.14). If desired, we may estimate b_0 from b_1 and the estimate $\bar{y}$ of γ_0. The expression we would use for this purpose would be the same as (1.11).

Computer packages exist for carrying out the computations of b_0 and b_1, and one would frequently depend on them. Application of these formulae to the simple regression example in Section 1.2 yields

$$b_0 = 6.3797, \text{ and } b_1 = -0.02777. \tag{1.16}$$

In terms of the b_0's and b_1's, the predicted values $\hat{y}_i$ are

$$\hat{y}_i = b_0 + b_1 x_{i1} = \bar{y} + b_1(x_{i1} - \bar{x}_1) \tag{1.17}$$

and the residuals e_i are

$$e_i = y_i - \hat{y}_i = y_i - \bar{y} - b_1(x_{i1} - \bar{x}_1). \tag{1.18}$$

It may be noted in passing that least squares residuals for the model (1.5) have the property

$$\sum_{i=1}^{n} e_i = \sum_{i=1}^{n}(y_i - \bar{y}) - b_1 \sum_{i=1}^{n}(x_{i1} - \bar{x}) = 0. \tag{1.19}$$

The method of least squares was apparently first published by the French mathematician Legendre in 1805. However, Carl Friedrich Gauss, who published it in 1809, claimed that he had been using the procedure since 1795, leading to one of the more famous disputes in the history of science. The reader who is interested in pursuing this dispute further may wish to consult Plackett (1972) and Stigler (1981).

---

[1] This is an appendix to this chapter which is at the end of it. Appendices to individual chapters have the chapter number in front of the 'A', while the appendix at the end of the book and sections in it start with letters.

1.5 Another Example and a Special Case

Example 1.2
Consider the data of Exhibit 1.7 on the population of zones and the number of telephones (household mains). We wish to see how population size affects the number of telephones. (Models connecting these two variables have been used to estimate population in small areas for non-census years.)

# of Residents	4041	2200	30148	60324	65468	30988
# of Household Mains	1332	690	11476	18368	22044	10686

EXHIBIT 1.7: Data on Population and Household Mains
SOURCE: Prof. Edwin Thomas, Department of Geography, University of Illinois at Chicago.

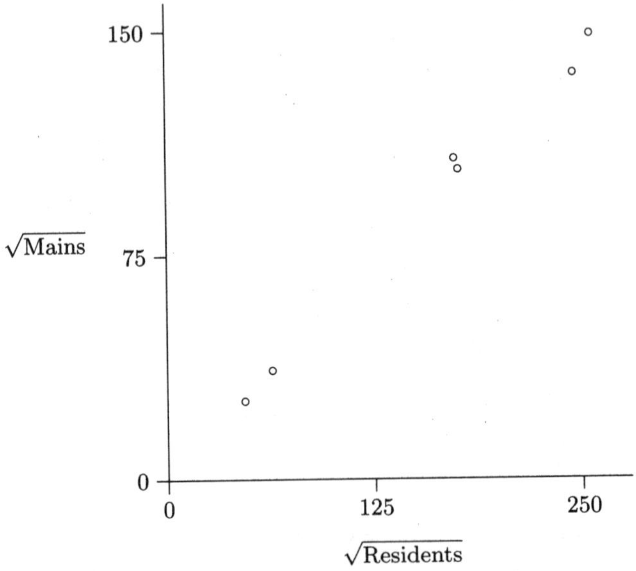

EXHIBIT 1.8: Plot of Square Root of Household Mains Against Square Root of Residents

Again for reasons that we shall have to defer for the moment, we prefer to take square roots of both variables and set

$$y_i = (\text{number of telephones})^{1/2}, \; x_{i1} = (\text{population size})^{1/2}.$$

A plot is shown in Exhibit 1.8. It appears to indicate a linear relationship with the line passing through the point (0, 0), which is perfectly reasonable since if there were no people in an area, there would usually be no household phones! Thus we would postulate a model of the form

$$y_i = \beta_1 x_{i1} + \epsilon_i, \tag{1.20}$$

i.e., the constant or intercept term β_0 would be missing. ■

Model (1.20) can be handled using least squares quite easily. Now we would minimize

$$\mathcal{S} = \sum_{i=1}^{n} (y_i - \beta_1 x_{i1})^2 \tag{1.21}$$

with respect to β_1. Then, from equating to zero the derivative

$$d\mathcal{S}/d\beta_1 = -2\sum_{i=1}^{n} (y_i - \beta_1 x_{i1}) x_{i1},$$

we get the least squares estimate b_1 of β_1 to be

$$b_1 = \sum_{i=1}^{n} y_i x_{i1} / \sum_{i=1}^{n} x_{i1}^2. \tag{1.22}$$

Since there is no β_0, the residuals are

$$e_i = y_i - \hat{y}_i = y_i - b_1 x_{i1},$$

but here $\sum_{i=1}^{n} e_i$ is not usually zero. For the data in Exhibit 1.7, b_1 turns out to be 0.578.

1.6 When Is Least Squares a Good Method? The Gauss-Markov Conditions

Perfectly reasonable questions to ask at this stage are: How good a procedure is least squares and does it always estimate the β_j's well? We shall have to defer a better answer to this question until Chapter 2, but for the moment we can say that least squares gives good predictions if certain conditions (called Gauss-Markov conditions) are met. In order to show the need for these conditions let us consider situations where good estimates would be difficult to get.

Exhibit 1.9a illustrates a case where a straight line is inappropriate and as a result we are not likely to get good predictions. In order to exclude such situations (i.e., force us to use models that are appropriate), we make the condition that

$$\text{E}(\epsilon_i) = 0, \quad \text{for all i.} \tag{1.23}$$

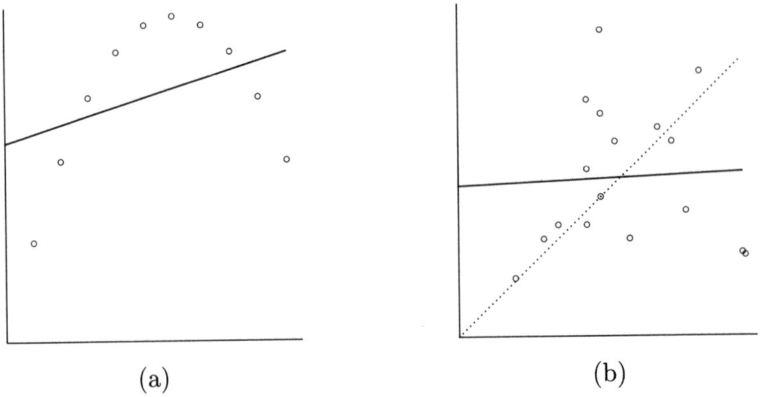

EXHIBIT 1.9: Violations of Some Gauss-Markov Conditions

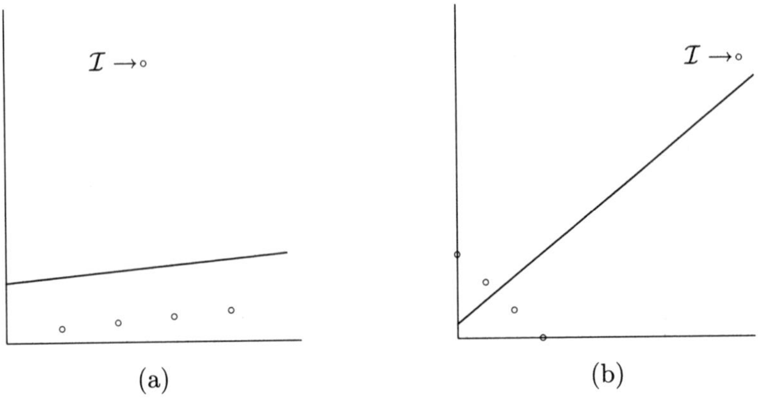

EXHIBIT 1.10: Examples of Influential Points

This implies that the expectation $\mathrm{E}(y_i)$ of y_i actually is $\beta_0 + \beta_1 x_{i1}$ in the simple regression case or $\beta_0 + \beta_1 x_{i1} + \cdots + \beta_k x_{ik}$ in the multiple regression case. As we shall see in Chapter 11, (1.23) can also be violated if necessary independent variables are left out of the model.

Another type of problem we shall need to guard against is shown by Exhibit 1.9b. Here assume that the true model is given by the dotted line and (1.23) holds, but the variance, var (ϵ_i) of ϵ_i, increases with x_i. The few points far from the dotted line can cause the least squares line, as shown by a continuous line in Exhibit 1.9b, to be quite bad (for much the same reasons as described in the next paragraph). This type of situation, often called heteroscedasticity, is prevented by imposing the condition

$$\mathrm{var}\,(\epsilon_i) = \mathrm{E}(\epsilon_i - \mathrm{E}(\epsilon_i))^2 = \mathrm{E}(\epsilon_i^2) = \sigma^2 \tag{1.24}$$

(i.e., $\mathrm{E}(\epsilon_i^2)$ is a constant) for all i.

Sometimes only a single point or a small number of points may violate (1.23) and/or (1.24). In Exhibit 1.10a, if the point '$\mathcal{I}$' had been absent, the regression line would have gone more or less through the remaining four points. However, with the presence of '$\mathcal{I}$' the regression line is as shown. Many procedures would be affected by such a point but least squares estimates are particularly affected. That is because the spacing between squares of equispaced numbers increases with the size of the numbers (e.g., $10^2-9^2 = 19$ while $2^2-1^2 = 3$). Points such as '$\mathcal{I}$' are called outliers because they are far removed from the regression line and, because they have a large effect, they are also called influential points. They deserve significant attention because frequently they represent a violation of (1.23). And when they do, they do not belong in the analysis and, as we have already seen, can hurt our analysis.

An even more potentially deadly situation is shown by Exhibit 1.10b. For much the same reason as before, the point '$\mathcal{I}$' can alter the entire direction of the line and, what is worse, '$\mathcal{I}$' would not even have a large residual to draw attention to itself. Such a point is an influential point but is not an outlier. Chapter 8 will be devoted to outliers and influential observations.

Yet another type of problem is perhaps best illustrated by an extreme example. If we had only two observations we could draw a straight line fitting them perfectly, but normally we would be reluctant to make a prediction based on them alone. Suppose we made 20 copies of each of the data points. We now have 40 'observations', but are certainly no better off. This is because our observations are related (to say the least!). We therefore require our observations to be uncorrelated:

$$\mathrm{E}(\epsilon_i \epsilon_j) = 0 \quad \text{for all} \quad i \neq j. \tag{1.25}$$

Conditions (1.23) , (1.24) and (1.25) are called the *Gauss-Markov conditions* and it is gratifying to note that they assure that an appropriate prediction made by a least squares fitted equation is good. 'Goodness' will be defined in Chapter 2, where a proof of the fact will also be given. Throughout the book, the Gauss-Markov conditions will be used as a benchmark. When they hold, least squares estimates are good and when they do not, we shall need to make appropriate changes which would cause approximate compliance with the conditions.

1.7 A Measure of Fit for Simple Regression

As already noted, when we have a good fit the residuals are small. Thus we can measure the quality of fit by the sum of squares of residuals $\sum_{i=1}^n e_i^2$.

However, this quantity is dependent on the units in which y_i's are measured. Thus, when $\beta_0 \neq 0$, a good measure of fit is

$$R^2 = 1 - \sum_{i=1}^{n} e_i^2 / \sum_{i=1}^{n} (y_i - \bar{y})^2. \tag{1.26}$$

This number lies between 0 and 1 (as will be shown in Section 2.8, p. 39) and the closer it is to 1 the better the fit.

When $\beta_0 = 0$ as in Section 1.5 a measure of fit is

$$R^2 = 1 - \sum_{i=1}^{n} e_i^2 / \sum_{i=1}^{n} y_i^2. \tag{1.27}$$

Since $\sum_{i=1}^{n} y_i^2$ is usually much larger than $\sum_{i=1}^{n}(y_i - \bar{y})^2$, this definition of R^2 is quite different from that in (1.26). *Therefore, models with β_0 cannot be compared with those without β_0 on the basis of R^2.*

Example 1.3
Running a regression of violent crimes (VIOL) against population (POP) using the data of Exhibit 1.11, we get the regression equation

$$\text{VIOL} = 433.6 + .00011\,\text{POP} \quad [R^2 = .486].$$

However, if New York is deleted, we get a substantial decline in R^2:

$$\text{VIOL} = 447.9 + .000085\,\text{POP} \quad [R^2 = .087].$$

This example serves to reminds us that R^2 depends not only on $\sum_{i=1}^{n} e_i^2$, as we would wish, but also on $\sum_{i=1}^{n}(y_i - \bar{y})^2$, and an increase in the value of the latter can increase R^2. The plot of Exhibit 1.12 illustrates the situation and shows a picture somewhat reminiscent of Exhibit 1.10b. ■

1.8 Mean and Variance of b_0 and b_1 Under Gauss-Markov Conditions

Since b_0 and b_1 depend on the y_i's, which are random variables, b_0 and b_1 are random variables. Their means and variances are given by

$$\begin{aligned} \mathrm{E}[b_0] &= \beta_0, \quad \mathrm{var}\,[b_0] = \sigma^2 \left[n^{-1} + \frac{\bar{x}_1^2}{\sum_{i=1}^{n}(x_{i1} - \bar{x}_1)^2} \right] \\ \mathrm{E}[b_1] &= \beta_1, \quad \mathrm{var}\,[b_1] = \sigma^2 / \sum_{i=1}^{n}(x_{i1} - \bar{x}_1)^2. \end{aligned} \tag{1.28}$$

Metro. Area	Violent Crimes	Property Crimes	Population
Allentown, PA	161.1	3162.5	636.7
Bakersfield, CA	776.6	7701.3	403.1
Boston, MA	648.2	5647.2	2763.4
Charleston, SC	851.7	5587.7	430.3
Corpus Christi, TX	611.5	6115.1	326.2
Elmira, NY	176.0	4693.5	97.7
Fort Lauderdale, FL	732.1	8044.4	1014.0
Greely, CO	434.9	5868.9	123.4
Jackson, MI	642.5	5402.4	151.5
La Crosse, WI	88.3	6261.3	91.1
Lexington, KY	338.2	4879.6	318.1
Madison, WI	177.4	659.2	323.5
Monroe, LA	472.6	3929.2	139.2
Norfolk, VA	500.9	5175.9	806.7
Peoria, IL	676.3	5146.1	365.9
Pueblo, CO	840.5	5709.1	126.0
Sacramento, CA	724.7	8087.4	1014.0
San Jose, CA	416.1	6280.4	1295.1
South Bend, IN	354.8	5328.8	280.8
Texarkana, TX	402.7	4225.2	127.0
Washington, DC	693.0	5895.4	3060.2
Youngstown, OH	356.4	3524.3	531.4
New York, NY	1469.9	6308.4	9119.7

EXHIBIT 1.11: Data on Violent and Property Crimes in 22 Metropolitan Areas
SOURCE: Dacey (1983, Ch. 3).

In the case where $\beta_0 = 0$, we have

$$E(b_1) = \beta_1 \text{ and } \operatorname{var}(b_1) = \sigma^2 / \sum_{i=1}^{n} x_{i1}^2. \tag{1.29}$$

These formulæ can be seen as special cases of Theorem 2.2 (p. 36) of Chapter 2, or can be proved directly as in the appendix to this chapter.

Since the expected value $E(b_0)$ of the estimate b_0 is β_0, b_0 is called an unbiased estimator of β_0; similarly, b_1 is an unbiased estimator of β_1. This is obviously pleasant.

In order to use the variances of b_0 and b_1 we encounter a slight problem. They depend on σ^2, which is not known. However (as will be shown in

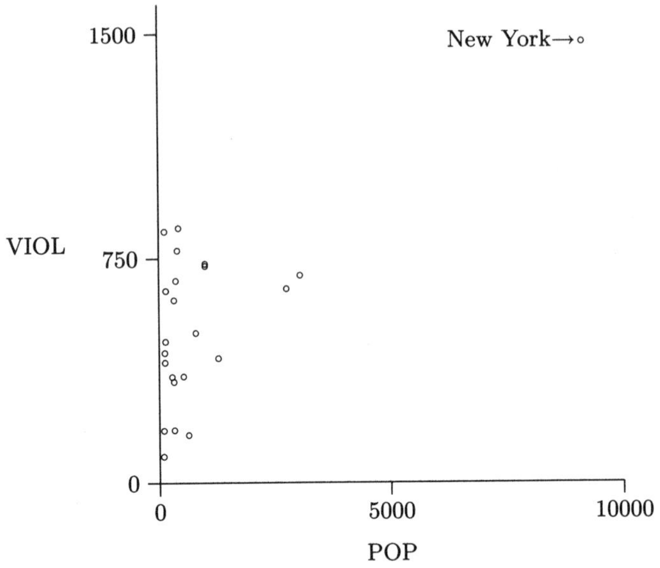

EXHIBIT 1.12: Plot of Violent Crimes Against Population

Chapter 2), an unbiased estimator of σ^2 is

$$s^2 = (n-2)^{-1} \sum_{i=1}^{n} e_i^2 \tag{1.30}$$

and if we replace σ^2 by s^2 in (1.28) and (1.29) we get estimates of var (b_0) and var (b_1). Square roots of these estimates are called *standard errors* and will be denoted by s.e.(b_0) and s.e.(b_1). Thus

$$\text{s.e.}(b_0) = s[n^{-1} + \bar{x}_1^2 / \sum_{i=1}^{n} (x_{i1} - \bar{x}_1)^2]^{1/2} \tag{1.31}$$

and

$$\text{s.e.}(b_1) = s/[\sum_{i=1}^{n} (x_{i1} - \bar{x}_1)^2]^{1/2} \tag{1.32}$$

when $\beta_0 \neq 0$; and when $\beta_0 = 0$,

$$\text{s.e.}(b_1) = s/[\sum_{i=1}^{n} x_{i1}^2]^{1/2} \text{ where } s^2 = (n-1)^{-1} \sum_{i=1}^{n} e_i^2. \tag{1.33}$$

These quantities are routinely provided by computer packages.

1.9 Confidence Intervals and Tests

Assume that the Gauss-Markov conditions (1.23), (1.24) and (1.25) hold, and let us now make the additional assumption that the ϵ_i's are normally distributed. Then the ϵ_i's are independently distributed as normal with mean 0 and variance σ^2. We shall denote this fact as $\epsilon_i \sim N(0, \sigma^2)$, i.e., '$\sim$' will stand for 'has the distribution'. It follows that $y_i \sim N(\beta_0 + \beta_1 x_{i1}, \sigma^2)$. Then, the b_j's, being linear combinations of y_i's, are also normal with means and variances as given in the last section. It may be shown that

$$(b_j - \beta_j)/\text{s.e.}(b_j) \sim t_{n-2} \tag{1.34}$$

for the simple regression case with $\beta_0 \neq 0$, where t_{n-2} is the Student's t distribution with $n-2$ degrees of freedom (Section B.5, p. 292, of Appendix B; a table is given on p. 320). From (1.34) we may obtain a $(1-\alpha)\times 100$ percent confidence interval (C.I.) for β_j as

$$b_j - \text{s.e.}(b_j)t_{n-2,\alpha/2} < \beta_j < b_j + \text{s.e.}(b_j)t_{n-2,\alpha/2} \tag{1.35}$$

where $j = 0$ or 1 and $t_{n-2,\alpha/2}$ denotes the upper $\alpha/2$ point of the t distribution with $n-2$ degrees of freedom.

Example 1.2 ctd.
Consider now the data on telephones from Section 1.5 and let us consider the model with $\beta_0 \neq 0$ (we shall test the hypothesis H: $\beta_0 = 0$ shortly). Exhibit 1.13 shows a portion of a typical output from a regression package. As we can see,

$$b_0 = 1.30, \quad b_1 = .571,$$

and

$$\text{s.e.}(b_0) = 4.28, \quad \text{s.e.}(b_1) = .024.$$

Since $t_{4,0.05} = 2.1318$, the 90 per cent confidence intervals for β_0 and β_1 are respectively

$$(-7.8241, 10.4241) \tag{1.36}$$

and

$$(.5198, .6221). \tag{1.37}$$

Since 0 is included in (1.36) we cannot reject the hypothesis H_0: $\beta_0 = 0$, but we can reject, say, H_0: $\beta_1 = .7$. Exhibit 1.14 shows an output for when the intercept term is missing. Now, using the fact $t_{5,.05} = 2.0151$, the C.I. for β_1 is seen to be (.5583, .5973).

The test for $\beta_j = 0$ is called for so often that most packages routinely carry it out. The values of

$$t(b_0) = b_0/\text{s.e.}(b_0) = 0.3037 \quad \text{and} \quad t(b_1) = b_1/\text{s.e.}(b_1) = 23.955$$

Variable	b_j	s.e.(b_j)	$t(b_j)$	P$[\|t\| > \|t(b_j)\|]$
Intercept	1.301	4.280	0.3037	0.7763
$[\text{Mains}]^{1/2}$	0.571	0.024	23.955	0.0001

$s = 4.714 \qquad R^2 = .9931$

EXHIBIT 1.13: Computer Output for Telephone Data

Variable	b_j	s.e.(b_j)	$t(b_j)$	P$[\|t\| > \|t(b_j)\|]$
$[\text{Mains}]^{1/2}$	0.578	0.0097	59.556	0.0001

$s = 4.264 \qquad R^2 = .9986$

EXHIBIT 1.14: Computer Output for Telephone Data When β_0 Is Missing

are also given in Exhibit 1.13. The probability that the value of a t distributed random variable would be numerically larger than $|t(b_0)| = 0.3037$ is .7763 and that of getting a t-value larger than $|t(b_1)| = 23.955$ is .0001. Thus we can reject $H : \beta_1 = 0$ at 5, 1 or even .1 per cent but cannot reject $H : \beta_0 = 0$ at any reasonable level of significance. ■

1.10 Predictions

One of the principal purposes of regression is to make predictions. Suppose x_{01} is a value of the independent variable x_1 for which we need to predict the dependent variable y. Obviously, such a prediction would be

$$\hat{y}_0 = b_0 + b_1 x_{01}. \tag{1.38}$$

Thus, using the output of Exhibit 1.13, if we wish to forecast the number of telephones for an area with 10000 people, we would have $y_0 = 1.301 + .571(100) = 58.4$ and the forecasted number of phones would be its square, which is 3411.

Since, as shown in Section 1.8, $\text{E}(b_0) = \beta_0$ and $\text{E}(b_1) = \beta_1$, we have

$$\text{E}(\hat{y}_0) = \beta_0 + \beta_1 x_{01}. \tag{1.39}$$

It has been shown in Appendix 1A (and may also be shown as a special case of the formulæ in Section 3.8.1, p. 71) that

$$\text{var}\,(\hat{y}_0) = \sigma^2[n^{-1} + (x_{01} - \bar{x}_1)^2 / \sum_{i=1}^{n} (x_{i1} - \bar{x}_1)^2]. \tag{1.40}$$

It might be noted in passing that $\text{var}\,(\hat{y}_0)$ obviously increases with $(x_{01} - \bar{x}_1)^2$, that is, $\text{var}\,(\hat{y}_0)$ gets larger the farther x_{01} is from $\bar{x}_1$.

Let y_0 be the observation, corresponding to x_{01}, that we would have got were we able to. Such a y_0 is called a *future observation*. Since we would be forecasting y_0 with $\hat{y}_0$ we would be interested in the difference

$$y_0 - \hat{y}_0. \tag{1.41}$$

In order to obtain the mean and variance of (1.41), we need to assume that y_0 is given by the same model as the observations on which the regression equation is built, i.e.,

$$y_0 = \beta_0 + \beta_1 x_{01} + \epsilon_0$$

where $\mathrm{E}(\epsilon_0) = 0$, $\mathrm{E}(\epsilon_0^2) = \sigma^2$, and $\mathrm{E}(\epsilon_0 \epsilon_i) = 0$ for $i = 1, \ldots, n$. Then, of course,

$$\mathrm{E}(y_0) = \beta_0 + \beta_1 x_{01}$$

and hence from (1.39)

$$\mathrm{E}(y_0 - \hat{y}_0) = 0. \tag{1.42}$$

The variance of (1.41) is

$$\begin{aligned} \operatorname{var}(y_0 - \hat{y}_0) &= \operatorname{var}(y_0) + \operatorname{var}(\hat{y}_0) \\ &= \sigma^2[1 + n^{-1} + (x_{01} - \bar{x}_1)^2 / \sum_{i=1}^{n} (x_{i1} - \bar{x}_1)^2]. \end{aligned} \tag{1.43}$$

Obviously, since σ^2 is not known, in practical applications we would normally replace σ^2 by s^2 in both (1.40) and (1.43). Using $s^2 = 22.22$ in Example 1.2, the standard error of $\hat{y}_0$ corresponding to a place with 10000 people turns out to be 2.39 and that of a future observation works out to 5.28.

If one is confused about the distinction between predicted value and future observation, one may wish to consider that the predicted value $\hat{y}_0$ is a point on the estimated regression line and its variance only reflects the fact that b_0 and b_1 are random variables. On the other hand, there is no reason to believe that a future observation will necessarily be on the true line $y = \beta_0 + \beta_1 x_1$ or on the fitted line. Indeed, $y_0 - \beta_0 - \beta_1 x_{01}$ has the variance σ^2. Thus the variance of $y_0 - \hat{y}_0$ would be due to a combination of both effects.

Appendix to Chapter 1

1A Some Derivations

Proof of (1.14): Since $\sum_{i=1}^n (y_i - \bar{y})\bar{x}_1 = \bar{x}_1 \sum_{i=1}^n (y_i - \bar{y}) = 0$,

$$\begin{aligned}\sum_{i=1}^n (y_i - \bar{y})(x_{i1} - \bar{x}_1) &= \sum_{i=1}^n (y_i x_{i1} - \bar{y} x_{i1}) - \bar{x}_1 \sum (y_i - \bar{y}) \\ &= \sum_{i=1}^n y_i x_{i1} - \bar{y} \sum_{i=1}^n x_{i1} = \sum_{i=1}^n y_i x_{i1} - n\bar{y}\bar{x}_1,\end{aligned} \tag{1.44}$$

and

$$\begin{aligned}\sum_{i=1}^n (x_{i1} - \bar{x}_1)^2 &= \sum_{i=1}^n x_{i1}^2 - 2\bar{x}_1 \sum_{i=1}^n x_{i1} + n\bar{x}_1^2 \\ &= \sum_{i=1}^n x_{i1}^2 - 2n\bar{x}_1^2 + n\bar{x}_1^2 = \sum_{i=1}^n x_{i1}^2 - n\bar{x}_1^2.\end{aligned} \tag{1.45}$$

Hence, the coefficient (1.13) may also be written as

$$b_1 = \sum_{i=1}^n (y_i - \bar{y})(x_{i1} - \bar{x}_1) / \sum_{i=1}^n (x_{i1} - \bar{x}_1)^2,$$

as stated on p. 8. □

Proof of (1.28): Since $\sum_{i=1}^n (x_{i1} - \bar{x}_1) = 0$ and

$$\sum_{i=1}^n (y_i - \bar{y})(x_{i1} - \bar{x}_1) = \sum_{i=1}^n y_i (x_{i1} - \bar{x}_1) - \bar{y} \sum_{i=1}^n (x_{i1} - \bar{x}_1) = \sum_{i=1}^n y_i (x_{i1} - \bar{x}_1),$$

it follows from (1.14) that

$$b_1 = \sum_{i=1}^n y_i (x_{i1} - \bar{x}_1) / \sum_{i=1}^n (x_{i1} - \bar{x}_1)^2 = \sum_{i=1}^n c_i y_i, \tag{1.46}$$

where $c_i = (x_{i1} - \bar{x}_1) / \sum_{i=1}^n (x_{i1} - \bar{x}_1)^2$. It is easy to verify that

$$\sum_{i=1}^n c_i = 0,$$

$$\sum_{i=1}^n c_i x_{i1} = \sum_{i=1}^n c_i x_{i1} - \bar{x}_1 \sum_{i=1}^n c_i = \sum_{i=1}^n c_i (x_{i1} - \bar{x}_1) = 1$$

and

$$\sum_{i=1}^n c_i^2 = [\sum_{i=1}^n (x_{i1} - \bar{x}_1)^2]^{-1}.$$

Hence, using standard results on means and variances of linear combinations of random variables (a summary of these is presented in the appendix to the book — see Section B.1.1, p. 284), we get

$$\begin{aligned} \mathrm{E}(b_1) &= \sum_{i=1}^{n} c_i \,\mathrm{E}(y_i) = \sum_{i=1}^{n} c_i\beta_0 + \beta_1 \sum_{i=1}^{n} c_i x_{i1} \\ &= \beta_0 \sum_{i=1}^{n} c_i + \beta_1 = \beta_1 \end{aligned} \tag{1.47}$$

and because $\mathrm{var}\,(y_i) = \mathrm{var}\,(\beta_0 + \beta_1 x_{i1} + \epsilon_i) = \mathrm{var}\,(\epsilon_i) = \sigma^2$, we have

$$\mathrm{var}\,(b_1) = \sum_{i=1}^{n} c_i^2 \mathrm{var}\,(y_i) = \sigma^2 \sum_{i=1}^{n} c_i^2 = \sigma^2 / \sum_{i=1}^{n} (x_{i1} - \bar{x}_1)^2. \tag{1.48}$$

Similarly, since

$$\mathrm{E}(\bar{y}) = n^{-1} \sum_{i=1}^{n} \mathrm{E}(y_i) = n^{-1} \sum_{i=1}^{n} (\beta_0 + \beta_1 x_{i1}) = \beta_0 + \beta_1 \bar{x}_1,$$

it follows that

$$\mathrm{E}(b_0) = \mathrm{E}(\bar{y} - b_1\bar{x}_1) = \beta_0 + \beta_1\bar{x}_1 - \bar{x}_1\,\mathrm{E}(b_1) = \beta_0 + \beta_1\bar{x}_1 - \bar{x}\beta_1 = \beta_0. \tag{1.49}$$

Now, from (1.46) we can write

$$b_0 = n^{-1} \sum_{i=1}^{n} y_i - \bar{x}_1 \sum_{i=1}^{n} c_i y_i = \sum_{i=1}^{n} (n^{-1} - \bar{x}_1 c_i) y_i. \tag{1.50}$$

Hence,

$$\begin{aligned} \mathrm{var}\,(b_0) &= \sum_{i=1}^{n} [n^{-1} - \bar{x}_1 c_i]^2 \mathrm{var}\,(y_i) \\ &= \sigma^2 \sum_{i=1}^{n} [n^{-2} - 2n^{-1}\bar{x}_1 c_i + \bar{x}_1^2 c_i^2] \\ &= \sigma^2 \left[n^{-1} + \frac{\bar{x}_1^2}{\sum_{i=1}^{n} (x_{i1} - \bar{x}_1)^2} \right]. \end{aligned} \tag{1.51}$$

This completes the proof. □

PROOF OF (1.29): In the case where $\beta_0 = 0$, we have from (1.22)

$$\begin{aligned} b_1 &= \sum_{i=1}^{n} y_i x_{i1} / \sum_{i=1}^{n} x_{i1}^2 = \beta_1 \sum_{i=1}^{n} x_{i1}^2 / \sum_{i=1}^{n} x_{i1}^2 + \sum_{i=1}^{n} \epsilon_i x_{i1} / \sum_{i=1}^{n} x_{i1}^2 \\ &= \beta_1 + \sum_{i=1}^{n} \epsilon_i x_{i1} / \sum_{i=1}^{n} x_{i1}^2. \end{aligned} \tag{1.52}$$

Hence,

$$\text{var}\,(b_1) = \sigma^2 \sum_{i=1}^{n} x_{i1}^2 / (\sum_{i=1}^{n} x_{i1}^2)^2 = \sigma^2 / \sum_{i=1}^{n} x_{i1}^2 \tag{1.53}$$

and $\text{E}(b_1) = \beta_1$. □

PROOF OF (1.40): Substituting the expression (1.11) for b_0 in (1.38) and using (1.46) we get

$$\hat{y}_0 = \bar{y} - b_1\bar{x}_1 + b_1 x_{01} = \bar{y} + b_1(x_{01} - \bar{x}_1) = \sum_{i=1}^{n}(n^{-1} + (x_{01} - \bar{x}_1)c_i)y_i.$$

Therefore,

$$\begin{aligned}\text{var}\,(\hat{y}_0) &= \sigma^2 \sum_{i=1}^{n}(n^{-1} + (x_{01} - \bar{x}_1)c_i)^2 \\ &= \sigma^2 \sum_{i=1}^{n}[n^{-2} + 2n^{-1}(x_{01} - \bar{x}_1)c_i + (x_{01} - \bar{x}_1)^2 c_i^2] \\ &= \sigma^2[n^{-1} + (x_{01} - \bar{x}_1)^2 / \sum_{i=1}^{n}(x_{i1} - \bar{x}_1)^2]\end{aligned}$$

on substituting for c_i from (1.46) and noting that $\sum_{i=1}^{n} c_i = 0$. □

Problems

Exercise 1.1: Show that for the estimator b_1 in (1.22), $\mathrm{E}[b_1] = \beta_1$.

Exercise 1.2: Prove (1.53).

Exercise 1.3: Let e_i be the residuals defined in (1.18). Find $\mathrm{var}(e_i)$.

Exercise 1.4: Let $\hat{y}_i$ be as defined as in (1.17). Find $\mathrm{E}(\hat{y}_i)$ and $\mathrm{var}(\hat{y}_i)$.

Exercise 1.5: Suppose in the model $y_i = \beta_0 + \beta_1 x_{i1} + \epsilon_i$, where $i = 1, \ldots, n$, $\mathrm{E}(\epsilon_i) = 0$, $\mathrm{E}(\epsilon_i^2) = \sigma^2$ and, for $i \neq j$, $\mathrm{E}(\epsilon_i \epsilon_j) = 0$, the measurements x_{i1} were in inches and we would like to write the model in centimeters, say, z_{i1}. If one inch is equal to c centimeters (c known), write the above model as $y_i = \beta_0^* + \beta_1^* z_{i1} + \epsilon_i$. Can you obtain the estimates of β_0^* and β_1^* from those of β_0 and β_1? Show that the value of R^2 remains the same for both models.

Exercise 1.6: Suppose $y_1, \ldots, y_n$ are independently distributed and $y_i = \mu + \epsilon_i$ for $i = 1, \ldots, n$. Find the least squares estimate of μ if $\mathrm{E}(\epsilon_i) = 0$ and $\mathrm{var}(\epsilon_i) = \sigma^2$. Give the variance of this estimate.

Exercise 1.7: Let $(y_1, x_1), \ldots, (y_n, x_n)$ and $(w_1, x_1), \ldots, (w_n, x_n)$ be two sets of independent observations where $x_1, \ldots, x_n$ are fixed constants. Suppose we fit the model $y_i = \alpha_1 + \beta x_i + \epsilon_i$ to the first data set and the model $w_i = \alpha_2 + \beta x_i + \eta_i$ to the second set. In each case $i = 1, \ldots, n$, and assume that all ϵ_i's and η_i's are independently distributed with zero means and common variance σ^2. Find the least squares estimates of α_1, α_2 and β.

Exercise 1.8: For $i = 1, \ldots, n$, let $y_i = \beta_0 + \beta_1 x_{i1} + \epsilon_i$ be the straight line regression model in which x_{i1}'s are such that $\sum_{i=1}^n x_{i1} = 0$, and ϵ_i's are independently distributed with mean zero and variance σ^2. What are the least squares estimators of β_0 and β_1? Find the mean and variance of these estimators.

Exercise 1.9: Stevens (1956) asked a number of subjects to compare notes of various decibel levels against a standard (80 decibels) and to assign them a loudness rating with the standard note being a 10. The data from this experiment are summarized in Exhibit 1.15. Run a regression using log y as a dependent variable and x as the independent variable.

Stimulus (x)	30	50	60	70	75	80	85	90	95	100
Median Response (y)	0.2	1.0	3.0	5.0	8.5	10.0	14.0	20.0	29.0	43.0
log(y)	-.70	.00	.48	.70	.93	1.00	1.15	1.30	1.46	1.63

EXHIBIT 1.15: Data from Stevens' Experiment
SOURCE: Dacey (1983, Ch.1) from Stevens (1956). Reproduced with permission from University of Illinois Press.

Exercise 1.10: Using the data shown in Exhibit 1.16, obtain an equation expressing stock prices as a function of earnings. At a 5 per cent level of significance, test the hypothesis that stock prices are unrelated to earnings, against the alternative that they are related.

Company	1972 Earnings per Share (in $'s)	Price (in $'s) in May, 1973
CROWN ZELLERBACH	1.83	28
GREAT NORTHERN NEKOOSA	3.35	45
HAMMERMILL PAPER	0.64	12
INTERNATIONAL PAPER	2.30	35
KIMBERLY-CLARK	2.39	45
MEAD	1.08	14
ST. REGIS PAPER	2.92	39
SCOTT PAPER	1.11	12
UNION CAMP	2.57	43
WESTVACO	1.22	23

EXHIBIT 1.16: Earnings and Prices of Selected Paper Company Stocks
SOURCE: Dacey (1983, Ch. 1) from Moody's Stock Survey, June 4, 1973, p. 610.

Exercise 1.11: Exhibit 1.17 gives data on population density (pd) and vehicle thefts (vtt) per thousand residents in 18 Chicago districts (D). District 1 represents downtown Chicago. Run a regression with vtt as the dependent variable and pd as the independent variable. Plot the residuals against pd. Do you notice an outlier? If so, can you explain why it is so? If appropriate, delete any outliers and re-estimate the model.

Now test the hypothesis that the slope is zero against the alternative that it is different from zero. Use 5 per cent as the level of significance.

D	pd	vtt	D	pd	vtt
1	3235	132.8	14	22919	13.3
2	24182	14.9	15	24534	15.1
3	20993	16.7	18	24987	16.2
6	15401	20.0	19	21675	12.5
7	19749	14.2	20	22315	11.8
10	19487	13.5	21	18402	19.6
11	19581	16.5	23	33445	10.5
12	14077	22.2	24	27345	10.1
13	18137	15.8	25	15358	19.0

EXHIBIT 1.17: Data on Population Density and Vehicle Thefts
SOURCE: Mark Buslik, Chicago Police Department.

Exercise 1.12: Exhibit 1.18 gives data on the number of marriages (ma) that occurred between residents of each of 8 annular zones and residents of Simsbury, Connecticut, for the period 1930–39. The number of residents of each zone is given as pop and the midpoint of distance between Simsbury and the band is given as d (e.g., the first annular zone represents a band 5–7 miles from Simsbury with a midpoint of 6). Run a regression of log[ma/pop] against d. Write a sentence explaining your findings to a non-technical audience.

d	pop	ma	d	pop	ma
6	3464	26	14	15207	7
8	4892	12	16	175458	49
10	2583	4	18	95179	18
12	39411	12	20	62076	7

EXHIBIT 1.18: Data on Simsbury Marriages
SOURCE: Dacey (1983, Ch. 4) from Ellsworth (1948).

Price	P	B	Price	P	B	Price	P	B	Price	P	B
10.25	112	p	24.50	146	c	24.75	158	c	30.50	276	c
14.25	260	p	19.75	212	c	16.50	322	p	22.75	264	c
29.25	250	c	30.25	292	c	12.50	188	p	17.75	378	p
17.50	382	p	16.25	340	p	16.75	240	p	29.50	251	c
12.00	175	p	29.00	252	c	17.50	425	p	27.50	202	c

EXHIBIT 1.19: Data on Book Prices, Pages and Type of Binding

Exercise 1.13: The data set given in Exhibit 1.19 was compiled by one of the authors from the Spring, 1988, catalogue of American Government books put out by a certain publisher. It lists prices, number of pages (P) and the binding (B; p stands for paperback and c for cloth). Fit a straight line to the paperback data, using price as the dependent variable and number of pages as the independent variable. What do the parameters say about the pricing policy of the publisher? Repeat the same exercise for cloth-bound books. Estimate the price of a paperback book of 100 pages and a 400-page cloth-bound book. Also estimate the prices of 250-page books with the two types of binding. In each case give the 95 per cent confidence interval within which the price of such a book when produced will lie.

Exercise 1.14: The data set in Exhibit 1.20 was given to us by Dr. T.N.K. Raju, Department of Neonatology, University of Illinois at Chicago. Regress each of the infant mortality rates (IMR) against the Physical Quality of Life Index (PQLI — which is an indicator of average wealth). In each case try taking logs of

State	PQLI Score	Comb- ined IMR	Rural Male IMR	Rural Female IMR	Urban Male IMR	Urban Female IMR
UTTAR PRAD.	17	167	159	187	110	111
MADHYA PRAD.	28	135	148	134	88	83
ORISSA	24	133	131	142	78	81
RAJASTHAN	29	129	135	142	55	77
GUJARAT	36	118	120	135	92	84
ANDHRA PRAD.	33	112	138	101	79	46
HARYANA	55	109	107	128	57	60
ASSAM	35	118	133	106	87	85
PUNJAB	62	103	115	108	58	73
TAMILNADU	43	103	125	115	67	59
KARNATAKA	52	75	92	70	51	59
MAHARASHTRA	60	75	95	72	50	62
KERALA	92	39	42	42	22	30

EXHIBIT 1.20: Data on Physical Quality of Life Index (PQLI) Scores and Infant Mortality Rates (IMR) for Selected Indian States

1. the independent variable only,
2. the dependent variable only, and
3. both independent and dependent variables.

Using suitable plots, visually judge which of 1, 2 or 3 above or the untransformed case gives the best fit.

L	D	L	D	L	D
1000	125	9000	920	9400	2750
2000	225	9250	1040	9375	3200
3000	325	9175	1320	9450	3750
4000	425	9150	1500	9500	4500
5000	525	9150	1600	9600	5000
6000	625	9000	1840	9700	6500
7000	725	9250	2160	9900	8000
8000	880	9125	2480	9900	9500

EXHIBIT 1.21: Data on Loads (L) and Deformation (D) of a Bar

Exercise 1.15: Exhibit 1.21 gives data on loads, in pounds, and corresponding deformation, in inches, of a mild steel bar (of length 8 ins. and average diameter .564 ins). The data were provided by M.R. Khavanin, Department of Mechanical Engineering, University of Illinois at Chicago.

1. Run a regression of the log of D against the log of L. Obtain a plot of the residuals against the predicted values. What does the plot tell you about the relationship between the two quantities?

2. Although, presumably, deformation depends on load, and not the other way around, run a regression of the log of L against the log of D. Plot residuals against the predicted and independent variables and take whatever action you think is warranted to get a good fit.

Do you have a physical explanation for what you observed in doing this exercise?

CHAPTER 2

Multiple Regression

2.1 Introduction

Formulae for multiple regression are much more compact in matrix notation. Therefore, we shall start off in the next section applying such notation first to simple regression, which we considered in Chapter 1, and then to multiple regression. After that we shall derive formulae for least squares estimates and present properties of these estimates. These properties will be derived under the Gauss-Markov conditions which were presented in Chapter 1 and are essentially restated in Section 2.5.

2.2 Regression Model in Matrix Notation

We begin this section by writing the familiar straight line case of Chapter 1 in matrix notation. Recall that the regression model then was:

$$\begin{aligned} y_1 &= \beta_0 + \beta_1 x_{11} + \epsilon_1 \\ &\cdots\cdots\cdots\cdots\cdots \\ &\cdots\cdots\cdots\cdots\cdots \\ y_n &= \beta_0 + \beta_1 x_{n1} + \epsilon_n. \end{aligned} \tag{2.1}$$

Now if we set

$$\boldsymbol{y} = \begin{pmatrix} y_1 \\ y_2 \\ \vdots \\ y_n \end{pmatrix}, \ \boldsymbol{\epsilon} = \begin{pmatrix} \epsilon_1 \\ \epsilon_2 \\ \vdots \\ \epsilon_n \end{pmatrix}, \ X = \begin{pmatrix} 1 & x_{11} \\ 1 & x_{21} \\ \vdots & \vdots \\ 1 & x_{n1} \end{pmatrix}, \ \boldsymbol{\beta} = \begin{pmatrix} \beta_0 \\ \beta_1 \end{pmatrix}, \tag{2.2}$$

then it is easy to verify that (2.1) may be written as

$$\boldsymbol{y} = X\boldsymbol{\beta} + \boldsymbol{\epsilon}. \tag{2.3}$$

Now let us consider the case of more than one independent variable. Suppose we have k independent variables $x_1, \ldots, x_k$; then the regression model is

$$\begin{aligned} y_1 &= \beta_0 + \beta_1 x_{11} + \beta_2 x_{12} + \cdots + \beta_k x_{1k} + \epsilon_1 \\ &\cdots\cdots\cdots\cdots\cdots\cdots\cdots\cdots\cdots\cdots \\ &\cdots\cdots\cdots\cdots\cdots\cdots\cdots\cdots\cdots\cdots \\ y_n &= \beta_0 + \beta_1 x_{n1} + \beta_2 x_{n2} + \cdots + \beta_k x_{nk} + \epsilon_n. \end{aligned} \tag{2.4}$$

Letting

$$X = \begin{pmatrix} 1 & x_{11} & \dots & x_{1k} \\ \multicolumn{4}{c}{\dotfill} \\ \multicolumn{4}{c}{\dotfill} \\ 1 & x_{n1} & \dots & x_{nk} \end{pmatrix} \text{ and } \boldsymbol{\beta} = \begin{pmatrix} \beta_0 \\ \vdots \\ \beta_k \end{pmatrix}, \tag{2.5}$$

the model (2.4), called the multiple regression model, may also be written in the form (2.3).

The matrix X is called a *design matrix.* As in simple regression, the β_0 term in (2.4) is often called the constant term or the intercept. Note that the first column of X, i.e., the column of 1's, corresponds to it. If for some reason we do not want to keep β_0 in the model, we would delete this column. As mentioned in the last chapter, the last k elements in the ith row of X constitute the ith design point of the model and an observation y_i together with its corresponding design point constitute the ith case or data point.

GPA (max=4)	3.95	3.84	3.68	3.59	3.57	3.49	3.47	3.40	3.08
Verbal SAT (SATV)	74	76	66	76	76	66	71	71	57
Math. SAT (SATM)	79	71	75	74	70	67	73	79	76

EXHIBIT 2.1: Data on Grade Point Average and SAT Scores.
SOURCE: Dacey (1983).

Example 2.1

For the data presented in Exhibit 2.1, we may write a multiple regression model to predict GPA on the basis of SATV and SATM as

$$\begin{aligned} y_1 &= 3.95 = \beta_0 + \beta_1(74) + \beta_2(79) + \epsilon_1 \\ y_2 &= 3.84 = \beta_0 + \beta_1(76) + \beta_2(71) + \epsilon_2 \\ &\dots\dots\dots\dots\dots\dots\dots\dots\dots\dots \\ y_9 &= 3.08 = \beta_0 + \beta_1(57) + \beta_2(76) + \epsilon_9. \end{aligned}$$

The values of y and X would be:

$$\boldsymbol{y} = \begin{pmatrix} 3.95 \\ 3.84 \\ 3.68 \\ 3.59 \\ 3.57 \\ 3.49 \\ 3.47 \\ 3.40 \\ 3.08 \end{pmatrix} \text{ and } X = \begin{pmatrix} 1 & 74 & 79 \\ 1 & 76 & 71 \\ 1 & 66 & 75 \\ 1 & 76 & 74 \\ 1 & 76 & 70 \\ 1 & 66 & 67 \\ 1 & 71 & 73 \\ 1 & 71 & 79 \\ 1 & 57 & 76 \end{pmatrix}.$$

■

2.3 Least Squares Estimates

We obtain the least squares estimate of $\boldsymbol{\beta}$ in the multiple regression model by minimizing

$$\begin{aligned} \mathcal{S} &= \sum_{i=1}^{n}(y_i - \beta_0 - \beta_1 x_{1i} \cdots - \beta_k x_{ik})^2 = (\boldsymbol{y} - X\boldsymbol{\beta})'(\boldsymbol{y} - X\boldsymbol{\beta}) \\ &= \boldsymbol{y}'\boldsymbol{y} - \boldsymbol{\beta}'X'\boldsymbol{y} - \boldsymbol{y}'X\boldsymbol{\beta} + \boldsymbol{\beta}'X'X\boldsymbol{\beta} \\ &= \boldsymbol{y}'\boldsymbol{y} - 2\boldsymbol{\beta}'(X'\boldsymbol{y}) + \boldsymbol{\beta}'(X'X)\boldsymbol{\beta}, \end{aligned} \tag{2.6}$$

since $\boldsymbol{y}'X\boldsymbol{\beta}$, being a scalar, equals $\boldsymbol{\beta}'(X'\boldsymbol{y})$. In order to minimize (2.6), we could differentiate it with respect to each β_j and set the derivative equal to zero. Or, equivalently, we can do it more compactly using matrix differentiation (see Appendix 2A):

$$\partial \mathcal{S}/\partial \boldsymbol{\beta} = -2X'\boldsymbol{y} + 2X'X\boldsymbol{\beta}. \tag{2.7}$$

Setting (2.7) equal to zero and replacing $\boldsymbol{\beta}$ by $\boldsymbol{b}$, we see that the least squares estimate $\boldsymbol{b}$ of $\boldsymbol{\beta}$ is given by

$$(X'X)\boldsymbol{b} = X'\boldsymbol{y}. \tag{2.8}$$

That this indeed gives a minimum will be shown at the end of this section. If $X'X$ is non-singular, (2.8) has a unique solution:

$$\boldsymbol{b} = (X'X)^{-1}X'\boldsymbol{y}. \tag{2.9}$$

When $(X'X)$ is singular (2.8) can still be solved by using generalized inverses (defined in Appendix A, Section A.12, p. 278). We get from Corollary A.1:

$$\boldsymbol{b} = (X'X)^{-}X'\boldsymbol{y} = X^{-}\boldsymbol{y}. \tag{2.10}$$

While this estimate is not unique, it follows from Corollary A.1(iii) that $X(X'X)^{-}X'$ is unique, and consequently, $X\boldsymbol{b}$ is unique. It is a simple matter to see that if β_0 is absent and the column of 1's deleted from X, (2.8), (2.9) and (2.10) continue to hold.

As for the simple regression case, we define residuals e_i by

$$\boldsymbol{e} = \boldsymbol{y} - \hat{\boldsymbol{y}}, \tag{2.11}$$

where

$$\boldsymbol{e} = \begin{pmatrix} e_1 \\ \vdots \\ e_n \end{pmatrix}, \quad \hat{\boldsymbol{y}} = \begin{pmatrix} \hat{y}_1 \\ \vdots \\ \hat{y}_n \end{pmatrix} = X\boldsymbol{b} = X(X'X)^{-1}X'\boldsymbol{y} = H\boldsymbol{y}, \tag{2.12}$$

and $H = X(X'X)^{-1}X'$. Let $M = I - H$. Then

$$MX = (I - H)X = X - X(X'X)^{-1}X'X = X - X = 0.$$

Using M and (2.3) we can express $\boldsymbol{e}$ in terms of $\boldsymbol{y}$ or $\boldsymbol{\epsilon}$ as follows:

$$\begin{aligned} \boldsymbol{e} &= \boldsymbol{y} - H\boldsymbol{y} = M\boldsymbol{y} \\ &= MX\boldsymbol{\beta} + M\boldsymbol{\epsilon} = M\boldsymbol{\epsilon}. \end{aligned} \tag{2.13}$$

Theorem 2.1 *The residuals are orthogonal to the predicted values as well as the design matrix X in the model $\boldsymbol{y} = X\boldsymbol{\beta} + \boldsymbol{\epsilon}$.*

PROOF: Since

$$X'\boldsymbol{e} = X'M\boldsymbol{\epsilon} = 0\boldsymbol{\epsilon} = \mathbf{o}, \tag{2.14}$$

the null vector, it follows that

$$\hat{\boldsymbol{y}}'\boldsymbol{e} = \boldsymbol{b}'X'\boldsymbol{e} = 0, \tag{2.15}$$

which proves the theorem. □

It follows from Theorem 2.1 that if β_0 is in the model, and consequently the first column of X is $\mathbf{1} = (1, \dots, 1)'$, then $\sum_{i=1}^{n} e_i = \mathbf{1}'\boldsymbol{e} = 0$.

To conclude this section we now show that the minimum of $\mathcal{S} = (\boldsymbol{y} - X\boldsymbol{\beta})'(\boldsymbol{y} - X\boldsymbol{\beta})$ is indeed attained at $\boldsymbol{b} = \boldsymbol{\beta}$. Note that, from Theorem 2.1,

$$(\boldsymbol{b} - \boldsymbol{\beta})'X'(\boldsymbol{y} - X\boldsymbol{b}) = (\boldsymbol{y} - X\boldsymbol{b})'X(\boldsymbol{b} - \boldsymbol{\beta}) = \boldsymbol{e}'X(\boldsymbol{b} - \boldsymbol{\beta}) = 0. \tag{2.16}$$

Hence,

$$\begin{aligned} \mathcal{S} &= (\boldsymbol{y} - X\boldsymbol{b} + X\boldsymbol{b} - X\boldsymbol{\beta})'(\boldsymbol{y} - X\boldsymbol{b} + X\boldsymbol{b} - X\boldsymbol{\beta}) \\ &= (\boldsymbol{y} - X\boldsymbol{b})'(\boldsymbol{y} - X\boldsymbol{b}) + (\boldsymbol{b} - \boldsymbol{\beta})'(X'X)(\boldsymbol{b} - \boldsymbol{\beta}). \end{aligned}$$

Both expressions in the last line are quadratic forms and hence positive, and the first of these does not depend on $\boldsymbol{\beta}$. Therefore, $\mathcal{S}$ attains its minimum at $\boldsymbol{b} = \boldsymbol{\beta}$.

2.4 Examples

Example 2.2
Several computer packages are available for computing $\boldsymbol{b}$. From Exhibit 2.3, which illustrates the result of applying such a program to the data of Exhibit 2.2 (not all the variables shown in Exhibit 2.2 have been used), we get the regression equation (with somewhat greater rounding than in Exhibit 2.3)

$$\begin{aligned} \text{PRICE} = {} & 18.48 + .018\,\text{FLR} + 4.03\,\text{RMS} - 7.75\,\text{BDR} \\ & + 2.20\,\text{BTH} + 1.37\,\text{GAR} + .257\,\text{LOT} + 7.09\,\text{FP} + 10.96\,\text{ST}. \end{aligned} \tag{2.17}$$

Price	BDR	FLR	FP	RMS	ST	LOT	TAX	BTH	CON	GAR	CDN	L1	L2
53	2	967	0	5	0	39	652	1.5	1	0.0	0	1	0
55	2	815	1	5	0	33	1000	1.0	1	2.0	1	1	0
56	3	900	0	5	1	35	897	1.5	1	1.0	0	1	0
58	3	1007	0	6	1	24	964	1.5	0	2.0	0	1	0
64	3	1100	1	7	0	50	1099	1.5	1	1.5	0	1	0
44	4	897	0	7	0	25	960	2.0	0	1.0	0	1	0
49	5	1400	0	8	0	30	678	1.0	0	1.0	1	1	0
70	3	2261	0	6	0	29	2700	1.0	0	2.0	0	1	0
72	4	1290	0	8	1	33	800	1.5	1	1.5	0	1	0
82	4	2104	0	9	0	40	1038	2.5	1	1.0	1	1	0
85	8	2240	1	12	1	50	1200	3.0	0	2.0	0	1	0
45	2	641	0	5	0	25	860	1.0	0	0.0	0	0	1
47	3	862	0	6	0	25	600	1.0	1	0.0	0	0	1
49	4	1043	0	7	0	30	676	1.5	0	0.0	0	0	1
56	4	1325	0	8	0	50	1287	1.5	0	0.0	0	0	1
60	2	782	0	5	1	25	834	1.0	0	0.0	0	0	1
62	3	1126	0	7	1	30	734	2.0	1	0.0	1	0	1
64	4	1226	0	8	0	37	551	2.0	0	2.0	0	0	1
66	2	929	1	5	0	30	1355	1.0	1	1.0	0	0	1
35	4	1137	0	7	0	25	561	1.5	0	0.0	0	0	0
38	3	743	0	6	0	25	489	1.0	1	0.0	0	0	0
43	3	596	0	5	0	50	752	1.0	0	0.0	0	0	0
46	2	803	0	5	0	27	774	1.0	1	0.0	1	0	0
46	2	696	0	4	0	30	440	2.0	1	1.0	0	0	0
50	2	691	0	6	0	30	549	1.0	0	2.0	1	0	0
65	3	1023	0	7	1	30	900	2.0	1	1.0	0	1	0

PRICE = Selling price of house in thousands of dollars
BDR = Number of bedrooms
FLR = Floor space in sq.ft. (computed from dimensions of each room and then augmented by 10%)
FP = Number of fireplaces
RMS = Number of rooms
ST = Storm windows (1 if present, 0 if absent)
LOT = Front footage of lot in feet
TAX = Annual taxes
BTH = Number of bathrooms
CON = Construction (0 if frame, 1 if brick)
GAR = Garage size (0 = no garage, 1= one-car garage, etc.)
CDN = Condition (1='needs work', 0 otherwise)
L1 = Location (L1=1 if property is in zone A, L1=0 otherwise)
L2 = Location (L2=1 if property is in zone B, L2=0 otherwise)

EXHIBIT 2.2: Data on House Prices
SOURCE: Ms. Terry Tasch of Long-Kogan Realty, Chicago.

From this equation we can estimate the selling price of a house with 1000 square feet of floor area, 8 rooms, 4 bedrooms, 2 baths, storm windows, no fireplaces, 40 foot frontage and a 1 car garage to be about \$65,000, as follows:

$$\begin{aligned} 18.48 + .018(1000) + 4.03(8) - 7.75(4) + 2.2(2) + \\ 1.37(1) + .257(40) + 7.09(0) + 10.96(1) = 64.73. \end{aligned} \tag{2.18}$$

We can also see that, according to (2.17), adding a space for an additional car in a garage would raise the price by about \$1370, every square foot increase in floor area would increase the price by about \$18, etc. In each case the assumption is that each of the above mentioned changes is marginal, i.e., nothing else changes. Another way of putting this is to say that the effect of adding an extra garage space is \$1370, other things being equal, i.e., for two houses which are otherwise identical (in terms of the other variables in the model) but with one having space for an extra car in the garage, the price difference would be estimated to be about \$1370.

If the reader is disturbed by the negative sign associated with BDR, he/she should note that the estimated loss of price occurs if we increase the number of bedrooms without increasing the number of rooms or floor area. If we also increased the number of rooms by one and added a bathroom and some floor area to account for these additions, then the estimate of price would go up. In a situation where there are several related variables, signs which at first glance would appear counter-intuitive are not uncommon. Here (as in many other of these cases) further thought shows that the sign may be plausible. In addition, the reader should bear in mind that these estimates are random variables, and even more importantly, that we may not have considered important variables and may in other ways have fallen far short of perfection. It is also too true that a perfect model is seldom possible!

We draw the reader's attention to the variable ST which takes only two values, 0 and 1, and as such is called a dichotomous variable. Such variables, which are frequently used to indicate the presence or absence of an attribute, are also called dummy variables or indicator variables and fall in the general category of qualitative variables. The interpretation of the coefficients of such variables is about the same as for any other coefficient. We shall consider indicator variables further in Chapter 4. Suffice it to say for the moment that, that other features being equal, the presence of storm windows seems to enhance prices by a whopping ten thousand dollars! ■

Example 2.3 (Continuation of Example 1.1, Page 2)

As another example consider the data of Section 1.2 on speed of vehicles on a road and density of vehicles. In that section we fitted an equation of the form

$$(\text{Speed})^{1/2} = \beta_0 + \beta_1(\text{density}) + \beta_2(\text{density})^2 + \epsilon. \tag{2.19}$$

Variable	b_j	s.e.(b_j)	$t(b_j)$	$P[\lvert t\rvert > \lvert t(b_j)\rvert]$
Intercept	18.48	5.310	3.48	.0029
FLR	0.018	.0032	5.46	.0001
RMS	4.026	1.637	2.46	.0249
BDR	-7.752	1.872	-4.14	.0007
BTH	2.205	2.646	.833	.4163
GAR	1.372	1.453	.944	.3584
LOT	0.257	0.137	1.88	.0775
FP	7.091	3.190	2.23	.0401
ST	10.96	2.323	4.72	.0002

$$R^2 = .9027 \quad R_a^2 = .8568 \quad s = 4.780$$

EXHIBIT 2.3: Result of Applying a Least Squares Package Program on House Price Data

Here all we needed to do was to create a second independent variable each value of which is the square of each density. ■

Example 2.4

Let us return to the data of Exhibit 2.1, p. 29. On applying the least squares one gets

$$\text{GPA} = 1.22 + 0.029 \text{ SATV} + 0.00044 \text{ SATM}.$$

This would seem to indicate that SAT mathematics scores have very little effect on GPA. But we must not forget that when judging the effect of one variable we must bear in mind what other variables are already in the model. With SATM alone we get

$$\text{GPA} = 3.55 + .0001 \text{ SATM}.$$

Thus, at least for these students, SAT mathematics scores are not very good predictors of GPA. ■

Users of some computer packages will notice that nowadays they provide, in addition to the estimates we have mentioned, a statistic called *standardized coefficients* or *betas*. These are the regression coefficients b_j divided by the sample standard deviation of the dependent variable values and multiplied by the sample standard deviation of the corresponding independent variable values. This obviously renders the coefficients unit free. No further mention will be made in this book of standardized coefficients.

2.5 Gauss-Markov Conditions

In order for estimates of $\boldsymbol{\beta}$ to have some desirable statistical properties, we need the following assumptions, called Gauss-Markov (G-M) conditions, which have already been introduced in Chapter 1:

$$\begin{aligned} \mathrm{E}(\epsilon_i) &= 0 && (2.20)\\ \mathrm{E}(\epsilon_i^2) &= \sigma^2 && (2.21)\\ \mathrm{E}(\epsilon_i\epsilon_j) &= 0 \text{ when } i \neq j, && (2.22)\end{aligned}$$

for all $i, j = 1, \ldots, n$. We can also write these conditions in matrix notation as

$$\mathrm{E}(\boldsymbol{\epsilon}) = \mathbf{o}, \ \ \mathrm{E}(\boldsymbol{\epsilon}\boldsymbol{\epsilon}') = \sigma^2 I. \qquad (2.23)$$

Note that $\mathbf{o}$ is the vector of zeros. We shall use these conditions repeatedly in the sequel.

Note that G-M conditions imply that

$$\mathrm{E}(\boldsymbol{y}) = X\boldsymbol{\beta} \qquad (2.24)$$

and

$$\mathrm{cov}(\boldsymbol{y}) = \mathrm{E}[(\boldsymbol{y} - X\boldsymbol{\beta})(y - X\boldsymbol{\beta})'] = \mathrm{E}(\boldsymbol{\epsilon}\boldsymbol{\epsilon}') = \sigma^2 I. \qquad (2.25)$$

It also follows that (see (2.13))

$$E[\boldsymbol{e}\boldsymbol{e}'] = ME[\boldsymbol{\epsilon}\boldsymbol{\epsilon}']M = \sigma^2 M \qquad (2.26)$$

since M is idempotent. Therefore,

$$\mathrm{var}\,(e_i) = \sigma^2 m_{ii} = \sigma^2[1 - h_{ii}] \qquad (2.27)$$

where m_{ij} and h_{ij} are the ijth elements of M and H respectively. Because a variance is non-negative and a covariance matrix is at least positive semi-definite, it follows that $h_{ii} \leq 1$ and M is at least positive semi-definite.

2.6 Mean and Variance of Estimates Under G-M Conditions

Because of (2.24)

$$\mathrm{E}(\boldsymbol{b}) = \mathrm{E}[(X'X)^{-1}X'\boldsymbol{y}] = (X'X)^{-1}X'X\boldsymbol{\beta} = \boldsymbol{\beta}. \qquad (2.28)$$

As discussed in Section B.1.1, p. 284, if for any parameter θ, its estimate t has the property that $\mathrm{E}(t) = \theta$, then t is an *unbiased* estimator of θ. Thus under G-M conditions, $\boldsymbol{b}$ is an unbiased estimator of $\boldsymbol{\beta}$. Note that we only used the first of the G-M conditions to prove this. Therefore violation of conditions (2.21) and (2.22) will not lead to bias.

Theorem 2.2 *Under the first G-M condition (2.20), the least squares estimator* $\boldsymbol{b}$ *is an unbiased estimator of* $\boldsymbol{\beta}$. *Further, under the G-M conditions (2.20)–(2.22),*

$$\text{cov}(\boldsymbol{b}) = \sigma^2(X'X)^{-1}.$$

PROOF: We have already established unbiasedness. Let $A = (X'X)^{-1}X'$, then $\boldsymbol{b} = A\boldsymbol{y}$ and we get, from (2.25),

$$\begin{aligned}\text{cov}(\boldsymbol{b}) &= A\,\text{cov}(\boldsymbol{y})A' = \sigma^2 AIA' = \sigma^2 AA' \\ &= \sigma^2(X'X)^{-1}X'X(X'X)^{-1} = \sigma^2(X'X)^{-1},\end{aligned} \tag{2.29}$$

which completes the proof. □

Corollary 2.1 *If* $\text{tr}[(X'X)^{-1}] \to 0$ *as* $n \to \infty$, *then the estimator* $\boldsymbol{b}$ *is a consistent estimator of* $\boldsymbol{\beta}$.

The proof follows from the fact that when $(X'X)^{-1} \to 0$, $\text{cov}(\boldsymbol{b}) \to 0$ as $n \to \infty$. See Appendix 2A at the end of this chapter for a definition of consistency.

From (2.13), $\text{E}(\boldsymbol{e}) = \text{E}(M\boldsymbol{\epsilon}) = 0$. Therefore, it follows from (2.27) that

$$\text{cov}(\boldsymbol{e}) = \text{E}[\boldsymbol{e}\boldsymbol{e}'] = \sigma^2 MM' = \sigma^2 M. \tag{2.30}$$

Writing the ith equation in (2.4) as

$$y_i = \boldsymbol{x}_i'\boldsymbol{\beta} + \epsilon_i,$$

where $\boldsymbol{x}_i' = (1, x_{i1}, \ldots, x_{ik})$, the predicted value of y_i can be defined as

$$\hat{y}_i = \boldsymbol{x}_i'\boldsymbol{b}.$$

Consequently the predicted value of $\boldsymbol{y}$ is

$$\hat{\boldsymbol{y}} = X\boldsymbol{b}.$$

From Theorem 2.2 we get, on using the results on the covariance of a random vector (Section B.2, Page 286),

$$\text{var}\,(\hat{y}_i) = \boldsymbol{x}_i'\,\text{cov}(\boldsymbol{b})\boldsymbol{x}_i = \sigma^2\boldsymbol{x}_i'(X'X)^{-1}\boldsymbol{x}_i = \sigma^2 h_{ii}$$

and

$$\text{cov}(\hat{\boldsymbol{y}}) = X\,\text{cov}(b)X' = \sigma^2 X(X'X)^{-1}X' = \sigma^2 H.$$

Obviously the first of the two results follows from the second. It also follows that $h_{ii} \geq 0$ and that H is at least positive semi-definite.

2.7 Estimation of σ^2

To use several of the formulæ of the last section, we need σ^2, which is usually not known and needs to be estimated. This can be done using the residuals e_i. Because $M = (m_{ij})$ is a symmetric idempotent matrix,

$$\sum_{i=1}^{n} e_i^2 = \boldsymbol{e}'\boldsymbol{e} = \boldsymbol{\epsilon}'M'M\boldsymbol{\epsilon} = \boldsymbol{\epsilon}'M\boldsymbol{\epsilon} = \sum_{i=1}^{n} m_{ii}\epsilon_i^2 + \sum_{\substack{i,j=1 \\ i\neq j}}^{n} m_{ij}\epsilon_i\epsilon_j. \tag{2.31}$$

The quantity $\sum_{i=1}^{n} e_i^2$ is often called the *residual sum of squares* and is denoted as *RSS*. It follows from (2.31) that

$$\begin{aligned} \mathrm{E}(\sum_{i=1}^{n} e_i^2) &= \sum_{i=1}^{n} m_{ii}\,\mathrm{E}(\epsilon_i^2) + \sum_{\substack{i,j=1 \\ i\neq j}}^{n} m_{ij}\,\mathrm{E}(\epsilon_i\epsilon_j) \\ &= \sigma^2 \sum_{i=1}^{n} m_{ii} = \sigma^2\,\mathrm{tr}M = (n-k-1)\sigma^2 \end{aligned}$$

when an intercept is present and there are k independent variables, since then $\mathrm{tr}M = \mathrm{tr}I_n - \mathrm{tr}H = n-k-1$ — see (A.10) on p. 278. Therefore, if we let

$$s^2 = \sum_{i=1}^{n} e_i^2/(n-k-1) \tag{2.32}$$

we see that s^2 is an unbiased estimate of σ^2. When an intercept term is absent and there are k independent variables

$$s^2 = \sum_{i=1}^{n} e_i^2/(n-k) \tag{2.33}$$

is an unbiased estimator of σ^2. The divisor $n-k$ in the last formula and $n-k-1$ in (2.32) are the degrees of freedom.

Also, as shown in Appendix 2A, $s^2 \to \sigma^2$ in probability as $n \to \infty$; that is, s^2 is a consistent estimator of σ^2. Thus, when σ^2 is not known, a consistent and unbiased estimator of $\mathrm{cov}(\boldsymbol{b})$ is given by

$$\widehat{\mathrm{cov}(\boldsymbol{b})} = s^2(X'X)^{-1} = G = (g_{ij}), \quad \text{say.} \tag{2.34}$$

The matrix

$$(g_{ij}/g_{ii}^{1/2}g_{jj}^{1/2}) \tag{2.35}$$

is an estimate of the correlation matrix of $\boldsymbol{b}$. Matrices (2.34) and (2.35) are available from typical least squares computer packages. We can summarize the above results in the following theorem:

Theorem 2.3 *Under G-M conditions, s^2 is an unbiased and consistent estimator of σ^2, and $s^2(X'X)^{-1}$ is a consistent and unbiased estimator of* $\text{cov}(\boldsymbol{b})$.

The residuals and the residual sum of squares play a very important role in regression analysis. As we have just seen, the residual sum of squares when divided by $n-k-1$ gives an unbiased estimator of σ^2. In fact, under the assumption of normality of observations this is the best unbiased estimator in the sense that it has uniformly minimum variance among all unbiased estimators which are quadratic functions of the y_i's (i.e., estimators of the form $\boldsymbol{y}'A\boldsymbol{y}$, where A is a symmetric matrix; note that $s^2 = \boldsymbol{y}'M\boldsymbol{y}/[n-k-1]$). This result holds for even more general distributions (see Rao, 1952, and Hsu, 1938).

As we shall see later in the book, residuals are used to detect the presence of outliers and influential points, check for normality of the data, detect the adequacy of the model, etc. In short, they are used to determine if it is reasonable to assume that the Gauss-Markov conditions are being met. Residuals can, obviously, be used to determine the quality of fit of the regression equation. In the next section we give such a measure of fit. But first, we present the following theorem, which shows the connection between the residual sum of squares, total sum of squares $\sum_{i=1}^n y_i^2$ and the predicted sum of squares $\sum_{i=1}^n \hat{y}_i^2$.

Theorem 2.4 *Let $\bar{y} = n^{-1}\sum_{i=1}^n y_i$. Then,*

$$\sum_{i=1}^n e_i^2 = \sum_{i=1}^n y_i^2 - \sum_{i=1}^n \hat{y}_i^2 = \left(\sum_{i=1}^n y_i^2 - n\bar{y}^2\right) - \left(\sum_{i=1}^n \hat{y}_i^2 - n\bar{y}^2\right).$$

PROOF: Since, from Theorem 2.1, $\hat{\boldsymbol{y}}'\boldsymbol{e} = \sum_{i=1}^n e_i\hat{y}_i = 0$,

$$\begin{aligned}\sum_{i=1}^n y_i^2 &= \sum_{i=1}^n (y_i - \hat{y}_i + \hat{y}_i)^2 \\ &= \sum_{i=1}^n (y_i - \hat{y}_i)^2 + \sum_{i=1}^n \hat{y}_i^2 + 2\sum_{i=1}^n (y_i - \hat{y}_i)\hat{y}_i \\ &= \sum_{i=1}^n e_i^2 + \sum_{i=1}^n \hat{y}_i^2 + 2\sum_{i=1}^n e_i\hat{y}_i = \sum_{i=1}^n e_i^2 + \sum_{i=1}^n \hat{y}_i^2.\end{aligned}$$

The second part of the theorem is obvious. □

Corollary 2.2 *If there is an intercept in the model, that is, if there is a constant term β_0 in the model, then*

$$\sum_{i=1}^n e_i^2 = \sum_{i=1}^n (y_i - \bar{y})^2 - \sum_{i=1}^n (\hat{y}_i - \bar{y})^2$$

PROOF: Since there is a constant term in the model, we have, from Theorem 2.1, $\mathbf{1}'\boldsymbol{e} = \boldsymbol{e}'\mathbf{1} = \sum_{i=1}^n e_i = 0$. Hence, $\sum_{i=1}^n y_i = \sum_{i=1}^n \hat{y}_i$. Therefore, in this case the mean of the observations is the same as the mean of the predicted values. Hence, from Theorem 2.4

$$\sum_{i=1}^n e_i^2 = \left(\sum_{i=1}^n y_i^2 - n\bar{y}^2\right) - \left(\sum_{i=1}^n \hat{y}_i^2 - n\bar{y}^2\right) = \sum_{i=1}^n (y_i - \bar{y})^2 - \sum_{i=1}^n (\hat{y}_i - \bar{y})^2,$$

which proves the corollary. □

2.8 Measures of Fit

The measure of fit R^2, introduced in Section 1.7, p. 13, and given by (1.26) when there is an intercept term and by (1.27) when there is no intercept, is also appropriate for multiple regression. Its square root is the sample correlation between y_i's and $\hat{y}_i$'s, i.e., in the case of a regression model with an intercept,

$$R = \sum_{i=1}^n (y_i - \bar{y})(\hat{y}_i - \bar{y}) / [\sum_{i=1}^n (y_i - \bar{y})^2 \sum_{i=1}^n (\hat{y}_i - \bar{y})^2]^{1/2} \tag{2.36}$$

since, as we seen in the proof of Corollary 2.2, $n^{-1}\sum_{i=1}^n y_i = n^{-1}\sum_{i=1}^n \hat{y}_i$.

To see that the square of (2.36) is indeed (1.26), notice that because, by (2.15), $\sum_{i=1}^n e_i \hat{y}_i = 0$,

$$\begin{aligned}\sum_{i=1}^n (y_i - \bar{y})(\hat{y}_i - \bar{y}) &= \sum_{i=1}^n (y_i - \hat{y}_i + \hat{y}_i - \bar{y})(\hat{y}_i - \bar{y}) \\ &= \sum_{i=1}^n (y_i - \hat{y}_i)(\hat{y}_i - \bar{y}) + \sum_{i=1}^n (\hat{y}_i - \bar{y})^2 \\ &= \sum_{i=1}^n e_i \hat{y}_i - \bar{y}\sum_{i=1}^n e_i + \sum_{i=1}^n (\hat{y}_i - \bar{y})^2 = \sum_{i=1}^n (\hat{y}_i - \bar{y})^2.\end{aligned}$$

Therefore,

$$R = [\sum_{i=1}^n (\hat{y}_i - \bar{y})^2]^{1/2} / [\sum_{i=1}^n (y_i - \bar{y})^2]^{1/2}$$

and, because of Corollary 2.2, we get

$$R^2 = 1 - \frac{\sum_{i=1}^n (y_i - \hat{y}_i)^2}{\sum_{i=1}^n (y_i - \bar{y})^2}. \tag{2.37}$$

From (2.37), on using Corollary 2.2 again, it follows that R^2 lies between 0 and 1, and, because R is nonnegative, $0 \le R \le 1$. It is shown in Appendix

2A.3 at the end of this chapter that R^2 is the sample multiple correlation between the dependent variable y and the independent variables $x_1, \ldots, x_k$ in the sense given in Section B.7 (Page 295) of Appendix B.

Some analysts prefer to use an *adjusted* R^2 denoted by R_a^2 and given by

$$R_a^2 = 1 - \frac{[\sum_{i=1}^n (y_i - \hat{y}_i)^2/(n-k-1)]}{[\sum_{i=1}^n (y_i - \bar{y})^2/(n-1)]} = 1 - s^2/[\sum_{i=1}^n (y_i - \bar{y})^2/(n-1)].$$

R_a^2 adjusts for the sample size, since it is often felt that small sample sizes tend to unduly inflate R^2. However, R_a^2 can take negative values.

Alternatively, s^2 can also be used as a measure of fit, smaller values of s^2 indicating a good fit. A rough practical use of s^2 stems from the fact that when the number of observations n is large, $4s$ is the approximate width of the 95 per cent confidence interval for a future observation. When we are primarily interested in prediction, this provides an excellent indication of the quality of fit. For example, in the house price example, if the regression model were to be used to estimate the selling price of a house not in the sample, we could be off by as much as ±$9500 at a 95 per cent level. Sometimes alternative estimates of σ are available (see Chapter 6). Then, s provides an excellent means of discovering how close our fit is to a theoretical ideal. In most cases one needs to consider both R^2 (or R_a^2) and s^2 in order to assess the quality of fit.

When the regression equation does not have an intercept, that is, when $\beta_0 = 0$, we define the square of the sample correlation between y_i's and $\hat{y}_i$'s as

$$R^2 = \left[\sum_{i=1}^n y_i \hat{y}_i\right]^2 \Big/ \left[(\sum_{i=1}^n y_i^2)(\sum_{i=1}^n \hat{y}_i^2)\right]. \tag{2.38}$$

Since

$$\sum_{i=1}^n y_i \hat{y}_i = \boldsymbol{y}'\hat{\boldsymbol{y}} = \boldsymbol{y}'X\boldsymbol{b} = \boldsymbol{y}'X(X'X)^{-1}X'\boldsymbol{y}$$
$$= \boldsymbol{y}'X(X'X)^{-1}(X'X)(X'X)^{-1}X'\boldsymbol{y} = \boldsymbol{b}'X'X\boldsymbol{b} = \hat{\boldsymbol{y}}'\hat{\boldsymbol{y}} = \sum_{i=1}^n \hat{y}_i^2$$

and, from Theorem 2.4,

$$\sum_{i=1}^n y_i^2 = \sum_{i=1}^n e_i^2 + \sum_{i=1}^n \hat{y}_i^2 = \sum_{i=1}^n (y_i - \hat{y}_i)^2 + \sum_{i=1}^n \hat{y}_i^2,$$

(2.38) becomes

$$(\sum_{i=1}^n \hat{y}_i^2)/(\sum_{i=1}^n y_i^2) = 1 - \frac{\sum_{i=1}^n (y_i - \hat{y}_i)^2}{\sum_{i=1}^n y_i^2}.$$

Note that here too $0 \le R \le 1$ and $0 \le R^2 \le 1$.

2.9 The Gauss-Markov Theorem

In most applications of regression we are interested in estimates of some linear function $L\boldsymbol{\beta}$ or $\boldsymbol{\ell}'\boldsymbol{\beta}$ of $\boldsymbol{\beta}$, where $\boldsymbol{\ell}$ is a vector and L is a matrix. Estimates of this type include the predicteds $\hat{y}_i$, the estimate $\hat{y}_0$ of a future observation, $\hat{\boldsymbol{y}}$ and even $\boldsymbol{b}$ itself. We first consider below $\boldsymbol{\ell}'\boldsymbol{\beta}$; the more general vector function is considered subsequently.

Although there may be several possible estimators, we shall confine ourselves to linear estimators — i.e., an estimator which is a linear function of $y_1, \ldots, y_n$, say $\boldsymbol{c}'\boldsymbol{y}$. We also require that these linear functions be unbiased estimators of $\boldsymbol{\ell}'\boldsymbol{\beta}$ and assume that such linear unbiased estimators for $\boldsymbol{\ell}'\boldsymbol{\beta}$ exist; $\boldsymbol{\ell}'\boldsymbol{\beta}$ is then called estimable.

In the following theorem we show that among all linear unbiased estimators, the least squares estimator $\boldsymbol{\ell}'\boldsymbol{b} = \boldsymbol{\ell}'(X'X)^{-1}X'\boldsymbol{y}$, which is also a linear function of $y_1, \ldots, y_n$ and which in (2.28) has already been shown to be unbiased, has the smallest variance. That is, $\mathrm{var}\,(\boldsymbol{\ell}'\boldsymbol{b}) \le \mathrm{var}\,(\boldsymbol{c}'\boldsymbol{y})$ for all $\boldsymbol{c}$ such that $\mathrm{E}(\boldsymbol{c}'\boldsymbol{y}) = \boldsymbol{\ell}'\boldsymbol{\beta}$. Such an estimator is called a best linear unbiased estimator (BLUE).

Theorem 2.5 (Gauss-Markov) *Let* $\boldsymbol{b} = (X'X)^{-1}X'\boldsymbol{y}$ *and* $\boldsymbol{y} = X\boldsymbol{\beta} + \boldsymbol{\epsilon}$. *Then under G-M conditions, the estimator* $\boldsymbol{\ell}'\boldsymbol{b}$ *of the estimable function* $\boldsymbol{\ell}'\boldsymbol{\beta}$ *is BLUE.*

PROOF: Let $\boldsymbol{c}'\boldsymbol{y}$ be another linear unbiased estimator of (estimable) $\boldsymbol{\ell}'\boldsymbol{\beta}$. Since $\boldsymbol{c}'\boldsymbol{y}$ is an unbiased estimator of $\boldsymbol{\ell}'\boldsymbol{\beta}$, $\boldsymbol{\ell}'\boldsymbol{\beta} = \mathrm{E}(\boldsymbol{c}'\boldsymbol{y}) = \boldsymbol{c}'X\boldsymbol{\beta}$ for all $\boldsymbol{\beta}$ and hence we have

$$\boldsymbol{c}'X = \boldsymbol{\ell}'. \tag{2.39}$$

Now,

$$\mathrm{var}\,(\boldsymbol{c}'\boldsymbol{y}) = \boldsymbol{c}'\,\mathrm{cov}(\boldsymbol{y})\boldsymbol{c} = \boldsymbol{c}'(\sigma^2 I)\boldsymbol{c} = \sigma^2\boldsymbol{c}'\boldsymbol{c},$$

and

$$\mathrm{var}\,(\boldsymbol{\ell}'\boldsymbol{b}) = \boldsymbol{\ell}'\,\mathrm{cov}(\boldsymbol{b})\boldsymbol{\ell} = \sigma^2\boldsymbol{\ell}'(X'X)^{-1}\boldsymbol{\ell} = \sigma^2\boldsymbol{c}'X(X'X)^{-1}X'\boldsymbol{c},$$

from (2.29) and (2.39). Therefore

$$\begin{aligned}\mathrm{var}\,(\boldsymbol{c}'\boldsymbol{y}) - \mathrm{var}\,(\boldsymbol{\ell}'\boldsymbol{b}) &= \sigma^2[\boldsymbol{c}'\boldsymbol{c} - \boldsymbol{c}'X(X'X)^{-1}X'\boldsymbol{c}] \\ &= \sigma^2\boldsymbol{c}'[I - X(X'X)^{-1}X']\boldsymbol{c} \ge 0,\end{aligned}$$

since $I - X(X'X)^{-1}X' = M$ is positive semi-definite (see end of Section 2.5). This proves the theorem. □

A slight generalization of the Gauss-Markov theorem is the following:

Theorem 2.6 *Under G-M conditions, the estimator* $L\boldsymbol{b}$ *of the estimable function* $L\boldsymbol{\beta}$ *is BLUE in the sense that*

$$\mathrm{cov}(C\boldsymbol{y}) - \mathrm{cov}(L\boldsymbol{b})$$

is positive semi-definite, where L is an arbitrary matrix and $C\boldsymbol{y}$ is another unbiased linear estimator of $L\boldsymbol{\beta}$.

This theorem implies that if we wish to estimate several (possibly) related linear functions of the β_j's, we cannot do better (in a BLUE sense) than use least squares estimates.

PROOF: As in the proof of the Gauss-Markov theorem, the unbiasedness of $C\boldsymbol{y}$ yields $L\boldsymbol{\beta} = C\,\mathrm{E}(\boldsymbol{y}) = CX\boldsymbol{\beta}$ for all $\boldsymbol{\beta}$, whence $L = CX$, and since $\mathrm{cov}(C\boldsymbol{y}) = \sigma^2 CC'$ and

$$\mathrm{cov}(L\boldsymbol{b}) = \sigma^2 L(X'X)^{-1}L' = \sigma^2 CX(X'X)^{-1}X'C',$$

it follows that

$$\mathrm{cov}(C\boldsymbol{y}) - \mathrm{cov}(L\boldsymbol{b}) = \sigma^2 C[I - X(X'X)^{-1}X']C',$$

which is positive semi-definite, since, as shown at the end of Section 2.5, the matrix $[I - X(X'X)^{-1}X'] = M$ is at least positive semi-definite. □

If, in addition to the Gauss-Markov conditions, we make the further assumption that y_i's are normally distributed, then it may be shown that $\boldsymbol{\ell}'\boldsymbol{b}$ has minimum variance in the entire class of unbiased estimators, and that s^2 has minimum variance among all unbiased estimators of σ^2 (see Rao, 1973, p. 319).

2.10 The Centered Model

Let $\bar{x}_j = n^{-1}\sum_{i=1}^n x_{ij}$ and $z_{ij} = x_{ij} - \bar{x}_j$ for all $i = 1, \ldots, n$ and $j = 1, \ldots, k$. Further, let Z be the matrix (z_{ij}) of the z_{ij}'s and set $\gamma_0 = \beta_0 + \beta_1\bar{x}_1 + \cdots + \beta_k\bar{x}_k$ and $\boldsymbol{\beta}_{(0)} = (\beta_1, \ldots, \beta_k)'$. Using this notation, the regression model (2.3) can be written as

$$\boldsymbol{y} = \gamma_0 \boldsymbol{1} + Z\boldsymbol{\beta}_{(0)} + \boldsymbol{\epsilon} = \begin{pmatrix} \boldsymbol{1} & Z \end{pmatrix} \begin{pmatrix} \gamma_0 \\ \boldsymbol{\beta}_{(0)} \end{pmatrix} + \boldsymbol{\epsilon}. \tag{2.40}$$

The model (2.40) is called a centered model or, more specifically, the centered version of (2.3). The two models (2.3) and (2.40) are equivalent. From γ_0 and $\boldsymbol{\beta}_{(0)}$ we can get $\boldsymbol{\beta}$ of (2.3) and *vice versa*. Viewing least squares estimation as a problem in minimization, we see that this correspondence extends to the estimates as well. In particular, the estimate $\boldsymbol{b}_{(0)}$ of $\boldsymbol{\beta}_{(0)}$ consists exactly of the last k components of the estimate $\boldsymbol{b}$ of $\boldsymbol{\beta}$ of (2.3).

Since, obviously, $\sum_{i=1}^n z_{ij} = 0$ for all j, it follows that $\boldsymbol{1}'Z = \boldsymbol{o}'$ and $Z'\boldsymbol{1} = \boldsymbol{o}$. Hence, applying (2.9) to the model (2.40), we see that the least squares estimates $\hat{\gamma}_0$ of γ_0 and $\boldsymbol{b}_0$ of $\boldsymbol{\beta}_{(0)}$ are given by

$$\begin{pmatrix} \hat{\gamma}_0 \\ \boldsymbol{b}_{(0)} \end{pmatrix} = \begin{pmatrix} n & \boldsymbol{o}' \\ \boldsymbol{o} & Z'Z \end{pmatrix}^{-1} \begin{pmatrix} \boldsymbol{1}' \\ Z' \end{pmatrix} \boldsymbol{y} = \begin{pmatrix} n^{-1} & \boldsymbol{o}' \\ \boldsymbol{o} & (Z'Z)^{-1} \end{pmatrix} \begin{pmatrix} n\bar{y} \\ Z'\boldsymbol{y} \end{pmatrix}.$$

Thus

$$\begin{aligned} \hat{\gamma}_0 &= \bar{y} \\ \text{and } \boldsymbol{b}_{(0)} &= (Z'Z)^{-1}Z'\boldsymbol{y} = (Z'Z)^{-1}Z'(\boldsymbol{y} - \mathbf{1}\bar{y}) \end{aligned} \tag{2.41}$$

since, as noted before, $Z'\mathbf{1} = \mathbf{o}$. Clearly, $b_0 = \bar{y} - b_1\bar{x}_1 - \cdots - b_k\bar{x}_k = \bar{y} - (\bar{x}_1, \ldots, \bar{x}_k)\boldsymbol{b}_{(0)}$. The covariance matrix of the least squares estimates γ_0 and $\boldsymbol{b}_{(0)}$ is

$$\operatorname{cov}\begin{pmatrix} \hat{\gamma}_0 \\ \boldsymbol{b}_{(0)} \end{pmatrix} = \sigma^2 \begin{pmatrix} n & \mathbf{o}' \\ \mathbf{o} & Z'Z \end{pmatrix}^{-1}.$$

Thus $\hat{\gamma}_0$ and $\boldsymbol{b}_{(0)}$ are uncorrelated, which implies that they would be independent when they are normal. This makes the centered model very convenient to use in tests involving the regression parameters $(\beta_1, \ldots, \beta_k)$, as we shall see in the next chapter.

In fact, often it is even more convenient to center the y_i's as well and obtain the least squares estimates from the model

$$y_{(0)i} = y_i - \bar{y} = \beta_1(x_{i1} - \bar{x}_1) + \cdots + \beta_k(x_{ik} - \bar{x}_k) + (\epsilon_i - \bar{\epsilon}) \tag{2.42}$$

where $\bar{\epsilon} = n^{-1}\sum_{i=1}^{n} \epsilon_i$ and $i = 1, \ldots, n$. Although, since the covariance matrix of $(\epsilon_1 - \bar{\epsilon}, \ldots, \epsilon_n - \bar{\epsilon})$ is $I - n^{-1}\mathbf{1}\mathbf{1}'$, the $\epsilon_i - \bar{\epsilon}$'s are not uncorrelated, it turns out that *in this case* proceeding as if the $(\epsilon_i - \bar{\epsilon})$'s were uncorrelated and applying the least squares formula (2.9) yields estimates that are BLUE (see Exercise 7.2, p. 146).

It is easily verified that the residual sum of squares for this estimate is

$$\boldsymbol{y}'\left[I - \begin{pmatrix} \mathbf{1} & Z \end{pmatrix} \begin{pmatrix} n & \mathbf{o}' \\ \mathbf{o} & Z'Z \end{pmatrix}^{-1} \begin{pmatrix} \mathbf{1}' \\ Z' \end{pmatrix}\right]\boldsymbol{y}$$

which simplifies to

$$\begin{aligned} &\boldsymbol{y}'[I - n^{-1}\mathbf{1}\mathbf{1}' - Z(Z'Z)^{-1}Z']\boldsymbol{y} \\ &\quad = [\boldsymbol{y}'\boldsymbol{y} - n\bar{y}^2] - \boldsymbol{y}'Z(Z'Z)^{-1}Z'\boldsymbol{y}. \end{aligned} \tag{2.43}$$

From (2.31) and (2.13) it follows that the residual sum of squares for the uncentered model is

$$\boldsymbol{y}'M\boldsymbol{y} = \boldsymbol{y}'\boldsymbol{y} - \boldsymbol{y}'X(X'X)^{-1}X'\boldsymbol{y}.$$

Since the residual sum of squares given by the two methods must be the same, we get

$$\boldsymbol{y}'X(X'X)^{-1}X'\boldsymbol{y} = \boldsymbol{y}'Z(Z'Z)^{-1}Z'\boldsymbol{y} + n\bar{y}^2. \tag{2.44}$$

Hence

$$\boldsymbol{y}'M\boldsymbol{y} = (\boldsymbol{y}'\boldsymbol{y} - n\bar{y}^2) - \boldsymbol{y}'Z(Z'Z)^{-1}Z'\boldsymbol{y}. \tag{2.45}$$

The first term in the last line of (2.43) is called the *corrected total sum of squares* and the second term is called the *sum of squares due to* $\beta_1, \ldots, \beta_k$. It is also sometimes called the *sum of squares due to regression* or the *model sum of squares*. The second term on the right side of (2.44) may be called the *sum of squares due to the intercept* or the *sum of squares due to* β_0.

2.11 Centering and Scaling

Let $s_{jj} = \sum_{i=1}^{n} z_{ij}^2$ and $D_{(0s)} = \text{diag}\,(s_{11}^{1/2}, \ldots, s_{kk}^{1/2})$. Define

$$\boldsymbol{\delta} = D_{(0s)}\boldsymbol{\beta}_{(0)} \text{ and } Z_{(s)} = ZD_{(0s)}^{-1}.$$

It is easily verified that the model (2.40) is equivalent to

$$\boldsymbol{y} = \begin{pmatrix} \mathbf{1} & Z_{(s)} \end{pmatrix} \begin{pmatrix} \gamma_0 \\ \boldsymbol{\delta} \end{pmatrix} + \boldsymbol{\epsilon}$$

which is called a *centered and scaled* model or the *centered and scaled version* of (2.3). It might be mentioned that if instead we had replaced Z and z_{ij}'s uniformly by X and x_{ij}'s, we would have obtained simply a *scaled* model. Notice that the diagonal elements of $\mathcal{R} = Z_{(s)}'Z_{(s)} = D_{(0s)}^{-1}Z'ZD_{(0s)}^{-1}$ are ones and the off-diagonal elements are the sample correlation coefficients (Section B.1.3, p. 285) between columns of X. Consequently $Z_{(s)}'Z_{(s)}$ is sometimes called the correlation matrix of the independent variables.

The least squares estimate of γ_0 in the centered and scaled model is $\bar{y}$ and that of $\boldsymbol{\delta}$ is

$$\boldsymbol{d} = [Z_{(s)}'Z_{(s)}]^{-1}Z_{(s)}'\boldsymbol{y}.$$

As in the last section it may be shown that $\boldsymbol{d}$ remains unchanged if we center the y_i's as well, and we could obtain estimates that are BLUE from the model

$$\boldsymbol{y}_{(0)} = Z_{(s)}\boldsymbol{\delta} + \boldsymbol{\epsilon}_{(0)} \tag{2.46}$$

where $\boldsymbol{y}_{(0)} = \boldsymbol{y} - \mathbf{1}\bar{y}$ and $\boldsymbol{\epsilon}_{(0)} = \boldsymbol{\epsilon} - \mathbf{1}\bar{\epsilon}$, by applying least squares and treating the components $\epsilon_{(0)i}$'s of $\boldsymbol{\epsilon}_{(0)}$ as if they were independently distributed. The covariance matrix of $\boldsymbol{d}$ is

$$\text{cov}(\boldsymbol{d}) = \sigma^2[Z_{(s)}'Z_{(s)}]^{-1}$$

and, as in Section 2.10, $\hat{\gamma}_0$ and $\boldsymbol{d}$ are uncorrelated.

2.12 *Constrained Least Squares

In this section we consider the case of computing least squares estimates when they are subject to a linear equality constraint of the form

$$C\boldsymbol{\beta} - \boldsymbol{d} = 0 \tag{2.47}$$

where C is an $m \times (k+1)$ matrix of rank m with $m < (k+1)$. For inequality constraints, see Judge and Takayama (1966).

We still minimize $\mathcal{S} = (\boldsymbol{y} - X\boldsymbol{\beta})'(y - X\boldsymbol{\beta})$ but now we do it under the constraint (2.47). Therefore, using the Lagrange multiplier $\boldsymbol{\lambda} = (\lambda_1, \ldots, \lambda_m)'$ and differentiating with respect to $\boldsymbol{\lambda}$ and $\boldsymbol{\beta}$, we have

$$\frac{\partial}{\partial \boldsymbol{\lambda}}[\mathcal{S} + \boldsymbol{\lambda}'(\boldsymbol{d} - C\boldsymbol{\beta})] = \boldsymbol{d} - C\boldsymbol{\beta} \tag{2.48}$$

and

$$\frac{\partial}{\partial \boldsymbol{\beta}}[\mathcal{S} + \boldsymbol{\lambda}'(\boldsymbol{d} - C\boldsymbol{\beta})] = 2(X'X)\boldsymbol{\beta} - 2X'\boldsymbol{y} - C'\boldsymbol{\lambda}. \tag{2.49}$$

Replacing $\boldsymbol{\beta}$ and $\boldsymbol{\lambda}$ by $\hat{\boldsymbol{\beta}}$ and $\hat{\boldsymbol{\lambda}}$ in (2.48) and (2.49) and equating the resulting expressions to zero we get (assuming $X'X$ to be non-singular) $C\hat{\boldsymbol{\beta}} = \boldsymbol{d}$ and

$$\hat{\boldsymbol{\beta}} = (X'X)^{-1}[X'\boldsymbol{y} + \frac{1}{2}C'\hat{\boldsymbol{\lambda}}] = \boldsymbol{b} + \frac{1}{2}(X'X)^{-1}C'\hat{\boldsymbol{\lambda}}. \tag{2.50}$$

It follows that

$$\boldsymbol{d} = C\hat{\boldsymbol{\beta}} = C\boldsymbol{b} + \frac{1}{2}C(X'X)^{-1}C'\hat{\boldsymbol{\lambda}}. \tag{2.51}$$

Substituting the solution for $\hat{\boldsymbol{\lambda}}$ from (2.51) into (2.50), we get

$$\hat{\boldsymbol{\beta}} = \boldsymbol{b} + (X'X)^{-1}C'[C(X'X)^{-1}C']^{-1}(\boldsymbol{d} - C\boldsymbol{b}). \tag{2.52}$$

Appendix to Chapter 2

2A.1 Matrix Differentiation

Let $\boldsymbol{\theta} = (\theta_1, \ldots, \theta_k)'$ and let $f(\boldsymbol{\theta})$ be a real-valued function of k variables $\theta_1, \ldots, \theta_k$. Then, the partial derivatives of $f(\boldsymbol{\theta})$ with respect to $\boldsymbol{\theta}$ are

$$\frac{\partial f(\boldsymbol{\theta})}{\partial \boldsymbol{\theta}} = \begin{pmatrix} \partial f(\boldsymbol{\theta})/\partial \theta_1 \\ \vdots \\ \partial f(\boldsymbol{\theta})/\partial \theta_k \end{pmatrix}. \tag{2.53}$$

Lemma 2.1 *Let* $\mathbf{c} = (c_1, \cdots, c_k)'$, $\boldsymbol{\beta} = (\beta_1, \cdots, \beta_k)'$ *and* $f(\boldsymbol{\beta}) = \mathbf{c}'\boldsymbol{\beta}$. *Then*

$$\frac{\partial f(\boldsymbol{\beta})}{\partial \boldsymbol{\beta}} = \mathbf{c}.$$

Proof: Since $f(\boldsymbol{\beta}) = \mathbf{c}'\boldsymbol{\beta} = \sum_{i=1}^k c_i\beta_i$ we find that $\partial f(\boldsymbol{\beta})/\partial\beta_i = c_i$. Hence

$$\frac{\partial f(\boldsymbol{\beta})}{\partial \boldsymbol{\beta}} = \begin{pmatrix} c_1 \\ \vdots \\ c_k \end{pmatrix} = \mathbf{c}.$$

□

Lemma 2.2 *Let* $A = (a_{ij}) = (\boldsymbol{a}_1, \ldots, \boldsymbol{a}_k)$ *be a* $k \times k$ *symmetric matrix where* $\boldsymbol{a}_i' = (a_{i1}, \ldots, a_{ik})$, *the* i*th row of the matrix* A. *Let* $\boldsymbol{\beta} = (\beta_1, \cdots, \beta_k)'$ *and* $f(\boldsymbol{\beta}) = \boldsymbol{\beta}' A\boldsymbol{\beta}$. *Then*

$$\frac{\partial f(\boldsymbol{\beta})}{\partial \boldsymbol{\beta}} = 2A\boldsymbol{\beta}.$$

Proof: Since

$$\boldsymbol{\beta}' A\boldsymbol{\beta} = \sum_{i=1}^k a_{ii}\beta_i^2 + 2\sum_{i<j}^k a_{ij}\beta_i\beta_j,$$

we get

$$\frac{\partial(\boldsymbol{\beta}' A\boldsymbol{\beta})}{\partial \beta_1} = 2a_{11}\beta_1 + 2\sum_{j=2}^k a_{1j}\beta_j = 2\sum_{j=1}^k a_{1j}\beta_j = 2\boldsymbol{a}_1'\boldsymbol{\beta}.$$

Hence

$$\frac{\partial(\boldsymbol{\beta}' A\boldsymbol{\beta})}{\partial \boldsymbol{\beta}} = 2\begin{pmatrix} \boldsymbol{a}_1'\boldsymbol{\beta} \\ \vdots \\ \boldsymbol{a}_k'\boldsymbol{\beta} \end{pmatrix} = 2\begin{pmatrix} \boldsymbol{a}_1' \\ \vdots \\ \boldsymbol{a}_k' \end{pmatrix}\boldsymbol{\beta} = 2A\boldsymbol{\beta}.$$

□

2A.2 Consistency of s^2

Obviously, it is most desirable that, as the sample size n gets large, an estimate t of a parameter θ be close to θ. A sequence $z_1, z_2, \ldots$ of random variables is said to converge in probability to a constant c if for every $\epsilon > 0$,

$$P(|z_n - c| > \epsilon) \to 0 \text{ as } n \to \infty.$$

In the sequel we shall denote this fact as $z_n \xrightarrow{P} c$, or where the context makes the meaning obvious, as $z \xrightarrow{P} c$.

The estimate t is said to be a consistent estimator of θ if $t \xrightarrow{P} \theta$, i.e., if for every $\epsilon > 0$,

$$P[|t - \theta| > \epsilon] \to 0,$$

as the sample size goes to infinity. Typically, consistency is established by using Chebyshev's inequality (see Rao, 1973, p.95) or by using Markov's inequality:

$$P[|X| \geq a] \leq \frac{E|X|^r}{a^r}, \text{ for } r \geq 1.$$

We shall use the latter to show the consistency of s^2. From (2.32) and (2.13) (M and H are defined just below equation (2.12)),

$$s^2 = (n - k - 1)^{-1} \boldsymbol{e}'\boldsymbol{e} = (n - k - 1)^{-1} \boldsymbol{\epsilon}'[I - H]\boldsymbol{\epsilon}.$$

Because $\boldsymbol{\epsilon}' H \boldsymbol{\epsilon}$ is a scalar, its trace is itself. Therefore, we get, on using Property 3 of Section A.6, p. 271,

$$\begin{aligned} E(\boldsymbol{\epsilon}' H \boldsymbol{\epsilon}) &= E[\operatorname{tr}(\boldsymbol{\epsilon}' H \boldsymbol{\epsilon})] \\ &= E[\operatorname{tr}(H \boldsymbol{\epsilon}\boldsymbol{\epsilon}')] = \operatorname{tr}[H\, E(\boldsymbol{\epsilon}\boldsymbol{\epsilon}')] = \sigma^2 \operatorname{tr}(H) = \sigma^2(k+1) \end{aligned}$$

(see Example A.11, p. 277). Hence, we get from Markov's inequality

$$P[n^{-1} \boldsymbol{\epsilon}' H \boldsymbol{\epsilon} > \eta] \leq \frac{E(\boldsymbol{\epsilon}' H \boldsymbol{\epsilon})}{n\eta} = \frac{\sigma^2(k+1)}{n\eta} \to 0$$

as $n \to \infty$. From the law of large numbers $n^{-1} \boldsymbol{\epsilon}'\boldsymbol{\epsilon} \to \sigma^2$ in probability. Hence, since $n(n - k - 1)^{-1} \to 1$, $s^2 \to \sigma^2$ in probability.

A random vector $\boldsymbol{t}$ is said to be a consistent estimator of $\boldsymbol{\theta}$ if $\boldsymbol{x} \xrightarrow{P} \boldsymbol{\theta}$, i.e., if for every $\delta > 0$,

$$P[(\boldsymbol{x} - \boldsymbol{\theta})'(\boldsymbol{x} - \boldsymbol{\theta}) > \delta] \to 0.$$

2A.3 R^2 as Sample Multiple Correlation

We may write (2.37) in matrix notation as

$$R^2 = 1 - \boldsymbol{y}' M \boldsymbol{y} / [\boldsymbol{y}'\boldsymbol{y} - n\bar{y}]^2.$$

It, therefore, follows from (2.45) that

$$R^2 = \boldsymbol{y}'Z(Z'Z)^{-1}Z'\boldsymbol{y}/[\boldsymbol{y}'\boldsymbol{y} - n\bar{y}^2]$$

which is the sample correlation between y and $x_1, \ldots, x_k$ as defined in Section B.7 (p. 295).

Problems

Exercise 2.1: Show that $\text{var}(e_i) = \sigma^2 - \text{var}(\hat{y}_i)$.

Exercise 2.2: Suppose $\boldsymbol{y} = X\boldsymbol{\beta} + \boldsymbol{\epsilon}$, where $\text{E}(\boldsymbol{\epsilon}) = 0$, $\text{cov}(\boldsymbol{\epsilon}) = \sigma^2 I_n$, the matrix X of dimension $n \times k$ has rank $k \leq n$, and $\boldsymbol{\beta}$ is a k-vector of regression parameters. Suppose, further, that we wish to predict the '$(n+1)$st observation' y_{n+1} at $\boldsymbol{x}'_{n+1} = (x_{n+1,1}, \ldots, x_{n+1,k})$; i.e., $y_{n+1} = \boldsymbol{x}'_{n+1}\boldsymbol{\beta} + \epsilon_{n+1}$ where ϵ_{n+1} has the same distribution as the other ϵ_i's and is independent of them. The predictor based on the least squares estimate of $\boldsymbol{\beta}$ is given by $\hat{y}_{n+1} = \boldsymbol{x}'_{n+1}\boldsymbol{b}$, where $\boldsymbol{b} = (X'X)^{-1}X'\boldsymbol{y}$.

1. Show that $\hat{y}_{n+1}$ is a linear function of $y_1, \ldots, y_n$ such that $\text{E}(\hat{y}_{n+1} - y_{n+1}) = 0$.

2. Suppose $\tilde{y}_{n+1} = \boldsymbol{a}'\boldsymbol{y}$ is another predictor of y_{n+1} such that $\text{E}(\tilde{y}_{n+1} - y_{n+1}) = 0$. Show that $\boldsymbol{a}$ must satisfy $\boldsymbol{a}'X = \boldsymbol{x}'_{n+1}$.

3. Find $\text{var}(\hat{y}_{n+1})$ and $\text{var}(\tilde{y}_{n+1})$.

4. Show that $\text{var}(\hat{y}_{n+1}) \leq \text{var}(\tilde{y}_{n+1})$.

Exercise 2.3: Let $y_i = \boldsymbol{x}'_i\boldsymbol{\beta} + \epsilon_i$ with $i = 1, \ldots, n$ be a regression model where $\text{E}(\epsilon_i) = 0$, $\text{var}(\epsilon_i) = \sigma^2$ and $\text{cov}(\epsilon_i, \epsilon_j) = 0$ when $i \neq j$. Suppose $e_i = y_i - \hat{y}_i$, where $\hat{y}_i = \boldsymbol{x}'_i\boldsymbol{b}$ and $\boldsymbol{b}$ is the least squares estimator of $\boldsymbol{\beta}$. Let $X' = (\boldsymbol{x}_1, \cdots, \boldsymbol{x}_n)$. Show that the variance of e_i is $[1 - \boldsymbol{x}'_i(X'X)^{-1}\boldsymbol{x}_i]\sigma^2$.

Exercise 2.4: In the model of Exercise 2.3, show that the $\hat{y}_i$ is a linear unbiased estimator of $\boldsymbol{x}'_i\boldsymbol{\beta}$ (that is, $\hat{y}_i$ is a linear function of $y_1, \ldots, y_n$ and $\text{E}(\hat{y}_i) = \boldsymbol{x}'_i\boldsymbol{\beta}$). What is the variance of $\hat{y}_i$? Does there exist any other linear unbiased estimator of $\boldsymbol{x}'_i\boldsymbol{\beta}$ with a smaller variance than the estimator $\hat{y}_i$?

Exercise 2.5: Explain what would happen to the estimate b_j of β_j and to its standard error if we express the jth independent variable x_j in meters instead of millimeters.

What would happen to b_j and $\text{var}(b_j)$ in a model which includes an intercept term if we replaced all the values of $x_{1j}, \ldots, x_{nj}$ of x_j by numbers which were nearly constants?
[**Hint:** For the first portion of the problem multiply X by a diagonal matrix containing .001 in the (j, j)th position and 1's in other diagonal positions. For the second portion center the model.]

Exercise 2.6: Consider the models $\boldsymbol{y} = X\boldsymbol{\beta} + \boldsymbol{\epsilon}$ and $\boldsymbol{y}^* = X^*\boldsymbol{\beta} + \boldsymbol{\epsilon}^*$ where $\text{E}(\boldsymbol{\epsilon}) = 0$, $\text{cov}(\boldsymbol{\epsilon}) = \sigma^2 I$, $\boldsymbol{y}^* = \Gamma\boldsymbol{y}$, $X^* = \Gamma X$, $\boldsymbol{\epsilon}^* = \Gamma\boldsymbol{\epsilon}$ and Γ is a known $n \times n$ orthogonal matrix. Show that:

1. $\text{E}(\boldsymbol{\epsilon}^*) = 0$, $\text{cov}(\boldsymbol{\epsilon}^*) = \sigma^2 I$

2. $\boldsymbol{b} = \boldsymbol{b}^*$ and $s^{*2} = s^2$, where $\boldsymbol{b}$ and $\boldsymbol{b}^*$ are the least squares estimates of $\boldsymbol{\beta}$ and s^2 and s^{*2} are the estimates of σ^2 obtained from the two models.

Exercise 2.7: Let

$$y_{ij} = \boldsymbol{x}_i'\boldsymbol{\beta} + \epsilon_{ij}$$

for $i = 1, \ldots, q$ and $j = 1, \ldots, n$, where ϵ_{ij}'s are independently and identically distributed as $N(0, \sigma^2)$, $\boldsymbol{\beta}$ is a k-vector and $\boldsymbol{x}_i$'s are k-vectors of constants. Find the least squares estimates of $\boldsymbol{\beta}$. Using the residual sum of squares, give an unbiased estimate of σ^2.

Exercise 2.8: Show that the residual sum of squares $\boldsymbol{e}'\boldsymbol{e} = \boldsymbol{y}'\boldsymbol{y} - \boldsymbol{b}'X'X\boldsymbol{b} = \boldsymbol{y}'\boldsymbol{y} - \boldsymbol{b}'X'\boldsymbol{y} = \boldsymbol{y}'\boldsymbol{y} - \boldsymbol{b}'X'\hat{\boldsymbol{y}}$.

Exercise 2.9: Consider a house with 1200 square foot floor area, 6 rooms, 3 bedrooms, 2 baths and a 40 foot lot but no garage, fireplace or storm windows. Estimate its selling price using Exhibit 2.3. Estimate the increase in price if a garage were added. How about a fireplace and storm windows?

If the owner decides to add on a 300 square foot extension which adds a bedroom and a bathroom, how much would that add to its price? Now can you shed more light on the negative sign on the coefficient of bedrooms?

Assume that the cases given in Exhibit 2.2 were actually randomly drawn from a much larger data set. If the regression were run on the larger data set, would you expect s to increase, decrease or stay about the same? Give reasons. Did you make any assumptions about the model?

Exercise 2.10: Obtain the least squares estimate of $\boldsymbol{\beta}$ in the model $\boldsymbol{y} = X\boldsymbol{\beta} + \boldsymbol{\epsilon}$ from the following information:

$$(X'X)^{-1} = \begin{pmatrix} 3.08651 & -.0365176 & -.0397747 & -.051785 \\ -0.03652 & .0013832 & -.0000994 & .000332 \\ -0.03977 & -.0000994 & .0018121 & -.000102 \\ -0.05179 & .0003319 & -.0001022 & .002013 \end{pmatrix}$$

and

$$X'\boldsymbol{y} = \begin{pmatrix} 660.1 \\ 13878.9 \\ 17274.5 \\ 15706.1 \end{pmatrix}.$$

If an unbiased estimate of the variance of the components y_i of $\boldsymbol{y}$ is 50.5, what are unbiased estimates of the variances of the components of the least squares estimate you have obtained?

Exercise 2.11: Suppose that you need to fit the model $y_i = \beta_0 + \beta_1 x_{i1} + \beta_2 x_{i2} + \epsilon_i$, where $\mathrm{E}(\epsilon_i) = 0$, $\mathrm{E}(\epsilon_i\epsilon_j) = 0$ for $i \neq j$ and $\mathrm{E}(\epsilon_i^2) = \sigma^2$, to the following data:

y	x_1	x_2	y	x_1	x_2
-43.6	27	34	5.9	40	30
3.3	33	30	-4.5	15	17
-12.4	27	33	22.7	26	12
7.6	24	11	-14.4	22	21
11.4	31	16	-28.3	23	27

It turns out that

$$(X'X)^{-1} = \begin{pmatrix} 1.97015 & -.056231 & -.0157213 \\ -0.05623 & .002886 & -.0009141 \\ -0.01572 & -.000914 & .0017411 \end{pmatrix}$$

and

$$X'\boldsymbol{y} = \begin{pmatrix} -52.3 \\ -1076.3 \\ -2220.2 \end{pmatrix},$$

where X and $\boldsymbol{y}$ have their usual meanings.

1. Find the least squares estimator of $\boldsymbol{\beta} = (\beta_0, \beta_1, \beta_2)'$ and its covariance matrix.

2. Compute the estimate s^2 of σ^2.

3. Find the predicted value $\hat{y}_1$ and its variance.

4. What is the estimated variance of e_1, the residual corresponding to the first case?

Exercise 2.12: The variables in Exhibit 2.4 are: cars per person (AO), population in millions (POP), population density (DEN), per capita income in U.S. dollars (GDP), gasoline price in U.S. cents per liter (PR), tonnes of gasoline consumed per car per year (CON) and thousands of passenger-kilometers per person of bus and rail use (TR). Obtain a linear model expressing AO in terms of the other variables.

Quite obviously, bus and rail use (TR) is affected by car ownership. Does this cause a bias in the model you have just estimated? Does this cause a violation of any of the Gauss-Markov conditions?

Exercise 2.13: Strips of photographic film were identically exposed (same camera, lens, lens aperture, exposure time and subject, which was a grey card) and then developed for different times in different identical developers at different temperatures. The light transmission (y) through each strip was then measured and the values are shown in Exhibit 2.5. The units for y are unimportant for the present, but as a point of interest, it might be mentioned that an increase of 30 units implies a halving of the light passing

Country	AO	POP	DEN	GDP	PR	CON	TR
AUSTRIA	.27	7.5	89	7.7	49	1.1	2.6
BELGIUM	.30	9.8	323	9.8	59	1.0	1.6
CANADA	.42	23.5	2	8.7	17	2.8	.1
DENMARK	.28	5.1	119	11.0	56	1.2	1.9
FINLAND	.24	4.8	16	7.1	49	1.2	2.2
FRANCE	.33	53.3	97	8.8	61	1.0	1.5
GERMANY	.35	61.3	247	10.4	49	1.1	1.7
GREECE	.08	9.4	71	3.4	56	1.7	.7
ICELAND	.34	.2	2	9.8	57	1.2	2.0
IRELAND	.20	3.2	46	3.8	40	1.5	.3
ITALY	.30	56.7	188	4.6	61	.6	1.8
JAPAN	.18	114.9	309	8.5	49	1.2	3.5
LUXEMBURG	.43	.4	138	9.8	44	1.6	.8
NETHERLANDS	.30	13.9	412	9.4	56	1.0	1.5
NEW ZEALAND	.40	3.1	12	5.9	34	1.3	.2
NORWAY	.28	4.1	13	9.8	61	1.0	1.7
PORTUGAL	.10	9.8	107	1.8	68	.7	.9
SPAIN	.18	36.8	73	4.0	44	.8	1.3
SWEDEN	.34	8.3	18	10.6	42	1.3	1.7
SWITZERLAND	.32	6.3	153	13.3	56	1.3	2.0
TURKEY	.014	42.7	55	1.2	36	3.3	.1
U.K.	.27	55.8	229	5.5	35	1.2	1.6
U.S.A.	.53	218.2	23	9.7	17	2.7	.3
YUGOSLAVIA	.09	22.0	86	2.1	40	1.1	2.1

EXHIBIT 2.4: International Car Ownership Data
SOURCE: OECD (1982). Reproduced with permission of the OECD. All data are for 1978.

		Time		
		4	6.5	9
Temp.	60	43	68	97
	68	73	101	130
	76	94	127	155

EXHIBIT 2.5: Values of y for different combinations of development times (in minutes) and temperatures (degrees Fahrenheit)

through. Without using a computer, fit a model expressing y as a linear function of development time and developer temperature.
[**Hint:** Center the model. Incidentally, if a wider range of times or temperature were taken, the relationship would not have been linear.]

Exercise 2.14: Exhibit 2.6 gives data on actual voltage V_a and the corresponding value V_c of voltage computed from the measured power output (using light output from electronic flash). A definition of efficiency (E) is the ratio V_c/V_a. Obtain a model which expresses E as a quadratic polynomial in V_a (i.e., a model in V_a and V_a^2). Examine the residuals. Is the fit satisfactory?

V_a	354	321	291	264	240	219	200	183	167	153
V_c	354	315	281	250	223	199	177	158	150	125

EXHIBIT 2.6: Voltage Data
SOURCE: Armin Lehning, Speedotron Corporation.

Exercise 2.15: Exhibit 2.7, obtained from KRIHS (1985), gives data on the number of cars per person (AO), per capita GNP (GNP), average car price (CP) and gasoline price after taxes (OP) in South Korea from 1974 to 1983. GNP and car prices are in 1000 Korean wons, while gasoline prices are in wons per liter. Let D before a variable name denote first differences, e.g., $\text{DAO}_t = \text{AO}_{t+1} - \text{AO}_t$ where AO_t is the value of AO in the tth year. Use DAO as the dependent variable and estimate parameters of models which have:

1. GNP, CP and OP as the independent variables,
2. DGNP, DCP and DOP as the independent variables,
3. DGNP, DCP and OP as independent variables, and
4. DGNP, CP and OP as independent variables.

Examine the models you get along with their R^2. Which of the models makes the best intuitive sense?
[**Hint:** It seems intuitively reasonable that rapid increases in auto ownership rates would depend on *increases* in income, rather than income itself. The high R^2 for Model 1 is possibly due to the fact that DAO is, more or less, increasing over t, and GNP is monotonically increasing.]

Exercise 2.16: Using the data of Exercise 1.12, run a regression of log[ma] against log[pop] and d. Explain why the value of R^2 is so different from that obtained in Exercise 1.12. Do you feel the fit is worse now?

Exercise 2.17: Redo Exercise 1.11, p. 24, using an additional independent variable which is 1 for the first district but zero otherwise. Plot residuals and compare with the corresponding plot from Exercise 1.11.

Exercise 2.18: Exhibit 2.8 gives data on per capita output in Chinese yuan, number (SI) of workers in the factory, land area (SP) of the factory in square meters per worker, and investment (I) in yuans per worker for 17 factories in Shanghai.

1. Using least squares, fit a model expressing output in terms of the other variables.

Year	AO	GNP	CP	OP
1974	.0022	183	2322	189
1975	.0024	238	2729	206
1976	.0027	319	3069	206
1977	.0035	408	2763	190
1978	.0050	540	2414	199
1979	.0064	676	2440	233
1980	.0065	785	2430	630
1981	.0069	944	2631	740
1982	.0078	1036	3155	740
1983	.0095	1171	3200	660

EXHIBIT 2.7: Korean Auto Ownership Data

Output	SI	SP	I	Output	SI	SP	I
12090	56	840	10.54	18800	919	2750	14.74
11360	133	2040	11.11	28340	1081	3870	29.19
12930	256	2410	10.73	30750	1181	4240	21.21
12590	382	2760	14.29	29660	1217	2840	12.45
16680	408	2520	11.19	20030	1388	3420	17.33
23090	572	2950	14.03	17420	1489	3200	24.40
16390	646	2480	18.76	11960	1508	3060	28.26
16180	772	2270	13.53	15700	1754	2910	19.52
17940	805	4040	16.71				

EXHIBIT 2.8: Data on Per Capita Output of Workers in Shanghai

2. In addition to the variables in part 1 use SI^2 and SP times I and obtain another model.

3. Using the model of part 2, find the values of SP, SI and I that maximize per capita output.

This problem was suggested to one of the authors by Prof. Zhang Tingwei of Tongji University, Shanghai, who also provided the data.

Exercise 2.19: Exhibit 2.9 gives information on capital, labor and value added for each of three economic sectors: Food and kindred products (20), electrical and electronic machinery, equipment and supplies (36) and transportation equipment (37). The data were supplied by Dr. Philip Israelovich of the Federal Reserve Bank, who also suggested the exercise. For each sector:

1. Consider the model

$$V_t = \alpha K_t^{\beta_1} L_t^{\beta_2} \eta_t,$$

where the subscript t indicates year, V_t is value added, K_t is capital, L_t is labor and η_t is an error term, with $\text{E}[\log(\eta_t)] = 0$ and

Yr	Capital '20'	Capital '36'	Capital '37'	Labor '20'	Labor '36'	Labor '37'	Real Value Added '20'	Real Value Added '36'	Real Value Added '37'
72	243462	291610	1209188	708014	881231	1259142	6496.96	6713.75	11150.0
73	252402	314728	1330372	699470	960917	1371795	5587.34	7551.68	12853.6
74	246243	278746	1157371	697628	899144	1263084	5521.32	6776.40	10450.8
75	263639	264050	1070860	674830	739485	1118226	5890.64	5554.89	9318.3
76	276938	286152	1233475	685836	791485	1274345	6548.57	6589.67	12097.7
77	290910	286584	1355769	678440	832818	1369877	6744.80	7232.56	12844.8
78	295616	280025	1351667	667951	851178	1451595	6694.19	7417.01	13309.9
79	301929	279806	1326248	675147	848950	1328683	6541.68	7425.69	13402.3
80	307346	258823	1089545	658027	779393	1077207	6587.33	6410.91	8571.0
81	302224	264913	1111942	627551	757462	1056231	6746.77	6263.26	8739.7
82	288805	247491	988165	609204	664834	947502	7278.30	5718.46	8140.0
83	291094	246028	1069651	604601	664249	1057159	7514.78	5936.93	10958.4
84	285601	256971	1191677	601688	717273	1169442	7539.93	6659.30	10838.9
85	292026	248237	1246536	584288	678155	1195255	8332.65	6632.67	10030.5
86	294777	261943	1281262	571454	670927	1171664	8506.37	6651.02	10836.5

EXHIBIT 2.9: Data on Capital, Labor and Value Added for Three Sectors

var $[\log(\eta_t)]$ a constant. Assuming that the errors are independent, and taking logs of both sides of the above model, estimate β_1 and β_2.

2. The model given in 1 above is said to be of the Cobb-Douglas form. It is easier to interpret if $\beta_1 + \beta_2 = 1$. Estimate β_1 and β_2 under this constraint.

3. Sometimes the model

$$V_t = \alpha\gamma^t K_t^{\beta_1} L_t^{\beta_2} \eta_t$$

is considered where γ^t is assumed to account for technological development. Estimate β_1 and β_2 for this model.

4. Estimate β_1 and β_2 in the model in 3, under the constraint $\beta_1 + \beta_2 = 1$.

Exercise 2.20: The data set given in Exhibit 2.10 and in Exhibit 2.11 was compiled by Prof. Siim Soot, Department of Geography, University of Illinois at Chicago, from *Statistical Abstract of the United States, 1981, U.S. Bureau of the Census, Washington, D.C.* The variables are (the data are for 1980 except as noted):

POP	Total population (1000's)
UR	Per mil (per 10^{-3}) of population living in urban areas
MV	Per mil who moved between 1965 and 1970
BL	Number of blacks (1000's)
SP	Number of Spanish speaking (1000's)
AI	Number of Native Americans (100's)
IN	Number of inmates of all institutions (correctional, mental, TB, etc.) in 1970 (1000's)

PR Number of inmates of correctional institutions in 1970 (100's)
MH Homes and schools for the mentally handicapped (100's)
B Births per 1000
HT Death rate from heart disease per 100,000 residents
S Suicide rate, 1978, per 100,000
DI Death rate from diabetes, 1978, per 100,000
MA Marriage rate, per 10,000
D Divorce rate per 10,000
DR Physicians per 100,000
DN Dentists per 100,000
HS Per mil high school grads
CR Crime rate per 100,000 population
M Murder rate per 100,000 population
PI Prison rate (Federal and State) per 100,000 residents
RP % voting for Republican candidate in presidential election
VT % voting for presidential candidate among voting age population
PH Telephones per 100 (1979)
INC Per capita income expressed in 1972 dollars
PL Per mil of population below poverty level

and the cases represent each of the states of the United States.

1. Run a regression with M as the dependent variable. Among the independent variables include MA, D, PL, S, B, HT, UR, CR and HS. Explain the results you get. Some of the independent variables included may have little effect on M or may be essentially measuring the same phenomenon. If you detect any such variables, delete it and rerun the model. On the other hand, if you think that some important variable has been left out, add it to the model. Write a short report (to a non-technical audience) explaining the relationships you have found. Discuss the pro's and con's of including POP as an additional variable in this model.

2. Now try M as the dependent variable against INC, PL, VT and UR. Compare the results with those from part 1 above. Can you offer any explanations for what you observe?

3. We know that doctors like to locate in cities. It is also likely that they would tend to locate disproportionately in high income areas and possibly where there would be more business for them (e.g., where B and HT are high). Set up an appropriate model with DR as the dependent variable. Run a regression and write a report.

4. Choosing your own independent variables, run a regression with MA as the dependent variable. If you included D as one of the predictors, you would find it to be quite an important one. Can you explain why?

5. Run a regression with S as the dependent variable and UR, D, CR, PL, POP and MV as independent variables. Add and subtract variables as you think fit. Write a short report on the final model you get.

6. Try any other models that you think might yield interesting results.

State	POP	UR	MV	BL	SP	AI	IN	PR	MH	B	HT	S	DI
ME	1125	475	417	3	5	41	11	7	10	149	358	139	160
NH	921	522	450	4	6	14	9	5	10	145	306	150	170
VT	512	338	451	1	3	10	5	4	6	152	341	148	136
MA	5237	838	409	221	141	77	77	55	82	122	357	89	164
RI	948	870	426	28	20	29	11	5	10	128	402	115	224
CT	3107	788	425	217	124	45	34	36	45	124	321	103	143
NY	17757	846	404	2402	1659	387	218	236	223	134	396	96	177
NJ	7364	890	424	925	492	84	60	88	70	132	377	72	183
PA	11867	693	363	1048	154	95	124	127	146	135	414	119	211
OH	10797	733	452	1077	120	122	107	142	89	156	356	126	183
IN	5491	642	462	415	87	78	51	82	71	162	334	119	203
IL	11419	830	460	1675	636	163	119	122	106	164	379	100	153
MI	9259	707	447	1199	162	400	86	129	122	157	321	123	169
WI	4706	642	422	183	63	295	58	47	68	155	349	124	158
MN	4077	668	432	53	32	350	50	32	51	161	307	102	119
IA	2914	586	429	42	26	55	37	23	25	161	380	118	157
MO	4917	681	479	514	52	123	48	68	37	157	380	122	168
ND	654	488	436	3	4	202	9	3	13	179	313	98	153
SD	690	464	429	2	4	451	11	5	11	189	354	117	148
NE	1570	627	461	48	28	92	22	17	31	167	348	92	134
KS	2363	667	489	126	63	154	31	48	15	165	345	108	165
DE	596	707	460	96	10	13	5	7	2	153	334	146	252
MD	4216	803	485	958	65	80	39	89	48	140	308	109	162
VA	5346	660	497	1008	80	93	46	113	37	148	295	141	121
WV	1950	362	396	65	13	16	14	23	1	159	431	140	208
NC	5874	480	462	1316	57	646	48	108	52	150	313	117	144
SC	3119	541	453	948	33	58	23	51	11	173	295	113	154
GA	5464	623	514	1465	61	76	46	144	19	172	301	136	130
FL	9739	843	559	1342	858	193	61	162	59	137	410	177	174
KY	3661	508	464	259	27	36	27	56	13	167	379	125	173
TN	4591	604	474	726	34	51	34	67	29	156	324	126	121
AL	3891	600	455	996	33	76	31	57	25	166	304	105	170
MS	2520	473	446	887	25	62	16	26	12	189	320	96	167
AR	2285	516	494	373	18	94	20	21	10	167	364	105	172
LA	4203	683	442	1237	99	121	32	77	34	197	317	118	186
OK	3026	673	523	205	57	1695	37	60	24	170	364	135	151
TX	14228	796	525	1710	2986	401	116	216	132	190	267	127	127
MT	787	529	496	2	10	373	8	6	10	179	279	155	145
ID	945	540	513	3	37	105	6	6	5	221	241	134	116
WY	471	628	519	3	24	71	4	3	0	217	233	176	139
CO	2890	806	571	102	339	181	22	30	45	170	229	178	105
NM	1299	722	503	24	476	1048	6	15	8	206	168	171	140
AZ	2718	838	587	75	441	1529	11	33	12	191	265	193	120
UT	1461	844	461	9	60	193	7	9	1	301	198	128	128
NV	799	853	638	51	54	133	3	12	0	176	246	248	99
WA	4130	736	550	106	120	608	38	59	46	164	284	142	129
OR	2632	679	553	37	66	273	24	26	22	165	296	155	119
CA	23669	913	565	1819	4544	2013	214	499	137	167	278	163	107
AK	400	645	731	14	9	640	1	3	1	224	76	148	27
HI	965	865	541	17	71	28	4	4	8	192	156	118	127

EXHIBIT 2.10: Demographic Data for the 50 States of the U.S.

State	MA	D	DR	DN	HS	CR	M	PI	RP	VT	PH	INC	PL
ME	109	56	146	45	678	4368	28	61	456	648	54	4430	120
NH	102	59	159	53	703	4680	25	35	577	578	58	5105	79
VT	105	46	211	58	697	4988	22	67	444	583	52	4372	135
MA	78	30	258	71	723	6079	41	56	419	593	58	5660	71
RI	79	39	206	56	617	5933	44	65	372	590	57	5281	87
CT	82	45	242	73	703	5882	47	68	482	612	64	6552	67
NY	81	37	261	74	662	6912	127	123	467	480	54	5736	94
NJ	75	32	184	66	664	6401	69	76	520	551	66	6107	81
PA	80	34	183	55	648	3736	68	68	496	520	62	5273	97
OH	93	55	157	49	677	5431	81	125	515	554	56	5289	94
IN	110	77	126	43	670	4930	89	114	560	577	57	4995	81
IL	97	46	182	54	661	5275	106	94	496	578	66	5881	105
MI	97	48	154	53	686	6676	102	163	490	598	60	5562	91
WI	84	36	151	58	703	4799	29	85	479	677	55	5225	77
MN	91	37	185	62	724	4799	26	49	425	704	57	5436	83
IA	96	39	122	50	723	4747	22	86	513	629	59	5232	79
MO	109	57	158	48	641	5433	111	112	512	589	58	5021	120
ND	92	32	126	47	676	2964	12	28	642	651	63	4891	106
SD	130	39	102	43	689	3243	7	88	605	674	56	4362	131
NE	89	40	145	61	743	4305	44	89	655	568	61	5234	96
KS	105	54	150	46	731	5379	69	106	579	570	61	5580	80
DE	75	53	160	46	695	6777	69	183	472	549	64	5779	82
MD	111	41	257	59	693	6630	95	183	442	502	62	5846	77
VA	113	45	170	49	642	4620	86	161	530	480	53	5250	105
WV	94	53	133	39	533	2552	71	64	452	528	44	4360	151
NC	80	49	150	38	553	4640	106	244	493	439	53	4371	147
SC	182	47	134	36	571	5439	114	238	494	407	50	4061	172
GA	134	65	144	42	587	5604	138	219	409	417	55	4512	180
FL	117	79	188	50	648	8402	145	208	555	496	59	5028	144
KY	96	45	134	42	533	3434	88	99	491	500	48	4255	177
TN	135	68	158	48	549	4498	108	153	487	489	53	4315	158
AL	129	70	124	35	555	4934	132	149	488	490	50	4186	164
MS	112	56	106	32	523	3417	145	132	494	521	47	3677	261
AR	119	93	119	33	562	3811	92	128	481	516	47	4062	185
LA	103	38	149	40	583	5454	157	211	512	537	52	4727	193
OK	154	79	128	42	656	5053	100	151	605	528	57	5095	138
TX	129	69	152	42	645	6143	169	210	553	456	55	5336	152
MT	104	65	127	57	725	5024	40	94	568	652	56	4769	115
ID	148	71	108	55	715	4782	31	87	665	685	54	4502	103
WY	144	78	107	49	753	4986	62	113	626	542	56	6089	87
CO	118	60	199	61	781	7333	69	96	551	568	57	5603	91
NM	131	80	147	41	657	5979	131	106	549	514	46	4384	193
AZ	121	82	187	49	725	8171	103	160	606	452	53	4915	138
UT	122	56	164	64	802	5881	38	64	728	655	53	4274	85
NV	1474	168	138	49	757	8854	200	230	625	413	63	5999	88
WA	120	69	178	68	763	6915	51	106	497	580	56	5762	85
OR	87	70	177	69	755	6687	51	120	483	616	55	5208	89
CA	88	61	226	63	740	7833	145	98	527	495	61	6114	104
AK	123	86	118	56	796	6210	97	143	543	583	36	7141	67
HI	128	55	203	68	730	7482	87	65	425	436	47	5645	79

EXHIBIT 2.11: Demographic Data for the 50 States of the U.S.*ctd.*

CHAPTER 3

Tests and Confidence Regions

3.1 Introduction

Consider the example of house prices that we saw in the last chapter. It is reasonable to ask one or more of the following questions:

(a) Is the selling price affected by the number of rooms in a house, given that the other independent variables (e.g., floor area, lot size) remain the same?

(b) Suppose the realtor says that adding a garage will add \$5000 to the selling price of the house. Can this be true?

(c) Do lot size and floor area affect the price equally?

(d) Does either lot size or floor area have any effect on prices?

(e) Can it be true that storm windows add \$6000 and a garage adds \$4000 to the price of a house?

For some data sets we may even be interested in seeing whether all the independent variables together have any effect on the response variable. In the first part of this chapter we consider such testing problems. In the second part of the chapter we shall study the related issue of determining confidence intervals and regions. In the last section we return to testing.

3.2 Linear Hypothesis

To formulate the above questions as formal tests let us recall the regression model we used in the last chapter:

$$y_i = \beta_0 + \beta_1 x_{i1} + \cdots + \beta_8 x_{i8} + \epsilon_i \tag{3.1}$$

where i indexes the cases and $x_{i1}, \ldots, x_{i8}$ are as in Example 2.2 of Section 2.4, p. 31. The variable x_2 was the number of rooms and if it had no effect on the selling price, β_2 would be zero. Thus, corresponding to (a) above, our hypothesis would be

$$H : \beta_2 = 0 \tag{3.2}$$

which we would test against the alternative $A : \beta_2 \neq 0$. In other situations we might have chosen as alternative $\beta_2 > 0$ (or, for that matter, $\beta_2 < 0$).

For (b) we would test

$$H : \beta_5 = 5 \text{ versus } A : \beta_5 \neq 5. \tag{3.3}$$

For (c) we have

$$H : \beta_1 - \beta_6 = 0 \text{ versus } A : \beta_1 - \beta_6 \neq 0. \tag{3.4}$$

For (d) the hypothesis would be H:$\beta_1 = 0, \beta_6 = 0$ and for (e) it would be H:$\beta_5 = 4, \beta_8 = 6$. If we were interested in testing whether all the independent variables have any effect on the selling price, our hypothesis would be

$$H : \beta_1 = 0, \beta_2 = 0, \ldots, \beta_8 = 0$$

considered simultaneously, i.e.,

$$H : \boldsymbol{\beta}_{(0)} = \mathbf{o} \tag{3.5}$$

and we would test it against $A : \boldsymbol{\beta}_{(0)} \neq \mathbf{o}$, where $\boldsymbol{\beta}_{(0)} = (\beta_1, \ldots, \beta_8)'$. We may have been also interested in whether some of the factors, say the last three, simultaneously affect the dependent variable. Then we would test the hypothesis

$$H : (\beta_6, \beta_7, \beta_8)' = \mathbf{o} \text{ versus } A : (\beta_6, \beta_7, \beta_8)' \neq \mathbf{o}. \tag{3.6}$$

All these hypotheses, namely (3.2) through (3.6), are special cases of the *general linear hypothesis*

$$H : C\boldsymbol{\beta} - \boldsymbol{\gamma} = 0. \tag{3.7}$$

For example, if we choose

$$C = (0, 0, 1, 0, 0, 0, 0, 0, 0) \quad \text{and} \quad \gamma = 0, \tag{3.8}$$

we get (3.2). Similarly (3.3), (3.4), (3.5) and (3.6) correspond respectively to

$$C = (0, 0, 0, 0, 0, 1, 0, 0, 0) \quad \text{and} \quad \gamma = 5, \tag{3.9}$$

$$C = (0, 1, 0, 0, 0, 0, -1, 0, 0) \quad \text{and} \quad \gamma = 0, \tag{3.10}$$

$$C = (0 \;\; I_8) \quad \text{and} \quad \boldsymbol{\gamma} = \mathbf{o} \tag{3.11}$$

$$C = (0 \;\; I_3) \quad \text{and} \quad \boldsymbol{\gamma} = \mathbf{o}. \tag{3.12}$$

In the last two equations, C is a partitioned matrix consisting of the null matrix 0 and the r dimensional identity matrix I_r.

Under the assumption that ϵ_i's are identically and independently distributed as $N(0, \sigma^2)$ (in the sequel we shall abbreviate this statement to

'ϵ_i iid $N(0,\sigma^2)$'), the hypothesis $H : C\boldsymbol{\beta} = \boldsymbol{\gamma}$, where C is an $m \times (k+1)$ matrix of rank m with $m < (k+1)$, is rejected if

$$\frac{m^{-1}(C\boldsymbol{b}-\boldsymbol{\gamma})'[C(X'X)^{-1}C']^{-1}(C\boldsymbol{b}-\boldsymbol{\gamma})}{s^2} \geq F_{m,n-k-1,\alpha} \tag{3.13}$$

where $F_{m,n-k-1,\alpha}$ is the upper $100 \times \alpha$ per cent point of the F distribution with $(m, n-k-1)$ degrees of freedom (Section B.5, p. 292, of Appendix B; tables are given on pp. 322 *et seq.*), and, as before,

$$\boldsymbol{b} = (X'X)^{-1}X'\boldsymbol{y}, \quad \text{and} \quad s^2 = (n-k-1)^{-1}\boldsymbol{y}'[I-X(X'X)^{-1}X']\boldsymbol{y}. \tag{3.14}$$

Many statistical packages include this test which is shown in the next section to be a likelihood ratio test. The distribution of the left side of (3.13) is derived in Section 3.4.

3.3 *Likelihood Ratio Test

Assume that the Gauss-Markov conditions hold and the y_i's are normally distributed. That is, the ϵ_i's are independent and identically distributed as $N(0,\sigma^2)$, i.e., ϵ_i iid $N(0,\sigma^2)$. Then the probability density function of $y_1, \ldots, y_n$ is given by

$$(2\pi\sigma^2)^{-n/2}\exp[-\frac{1}{2\sigma^2}(\boldsymbol{y}-X\boldsymbol{\beta})'(\boldsymbol{y}-X\boldsymbol{\beta})]. \tag{3.15}$$

The same probability density function, when considered as a function of $\boldsymbol{\beta}$ and σ^2, given the observations $y_1, \ldots, y_n$, is called the likelihood function and is denoted by $\mathcal{L}(\boldsymbol{\beta},\sigma^2|\boldsymbol{y})$. The *maximum likelihood estimates* of $\boldsymbol{\beta}$ and σ^2 are obtained by maximizing $\mathcal{L}(\boldsymbol{\beta},\sigma^2|\boldsymbol{y})$ with respect to $\boldsymbol{\beta}$ and σ^2. Since $\log[z]$ is an increasing function of z, the same maximum likelihood estimates can be found by maximizing the logarithm of $\mathcal{L}$.

Since maximizing (3.15) with respect to $\boldsymbol{\beta}$ is equivalent to minimizing $(\boldsymbol{y}-X\boldsymbol{\beta})'(\boldsymbol{y}-X\boldsymbol{\beta})$, the maximum likelihood estimate of $\boldsymbol{\beta}$ is the same as the least squares estimate; i.e., it is $\boldsymbol{b} = (X'X)^{-1}X'\boldsymbol{y}$. The maximum likelihood estimate of σ^2, obtained by equating to zero the derivative of the log of the likelihood function with respect to σ^2 after substituting $\boldsymbol{b}$ for $\boldsymbol{\beta}$, is (see Exercise 3.5)

$$\frac{1}{n}(\boldsymbol{y}-X\boldsymbol{b})'(\boldsymbol{y}-X\boldsymbol{b}) = \frac{1}{n}\boldsymbol{e}'\boldsymbol{e}. \tag{3.16}$$

To obtain the maximum likelihood estimate of $\boldsymbol{\beta}$ under the constraint $C\boldsymbol{\beta} = \boldsymbol{\gamma}$, we need to maximize (3.15) subject to $C\boldsymbol{\beta} = \boldsymbol{\gamma}$. This is equivalent to minimizing $(\boldsymbol{y}-X\boldsymbol{\beta})'(\boldsymbol{y}-X\boldsymbol{\beta})$ subject to $C\boldsymbol{\beta} - \boldsymbol{\gamma} = \mathbf{o}$. Thus, the maximum likelihood estimate $\hat{\boldsymbol{\beta}}_H$ of $\boldsymbol{\beta}$ under the hypothesis H is the same as the constrained least squares estimate obtained in Section 2.12, p. 44, i.e.,

$$\hat{\boldsymbol{\beta}}_H = \boldsymbol{b} + (X'X)^{-1}C'[C(X'X)^{-1}C']^{-1}(C\boldsymbol{b} - \boldsymbol{\gamma}).$$

The maximum likelihood estimate of σ^2 is given by

$$n\,\hat{\sigma}_H^2 = (\boldsymbol{y} - X\hat{\boldsymbol{\beta}}_H)'(\boldsymbol{y} - X\hat{\boldsymbol{\beta}}_H). \tag{3.17}$$

Since

$$(\boldsymbol{y} - X\hat{\boldsymbol{\beta}}_H) = \boldsymbol{y} - X\boldsymbol{b} - X(X'X)^{-1}C'[C(X'X)^{-1}C']^{-1}(C\boldsymbol{b} - \boldsymbol{\gamma})$$

and

$$\begin{aligned}(\boldsymbol{y} - X\boldsymbol{b})'X(X'X)^{-1}C'[C(X'X)^{-1}C']^{-1}(C\boldsymbol{b} - \boldsymbol{\gamma})\\ = \boldsymbol{y}'[I - X(X'X)^{-1}X']X(X'X)^{-1}C'[C(X'X)^{-1}C']^{-1}(C\boldsymbol{b} - \boldsymbol{\gamma}) = 0,\end{aligned}$$

we get

$$n\,\hat{\sigma}_H^2 = (\boldsymbol{y} - X\boldsymbol{b})'(\boldsymbol{y} - X\boldsymbol{b}) + (C\boldsymbol{b} - \boldsymbol{\gamma})'[C(X'X)^{-1}C']^{-1}(C\boldsymbol{b} - \boldsymbol{\gamma}). \tag{3.18}$$

The *likelihood ratio test* statistic for testing the hypothesis $H : C\boldsymbol{\beta} - \boldsymbol{\gamma} = \mathbf{o}$ versus $A : C\boldsymbol{\beta} - \boldsymbol{\gamma} \neq \mathbf{o}$ is given by

$$\lambda = \frac{\max_A \mathcal{L}(\boldsymbol{\beta}, \sigma^2|\boldsymbol{y})}{\max_H \mathcal{L}(\boldsymbol{\beta}, \sigma^2|\boldsymbol{y})}$$

where $\max_A \mathcal{L}(\boldsymbol{\beta}, \sigma^2|\boldsymbol{y})$ and $\max_H \mathcal{L}(\boldsymbol{\beta}, \sigma^2|\boldsymbol{y})$ are the maximum values of the likelihood function under the alternative A and the hypothesis H respectively. These maximum values are obtained by substituting the maximum likelihood estimates of $\boldsymbol{\beta}$ and σ^2 under the two hypotheses into the respective likelihood function. Making these substitutions, we get

$$\begin{aligned}\max_A \mathcal{L}(\boldsymbol{\beta}, \sigma^2|\boldsymbol{y}) &= (2\pi\hat{\sigma}_A^2)^{-n/2}\exp[-n/2],\\ \max_H \mathcal{L}(\boldsymbol{\beta}, \sigma^2|\boldsymbol{y}) &= (2\pi\hat{\sigma}_H^2)^{-n/2}\exp[-n/2].\end{aligned}$$

Thus,

$$\begin{aligned}\lambda &= \left(\frac{\hat{\sigma}_H^2}{\hat{\sigma}_A^2}\right)^{n/2}\\ &= \left(\frac{(\boldsymbol{y} - X\boldsymbol{b})'(\boldsymbol{y} - X\boldsymbol{b}) + (C\boldsymbol{b} - \boldsymbol{\gamma})'[C(X'X)^{-1}C']^{-1}(C\boldsymbol{b} - \boldsymbol{\gamma})}{(\boldsymbol{y} - X\boldsymbol{b})'(\boldsymbol{y} - X\boldsymbol{b})}\right)^{n/2}\\ &= \left(1 + \frac{(C\boldsymbol{b} - \boldsymbol{\gamma})'[C(X'X)^{-1}C']^{-1}(C\boldsymbol{b} - \boldsymbol{\gamma})}{(\boldsymbol{y} - X\boldsymbol{b})'(\boldsymbol{y} - X\boldsymbol{b})}\right)^{n/2}\end{aligned}$$

and the hypothesis H is rejected for large values of λ, or equivalently for large values of (3.13).

This test has the following important and pleasant property. Let

$$\delta^2 = (C\boldsymbol{\beta} - \boldsymbol{\gamma})'[C(X'X)^{-1}C']^{-1}(C\boldsymbol{\beta} - \boldsymbol{\gamma})/\sigma^2.$$

Then $\delta = 0$ if and only if $C\boldsymbol{\beta} - \boldsymbol{\gamma} = 0$. It can be shown (using the results in the next section) that this likelihood ratio test is uniformly most powerful for testing the hypothesis $\delta = 0$ versus $\delta > 0$; that is, no other test has a larger probability of rejecting the hypothesis when it is not true.

3.4 *Distribution of Test Statistic

In this section we show that under the assumption that the ϵ_i's are iid $N(0, \sigma^2)$, (3.13) has the F distribution. The proof consists of showing that the numerator and denominator are each σ^2 times averaged chi-square variables and are independent (see Section B.5, p. 292).

Since $\boldsymbol{b} = (X'X)^{-1}X'\boldsymbol{y}$ is a linear function of $y_1, \ldots, y_n$, it follows from (2.29) that $\boldsymbol{b} \sim N_{k+1}(\boldsymbol{\beta}, \sigma^2(X'X)^{-1})$. Hence, for an $m \times (k+1)$ matrix C and a vector $\boldsymbol{\gamma}$,

$$C\boldsymbol{b} - \boldsymbol{\gamma} \sim N_m(C\boldsymbol{\beta} - \boldsymbol{\gamma}, \sigma^2 C(X'X)^{-1}C')$$

(see Theorem B.3, p. 288) and, therefore, under $H : C\boldsymbol{\beta} - \boldsymbol{\gamma} = \mathbf{o}$,

$$[C(X'X)^{-1}C']^{-1/2}(C\boldsymbol{b} - \boldsymbol{\gamma}) \sim N_m(\mathbf{o}, \sigma^2 I).$$

It follows that

$$Q = (C\boldsymbol{b} - \boldsymbol{\gamma})'[C(X'X)^{-1}C']^{-1}(C\boldsymbol{b} - \boldsymbol{\gamma}) \sim \sigma^2\chi^2_m \tag{3.19}$$

where χ^2_m denotes a chi-square distribution with m degrees of freedom.

Now consider the denominator. By (2.13), $\boldsymbol{e} = M\boldsymbol{\epsilon}$, which is a linear function of $\epsilon_1, \ldots, \epsilon_n$. Hence, $\boldsymbol{e} \sim N_n(\mathbf{o}, \sigma^2 M)$. We have seen in Section 2.3 that M is symmetric and idempotent with rank $n - k - 1$. Thus, from Example B.6, p. 291,

$$\boldsymbol{e}'\boldsymbol{e} = \boldsymbol{\epsilon}' M \boldsymbol{\epsilon} \sim \sigma^2\chi^2_{n-k-1}.$$

Under the hypothesis $H : C\boldsymbol{\beta} = \boldsymbol{\gamma}$,

$$\begin{aligned} C\boldsymbol{b} - \boldsymbol{\gamma} &= C\boldsymbol{b} - C\boldsymbol{\beta} = C(X'X)^{-1}[X'\boldsymbol{y} - X'X\boldsymbol{\beta}] \\ &= C(X'X)^{-1}X'[\boldsymbol{y} - X\boldsymbol{\beta}] = C(X'X)^{-1}X'\boldsymbol{\epsilon}. \end{aligned}$$

Thus, if $P = X(X'X)^{-1}C'[C(X'X)^{-1}C']^{-1}C(X'X)^{-1}X'$, we get from (3.19), $Q = \boldsymbol{\epsilon}'P\boldsymbol{\epsilon}$. But since $X'M = 0$, we have $PM = 0$. Hence, from Theorem B.6, given on Page 292, Q and $\boldsymbol{e}'\boldsymbol{e}$ are independently distributed and under H (see Section B.5, Page 292)

$$\frac{n-k-1}{m}\frac{Q}{\boldsymbol{e}'\boldsymbol{e}} \sim F_{m,n-k-1}. \tag{3.20}$$

Similarly, it can be shown that under the alternative hypothesis $A : C\boldsymbol{\beta} - \boldsymbol{\gamma} \neq 0$, the expression $(n-k-1)m^{-1}Q/\boldsymbol{e}'\boldsymbol{e}$ has a noncentral F distribution with $(m, n-k-1)$ degrees of freedom and noncentrality parameter

$$\delta^2 = (C\boldsymbol{\beta} - \boldsymbol{\gamma})'[C(X'X)^{-1}C']^{-1}(C\boldsymbol{\beta} - \boldsymbol{\gamma})\sigma^{-2}.$$

Therefore, the power of the test can be calculated for specified values of δ^2, m, and $n-k-1$. Tabulated values of non-central F distributions are available from statistical tables (see, for example, Tiku, 1975).

3.5 Two Special Cases

The test for the hypothesis $\boldsymbol{\beta}_{(0)} = (\beta_1, \ldots, \beta_k)' = \mathbf{o}$ against the alternative $\boldsymbol{\beta}_{(0)} \neq 0$ is called for so often that several packages carry it out routinely. Obviously, this hypothesis implies that our model is of little value. In order to test it, we could have used (3.13) with $C = (\mathbf{o}, I_k)$, but it is more convenient to consider the centered model (2.40). The likelihood ratio test would reject the hypothesis for large values of

$$k^{-1}\boldsymbol{b}_{(0)}'(Z'Z)\boldsymbol{b}_{(0)}/[(n-k-1)^{-1}(\boldsymbol{y} - \bar{y}\mathbf{1} - Z\boldsymbol{b}_{(0)})'(\boldsymbol{y} - \bar{y}\mathbf{1} - Z\boldsymbol{b}_{(0)})]. \quad (3.21)$$

The denominator is clearly the residual sum of squares (sometimes called the *error sum of squares*) divided by $n-k-1$. Using (2.41), the numerator can be seen to be

$$k^{-1}\boldsymbol{b}_{(0)}'(Z'Z)\boldsymbol{b}_{(0)} = k^{-1}\boldsymbol{y}'Z(Z'Z)^{-1}Z'\boldsymbol{y},$$

which is k^{-1} times the sum of squares due to $\boldsymbol{\beta}_{(0)}$ — see (2.43). Since, obviously,

$$[I - n^{-1}\mathbf{1}\mathbf{1}' - Z(Z'Z)^{-1}Z'][Z(Z'Z)^{-1}Z'] = 0,$$

it follows from (2.43) and Theorem B.6 of Appendix B that the error sum of squares and the sum of squares due to $\boldsymbol{\beta}_{(0)}$ are independent under normality. Therefore, (3.21) has an F distribution with k and $n-k-1$ degrees of freedom when $\boldsymbol{\epsilon} \sim N(\mathbf{o}, \sigma^2 I)$ and $\boldsymbol{\beta}_{(0)} = \mathbf{o}$.

Because of (2.43) the computation of the sums of squares is relatively easy. The corrected total sum of squares $\sum_{i=1}^{n}(y_i - \bar{y})^2$ is easy to compute since, by (2.43), it is the sum of the error sum of squares and the sum of squares due to $\boldsymbol{\beta}_{(0)}$. The error sum of squares is just the sum of squares of the residuals. The sum of squares due to $\boldsymbol{\beta}_{(0)}$ can then be obtained by subtraction. This computational ease made tableaus such as that in Exhibit 3.1 useful computational devices in pre-computer days. However, even today, several package programs print such tableaus routinely, usually with an additional column. This column gives the minimum significance

Source	DF	Sum of Squares	Mean Square	F-value
$\boldsymbol{\beta}_{(0)}$ Error	k $n-k-1$	$Q_1 = \boldsymbol{b}_{(0)}{}'Z'Z\boldsymbol{b}_{(0)}$ $Q_2 = \boldsymbol{e}'\boldsymbol{e}$	$Q_1^* = Q_1/k$ $Q_2^* = Q_2/(n-k-1)$	Q_1^*/Q_2^*
C. Total	$n-1$	$\sum_{i=1}^n (y_i - \bar{y})^2$		

EXHIBIT 3.1: Tableau for Testing if All Coefficients Except the Intercept Are Zero's

level at which the hypothesis could be rejected. This level is often referred to as a *p-value.*

The test for each β_j being zero is also routinely carried out. Here $m = 1$ and $C(X'X)^{-1}C' = CGC'$ reduces to g_{jj} where $G = (X'X)^{-1} = (g_{ij})$. Since $sg_{jj}^{1/2}$ is the standard error of b_j, (3.13) reduces to the square of

$$b_j/\text{s.e.}(b_j). \tag{3.22}$$

The expression (3.22), called the *t-value* of b_j, has a t distribution with $n-k-1$ degrees of freedom (Section B.5, p. 292).

As for the F test just mentioned, the minimum level at which the hypothesis $\beta_j = 0$ can be rejected is also printed. This level is called the *p-value of* b_j. Notice that the probability is for a two-tailed test. If we are testing against a *one-sided alternative* (e.g., $\beta_1 > 0$ or $\beta_1 < 0$ instead of $\beta_1 \neq 0$) we would halve the probability. If its p-value is less than α, we say b_j is *significant* at level α. Words like 'quite significant', 'very significant', etc., essentially refer to the size of the p-value — obviously the smaller it is the more significant is the corresponding b_j.

3.6 Examples

Example 3.1 (Continuation of Example 2.2, Page 31)

Now let us return to some of the specific questions posed at the beginning of this chapter. The test statistic (3.13) can usually be obtained from computer packages for most C and γ. As mentioned above, for the two cases where we wish to test if a particular β_j is zero and where we wish to test whether all the β_j's, except β_0, are zeros, these tests are routinely carried out (without being asked for). The tests for individual β_j's being 0 have been carried out in Exhibit 2.3, p. 34. For example, the t-value corresponding to the hypothesis $H : \beta_1 = 0$ is 5.461 and the p-value is .0001.

Computations to test $H : \boldsymbol{\beta}_{(0)} = \mathbf{o}$ versus $A : \boldsymbol{\beta}_{(0)} \neq \mathbf{o}$ are illustrated in Exhibit 3.2. We do not need to actually look at the tables since the

Source	DF	Sum of Squares	Mean Square	F-value	p-value
MODEL	8	3602.32	450.29	19.7	.0001
ERROR	17	388.488	22.852		
C. TOTAL	25	3990.81			

EXHIBIT 3.2: Tableau for Testing if All Coefficients Except the Intercept Are Zero's for House Price Example

probability of getting a value of F larger than the one obtained under H is also given. Here it is .0001, so that we can safely reject the hypothesis

$$\beta_1 = 0, \beta_2 = 0, \ldots, \beta_8 = 0.$$

Hypoth.	NUM/DEN	DF	F-Value	p-value
$\beta_5 = 5$	142.5/22.85	1,17	6.235	.0231
$\beta_6 - \beta_1 = 0$	69.60/22.85	1,17	3.046	.0990
$\beta_2 = 0$	138.2/22.85	1,17	6.049	.0249
$\beta_5 = 5$ & $\beta_8 = 5$	132.18/22.85	2,17	5.784	.0121

EXHIBIT 3.3: Some Tests of Hypotheses

In other cases, the computation of (3.13) has to be asked for. Exhibit 3.3 shows results for a number of tests on the model of Example 2.2. For each test the hypothesis H is given in the first column and the alternative is simply its negation. NUM and DEN are the numerator and denominator of (3.13); the other terms are obvious. ∎

3.7 Comparison of Regression Equations

Sometimes we are faced with the problem of testing if regression equations obtained from different data sets are the same against the alternative that they are different. Such problems typically arise when data on the same independent and dependent variables have been gathered for different time periods, different places or different groups of people. We illustrate below how such problems might be handled when there are two such equations; more than two equations can be handled in the same way.

Let the two models be

$$\boldsymbol{y}_1 = X_1\boldsymbol{\beta}_1 + \boldsymbol{\epsilon}_1 \tag{3.23}$$

and

$$y_2 = X_2\beta_2 + \epsilon_2 \tag{3.24}$$

respectively, where X_1 is $n_1 \times p$, X_2 is $n_2 \times p$ and β_1 and β_2 are p-dimensional vectors. Suppose we are willing to assume that the first r components of β_1 and β_2 are the same. There is no loss of generality here since we can permute the elements of β_1 and β_2 by permuting the columns of X_1 and X_2. Write

$$\beta_1 = \begin{pmatrix} \beta^{(1)} \\ \beta_1{}^{(2)} \end{pmatrix} \text{ and } \beta_2 = \begin{pmatrix} \beta^{(1)} \\ \beta_2{}^{(2)} \end{pmatrix}$$

where $\beta^{(1)}$ is r-dimensional and, obviously, $\beta_1{}^{(2)}$ and $\beta_2{}^{(2)}$ are of dimension $(p-r)$. Partition X_1 and X_2 in a corresponding way:

$$X_1 = \begin{pmatrix} X_1^{(1)} & X_1^{(2)} \end{pmatrix}, \quad X_2 = \begin{pmatrix} X_2^{(1)} & X_2^{(2)} \end{pmatrix}$$

where $X_1^{(1)}$ and $X_2^{(1)}$ have r columns each. Then (3.23) and (3.24) may be written as

$$y_1 = \begin{pmatrix} X_1{}^{(1)} & X_1{}^{(2)} \end{pmatrix} \begin{pmatrix} \beta^{(1)} \\ \beta_1{}^{(2)} \end{pmatrix} + \epsilon_1 = X_1{}^{(1)}\beta^{(1)} + X_1{}^{(2)}\beta_1{}^{(2)} + \epsilon_1 \tag{3.25}$$

and

$$y_2 = \begin{pmatrix} X_2{}^{(1)} & X_2{}^{(2)} \end{pmatrix} \begin{pmatrix} \beta^{(1)} \\ \beta_2{}^{(2)} \end{pmatrix} + \epsilon_2 = X_2{}^{(1)}\beta^{(1)} + X_2{}^{(2)}\beta_2{}^{(2)} + \epsilon_2. \tag{3.26}$$

Define

$$y = \begin{pmatrix} y_1 \\ y_2 \end{pmatrix}, \quad \epsilon = \begin{pmatrix} \epsilon_1 \\ \epsilon_2 \end{pmatrix},$$

$$X = \begin{pmatrix} X_1^{(1)} & X_1^{(2)} & 0 \\ X_2^{(1)} & 0 & X_2^{(2)} \end{pmatrix} \text{ and } \beta = \begin{pmatrix} \beta^{(1)} \\ \beta_1{}^{(2)} \\ \beta_2{}^{(2)} \end{pmatrix}.$$

Then it may readily be verified by multiplication that

$$y = X\beta + \epsilon \tag{3.27}$$

is a combination of (3.25) and (3.26) Then, the hypothesis $H : \beta_1{}^{(2)} - \beta_2{}^{(2)} = \mathbf{o}$ is equivalent to $H : C\beta = \mathbf{o}$ where

$$C = \begin{pmatrix} 0_{p-r,r} & I_{p-r} & -I_{p-r} \end{pmatrix} : (p-r) \times (2p-r),$$

with the subscripts of the 0 and I denoting matrix dimension. Now the test can be carried out using (3.13). provided, of course, that $\epsilon \sim N(\mathbf{o}, \sigma^2 I)$.

Example 3.2
The data in Exhibits 3.4 and 3.5 show world record times in seconds for running different distances for men and women. Previous work has shown that a model of the form

$$\log(\text{time}) = \beta_0 + \beta_1 \log(\text{distance}) \tag{3.28}$$

fits such data reasonably well. We would like to examine how the β's compare for men and women athletes.

Dist. (m.)	Time (secs.)	Dist. (m.)	Time (secs.)	Dist. (m.)	Time (secs.)
100	9.9	1000	136.0	10000	1650.8
200	19.8	1500	213.1	20000	3464.4
400	43.8	2000	296.2	25000	4495.6
800	103.7	3000	457.6	30000	5490.4
		5000	793.0		

EXHIBIT 3.4: Men's World Record Times for Running and Corresponding Distances
SOURCE: Encyclopædia Britannica, 15th Edition, 1974, Micropædia, IX, p. 485. Reproduced with permission from Encyclopædia Britannica, Inc.

Dist. (m.)	Time (secs.)	Dist. (m.)	Time (secs.)	Dist. (m.)	Time (secs.)
60	7.2	200	22.1	800	117.0
100	10.8	400	51.0	1500	241.4

EXHIBIT 3.5: Women's World Record Times for Running and Corresponding Distances
SOURCE: Encyclopædia Britannica, 15th Edition, 1974, Micropædia, IX, p. 487 et seq. Reproduced with permission from Encyclopædia Britannica, Inc.

Before embarking on the test it is perhaps interesting to examine each data set independently first. On running regressions for the two sets individually we get

$$\begin{aligned}\log(\text{time}) &= \underset{(.06)}{-2.823} + \underset{(.0078)}{1.112\log(\text{dist.})} \qquad (R^2 = .9995, s_1 = .05),\end{aligned}$$

and

$$\begin{aligned}\log(\text{time}) &= \underset{(.183)}{-2.696} + \underset{(.0317)}{1.112\log(\text{dist.})} \qquad (R^2 = .9968, s_2 = .087)\end{aligned}$$

for men and women respectively, where the parenthetic quantities under the coefficients are corresponding standard errors. Notice that the coefficients of log(distance) are surprisingly similar. (This phenomenon has been noticed before in both track and swimming — see Riegel, 1981.) Now let us test if $\boldsymbol{\beta}$'s for men and women are the same.

The design matrix X and the $\boldsymbol{y}$ vector that we need are

$$\boldsymbol{y} = \begin{pmatrix} \log(9.9) \\ \log(19.8) \\ \vdots \\ \log(5490.4) \\ \log(7.2) \\ \vdots \\ \log(241.4) \end{pmatrix}, \quad X = \begin{pmatrix} 1 & \log(100) & 0 & 0 \\ 1 & \log(200) & 0 & 0 \\ \vdots & \vdots & \vdots & \vdots \\ 1 & \log(30000) & 0 & 0 \\ 0 & 0 & 1 & \log(60) \\ \vdots & \vdots & \vdots & \vdots \\ 0 & 0 & 1 & \log(1500) \end{pmatrix}.$$

Writing $\boldsymbol{\beta}_1 = (\beta_{10}, \beta_{11})'$ and $\boldsymbol{\beta}_2 = (\beta_{20}, \beta_{21})$, we get as our vector of parameters $\boldsymbol{\beta} = (\beta_{10}, \beta_{11}, \beta_{20}, \beta_{21})'$. Our model is $\boldsymbol{y} = X\boldsymbol{\beta} + \boldsymbol{\epsilon}$ and least squares yields

$$\begin{aligned} b_{10} &= -2.823(.0770) & \qquad b_{20} &= -2.696(.1299) \\ b_{11} &= 1.112(.0097) & \qquad b_{21} &= 1.112(.0224), \end{aligned}$$

where standard errors are given within parentheses, $R^2 = .9997$ and $s = .062$. We wish to test if $\beta_{10} = \beta_{20}$ and $\beta_{11} = \beta_{21}$; i.e.,

$$H : C\boldsymbol{\beta} = 0 \text{ versus } A : C\boldsymbol{\beta} \neq 0$$

where

$$C = \begin{pmatrix} 1 & 0 & -1 & 0 \\ 0 & 1 & 0 & -1 \end{pmatrix} = (I_2, -I_2).$$

This test yields an F-value of .5293, which is not even significant at a 40 per cent level — therefore, we are unable to reject the hypotheses that $\beta_{10} = \beta_{20}$ and $\beta_{11} = \beta_{21}$.

Simply testing $\beta_{10} = \beta_{20}$ yields an $F = .7130$, which also makes it difficult to reject $\beta_{10} = \beta_{20}$. In testing $\beta_{11} = \beta_{21}$ against $\beta_{11} \neq \beta_{21}$ we get an F-value of .0001, which should come as no surprise and which, of course, is very strong evidence in *favor* of the hypothesis!

Finally note that, although track times for men and women are usually different, here we could not reject their inequality (probably because of the small size of the data set for women, which resulted in large standard errors). This serves to demonstrate that if we are unable to reject a hypothesis it does not mean that it is true! ■

3.8 Confidence Intervals and Regions

Confidence intervals (C.I.) for individual β_j's can be obtained in a straightforward way from the distribution of (3.22). Some other confidence intervals and confidence regions that are often needed are presented in the sections below, the first two of which contain simple generalizations of the material of Section 1.10.

3.8.1 C.I. for the Expectation of a Predicted Value

Let $\boldsymbol{x}_0' = (x_{00}, x_{01}, \ldots, x_{0k})$ represent a set of values of the independent variables, where $x_{00} = 1$ if an intercept is present. Then the predicted value of y at the point $\boldsymbol{x}_0$ is $\hat{y}_0 = \boldsymbol{x}_0'\boldsymbol{b}$. We can easily verify from (2.28) and (2.29) that under the Gauss-Markov conditions,

$$\begin{aligned} \mathrm{E}(\hat{y}_0) &= \boldsymbol{x}_0'\boldsymbol{\beta} \\ \operatorname{var}(\hat{y}_0) &= \boldsymbol{x}_0' \operatorname{cov}(\boldsymbol{b})\boldsymbol{x}_0 = \sigma^2[\boldsymbol{x}_0'(X'X)^{-1}\boldsymbol{x}_0]. \end{aligned} \tag{3.29}$$

Hence, if $y_1, \ldots, y_n$ are normally distributed,

$$\hat{y}_0 - \boldsymbol{x}_0'\boldsymbol{\beta} \sim N(0, \sigma^2[\boldsymbol{x}_0'(X'X)^{-1}\boldsymbol{x}_0]),$$

and it follows that

$$\frac{\hat{y}_0 - \boldsymbol{x}_0'\boldsymbol{\beta}}{s[\boldsymbol{x}_0'(X'X)^{-1}\boldsymbol{x}_0]^{1/2}}$$

has a Student's t distribution with $n-k-1$ degrees of freedom. Therefore, an $(1-\alpha) \times 100$ per cent C.I. for the mean predicted value $\boldsymbol{x}_0'\boldsymbol{\beta}$ is given by

$$\hat{y}_0 \pm t_{r,\alpha/2}\, s[\boldsymbol{x}_0'(X'X)^{-1}\boldsymbol{x}_0]^{1/2} \tag{3.30}$$

where $r = n-k-1$, and $t_{r,\alpha/2}$ is the upper $(1/2)\,\alpha \times 100$ per cent point of the t distribution with r degrees of freedom.

3.8.2 C.I. for a Future Observation

We have already seen future observations, in the simple regression case, in Section 1.10, p. 18. Here we examine the multiple regression case. Let y_0 denote a future observation at the point $\boldsymbol{x}_0$ of the independent variables, i.e., $y_0 = \boldsymbol{x}_0'\boldsymbol{\beta} + \epsilon_0$. Such a y_0 is estimated by $\hat{y}_0 = \boldsymbol{x}_0'\boldsymbol{b}$. Typically, y_0 is independent of the observations $y_1, \ldots, y_n$. Making this assumption, we can readily compute

$$\begin{aligned} \operatorname{var}(\hat{y}_0 - y_0) &= \operatorname{var}(y_0) + \operatorname{var}(\hat{y}_0) \\ &= \sigma^2 + \sigma^2\boldsymbol{x}_0'(X'X)^{-1}\boldsymbol{x}_0 = \sigma^2[1 + \boldsymbol{x}_0'(X'X)^{-1}\boldsymbol{x}_0] \end{aligned} \tag{3.31}$$

from (3.29). Since σ^2 can be estimated by s^2 and since under the Gauss-Markov conditions,

$$\mathrm{E}(\hat{y}_0 - y_0) = \mathrm{E}(\hat{y}_0) - \mathrm{E}(y_0) = \boldsymbol{x}_0'\boldsymbol{\beta} - \boldsymbol{x}_0'\boldsymbol{\beta} = 0,$$

it follows that

$$\frac{\hat{y}_0 - y_0}{s[1 + \boldsymbol{x}_0'(X'X)^{-1}\boldsymbol{x}_0]^{1/2}}$$

has a t distribution with $r = n - k - 1$ degrees of freedom. Hence a $(1 - \alpha) \times 100$ per cent C.I. for y_0 is given by

$$\hat{y}_0 \pm t_{r,\alpha/2} s[1 + \boldsymbol{x}_0'(X'X)^{-1}\boldsymbol{x}_0]^{1/2}.$$

Example 3.3 (Continuation of Example 2.2, Page 31)
A value of $\hat{y}_0$ corresponding to $\boldsymbol{x}_0' = (1, 1000, \ldots, 1)$ has been computed in (2.18) on page 33. The variance of $\hat{y}_0$ and of $y_0 - \hat{y}_0$ can be obtained from most (not all) packages, from which confidence intervals can be obtained. In SAS[1] PROC REG (SAS, 1985b), approximate 95 per cent confidence intervals for $\hat{y}_0$ and y_0 can be found, using the CLM and the CLI options, by appending the appropriate independent variable values to the data and declaring the dependent variable value as missing. For our example the intervals are approximately (59.1, 70.3) and (52.9, 76.5). ■

3.8.3 *Confidence Region for Regression Parameters

Sometimes we wish to consider simultaneously several β_j's or several linear combinations of β_j's. In such cases, *confidence regions* can be found in a fairly straightforward way from (3.20) — assuming, of course, that the Gauss-Markov conditions hold and the ϵ_i's are normally distributed. Suppose we wish to find a confidence region for $\boldsymbol{\beta}$. Then setting $C = I_{k+1}$ in (3.20) and (3.19), and noting that $s^2 = (n - k - 1)^{-1}\boldsymbol{e}'\boldsymbol{e}$, we get

$$(k+1)^{-1}s^{-2}(\boldsymbol{b} - \boldsymbol{\beta})'(X'X)(\boldsymbol{b} - \boldsymbol{\beta}) \sim F_{k+1,n-k-1}.$$

Therefore, a $(1 - \alpha) \times 100$ per cent confidence region for $\boldsymbol{\beta}$ is an ellipsoidal region given by

$$R = \{\boldsymbol{\beta} : (\boldsymbol{b} - \boldsymbol{\beta})'(X'X)(\boldsymbol{b} - \boldsymbol{\beta}) \le (k+1)s^2 F_{k+1,n-k-1,\alpha}\}. \qquad (3.32)$$

The boundary for such a confidence region can be found by writing the equation corresponding to the inequality in (3.32) and solving it. Unfortunately, such regions, unless they are two dimensional, are difficult to

[1]SAS is a registered trademark of SAS Institute Inc., Cary, North Carolina

visualize. Consequently, it may be preferable on occasion to replace such a region by a less precise rectangular one using the Bonferroni inequality, as we discuss in the next section.

3.8.4 *C.I.'s for Linear Combinations of Coefficients

In this section we present expressions for simultaneous (or joint) confidence intervals for ℓ different linear combinations $\boldsymbol{a}_1'\boldsymbol{\beta}, \ldots, \boldsymbol{a}_\ell'\boldsymbol{\beta}$ of parameters $\boldsymbol{\beta} = (\beta_0, \ldots, \beta_k)'$. We give below two different expressions. The first is usually better when the number ℓ is small; the second when it is large. For each case we assume that the Gauss-Markov conditions hold and the ϵ_i's are normally distributed.

Using the Bonferroni inequality (see below) a simultaneous $(1-\alpha)\times 100$ per cent C.I. for $\boldsymbol{a}_1'\boldsymbol{\beta}, \ldots, a_\ell'\boldsymbol{\beta}$ may be shown to be given by

$$\boldsymbol{a}_i'\boldsymbol{b} \pm t_{r,\alpha/(2\ell)} s[\boldsymbol{a}_i'(X'X)^{-1}\boldsymbol{a}_i]^{1/2},$$

where $r = n - k - 1$ and $i = 1, \ldots, \ell$. (The Bonferroni inequality may be stated as follows: If $E_1, \ldots, E_k$ are k events, then

$$1 - P\{E_1 \cap \cdots \cap E_\ell\} \leq \sum_{i=1}^{\ell} P(E_i^c)$$

where E_i^c is the complement of E_i for $i = 1, \ldots, \ell$.)

When ℓ is large the rectangular confidence regions given above might become rather wide. If ℓ is such that $t_{r,\alpha/2\ell} \geq (k+1)^{1/2} F_{k+1,r,\alpha}^{1/2}$ (see (3.30)), it might be preferable to use the confidence region given by

$$\boldsymbol{a}_i'\boldsymbol{b} \pm s(k+1)^{1/2} F_{k+1,r,\alpha}^{1/2} [\boldsymbol{a}_i'(X'X)^{-1}\boldsymbol{a}_i]^{1/2}.$$

In practice, unless ℓ is quite small, it is desirable to compute both sets of intervals and then decide which is better.

Problems

Exercise 3.1: Consider the regression model

$$y_i = \beta_0 + \beta_1 x_{i1} + \cdots + \beta_6 x_{i6} + \epsilon_i, \text{ where } i = 1, \ldots, n$$

and ϵ_i's are identically and independently distributed as $N(0, \sigma^2)$. Define the matrix C and the vector γ, in the notation of Section 3.2, in order to test (against appropriate alternatives) each of the following hypotheses: (a) $\beta_1 = \cdots = \beta_6 = 0$, (b) $\beta_1 = 5\beta_6$, (c) $\beta_6 = 10$, (d) $\beta_1 = \beta_2$, (e) $\beta_3 = \beta_4 + \beta_5$.

Exercise 3.2: Consider the two sets of cases

$$(u_1, x_1), \ldots, (u_n, x_n) \text{ and } (v_1, z_1), \ldots, (v_n, z_n)$$

and the two models

$$u_i = \alpha_1 + \beta_1 x_i + \eta_i \text{ and } v_i = \alpha_2 + \beta_2 z_i + \delta_i$$

for $i = 1, \ldots, n$, where x_i's and z_i's are fixed constants and all η_i's and δ_i's are independently distributed with zero means and common variance σ^2. By defining $\boldsymbol{y}, X, \boldsymbol{\beta}$ and $\boldsymbol{\epsilon}$ appropriately, write the two equations given above as one regression model $\boldsymbol{y} = X\boldsymbol{\beta} + \boldsymbol{\epsilon}$.

Exercise 3.3: Consider two independent sets of observations. The first set of observations are taken on n subjects who were not given any medication and the model is assumed to be

$$y_{1i} = \beta_0 + \epsilon_{1i}, \text{ where } i = 1, \ldots, n$$

and ϵ_{i1}'s are iid $N(0, \sigma^2)$. For the second set of observations, the ith observation was taken on a subject who received a dose x_i of a medication, and the model is assumed to be

$$y_{2i} = \beta_0 + \beta_1 x_i + \epsilon_{2i} \text{ where } i = 1, \ldots, n$$

and ϵ_{2i}'s are iid $N(0, \sigma^2)$.

1. Find the least squares estimates of β_0 and β_1 and their variances.

2. Estimate σ^2.

3. Write the test statistic for testing the hypothesis $\beta_1 = 0$ against the alternative $\beta_1 > 0$.

4. If the first set had n_1 independent observations and the second set had $n_2 \neq n_1$ independent observations, what changes would you need to make?

Exercise 3.4: Suppose we need to compare the effects of two drugs each administered to n subjects. The model for the effect of the first drug is

$$y_{1i} = \beta_0 + \beta_1 x_{1i} + \epsilon_{1i}$$

while for the second drug it is

$$y_{2i} = \beta_0 + \beta_2 x_{2i} + \epsilon_{2i},$$

and in each case $i = 1, \ldots, n$ and $\bar{x}_1 = \bar{x}_2 = 0$. Assume that all observations are independent and that for each i both ϵ_{1i} and ϵ_{2i} are normally distributed with mean 0 and variance σ^2.

1. Obtain the least squares estimator for $\boldsymbol{\beta} = (\beta_0, \beta_1, \beta_2)'$ and its covariance matrix.

2. Estimate σ^2.

3. Write the test statistic for testing $\beta_1 = \beta_2$ against the alternative that $\beta_1 \neq \beta_2$.

Exercise 3.5: *For the likelihood function (3.15), find the maximum likelihood estimate of σ^2.

Exercise 3.6: Consider the two models $\boldsymbol{y}_1 = X_1\boldsymbol{\beta}_1 + \boldsymbol{\epsilon}_1$ and $\boldsymbol{y}_2 = X_2\boldsymbol{\beta}_2 + \boldsymbol{\epsilon}_2$ where the X_i's are $n_i \times p$ matrices. Suppose that $\boldsymbol{\epsilon}_i \sim N(\mathbf{o}, \sigma_i^2 I)$ where $i = 1, 2$ and that $\boldsymbol{\epsilon}_1$ and $\boldsymbol{\epsilon}_2$ are independent.

1. Assuming that the σ_i's are known, obtain a test for the hypothesis $\boldsymbol{\beta}_1 = \boldsymbol{\beta}_2$.

2. *Assume that $\sigma_1 = \sigma_2$ but they are unknown. Derive a test for the hypothesis $\boldsymbol{\beta}_1 = \boldsymbol{\beta}_2$.

Exercise 3.7: Let $\boldsymbol{y}_1 = X_1\boldsymbol{\beta}_1 + \boldsymbol{\epsilon}_1$ and $\boldsymbol{y}_2 = X_2\boldsymbol{\beta}_2 + \boldsymbol{\epsilon}_2$ where $\boldsymbol{y}_1$ and $\boldsymbol{y}_2$ are independently distributed, $\mathrm{E}(\boldsymbol{\epsilon}_1) = 0$, $\mathrm{cov}(\boldsymbol{\epsilon}_1) = \sigma^2 I_{n_1}$, $\mathrm{E}(\boldsymbol{\epsilon}_2) = 0$, $\mathrm{cov}(\boldsymbol{\epsilon}_2) = \sigma^2 I_{n_2}$, and $\boldsymbol{\beta}_1$ and $\boldsymbol{\beta}_2$ are p_1 and p_2 vectors respectively. Let

$$\boldsymbol{\beta}_1 = \begin{pmatrix} \boldsymbol{\beta}_1^{(1)} \\ \boldsymbol{\beta}_1^{(2)} \end{pmatrix} \quad \text{and} \quad \boldsymbol{\beta}_2 = \begin{pmatrix} \boldsymbol{\beta}_2^{(1)} \\ \boldsymbol{\beta}_2^{(2)} \end{pmatrix}$$

where $\boldsymbol{\beta}_1^{(1)}$ and $\boldsymbol{\beta}_2^{(1)}$ are r-vectors (which makes $\boldsymbol{\beta}_1^{(2)}$ a $(p_1 - r)$-vector and $\boldsymbol{\beta}_2^{(2)}$ a $(p_2 - r)$-vector). Write the two models above as one regression model and give a procedure for testing $H : \boldsymbol{\beta}_1^{(1)} = \boldsymbol{\beta}_2^{(1)}$.

Exercise 3.8: Consider the model in Exercise 3.6 and assume that $\sigma_1 = \sigma_2 = \sigma$ with σ unknown (notice that the situation is akin to that in Section 3.7 with $r = 0$).

1. Show that the test statistic for $H: \boldsymbol{\beta}_1 = \boldsymbol{\beta}_2$ obtained using (3.13) can be written as

$$(\hat{\boldsymbol{\beta}}_1 - \hat{\boldsymbol{\beta}}_2)'[(X_1'X_1)^{-1} + (X_2'X_2)^{-1}]^{-1}(\hat{\boldsymbol{\beta}}_1 - \hat{\boldsymbol{\beta}}_2)/ps^2,$$

where

$$(n_1 + n_2 - 2p)s^2 = \boldsymbol{y}_1'M_1\boldsymbol{y}_1 + \boldsymbol{y}_2'M_2\boldsymbol{y}_2,$$

and for $i = 1, 2$,

$$\hat{\boldsymbol{\beta}}_i = (X_i'X_i)^{-1}X_i'\boldsymbol{y}_i \text{ and } M_i = I_{n_i} - X_i(X_i'X_i)^{-1}X_i'.$$

2. Show that

$$[(X_1'X_1)^{-1} + (X_2'X_2)^{-1}]^{-1}$$
$$= X_1'X_1(\tilde{X}'\tilde{X})^{-1}X_2'X_2 = X_2'X_2(\tilde{X}'\tilde{X})^{-1}X_1'X_1,$$

where

$$\tilde{X} = \begin{pmatrix} X_1 \\ X_2 \end{pmatrix}.$$

3. Let $\boldsymbol{y} = (\boldsymbol{y}_1', \boldsymbol{y}_2')'$, $\hat{\boldsymbol{\beta}} = (\tilde{X}'\tilde{X})^{-1}\tilde{X}'\boldsymbol{y}$ and, for $i = 1, 2$, let $\hat{\boldsymbol{y}}_i = X_i\hat{\boldsymbol{\beta}}$. Show that

$$(\boldsymbol{y} - \tilde{X}\hat{\boldsymbol{\beta}})'\tilde{X} = 0 \text{ and } (\boldsymbol{y}_1 - \hat{\boldsymbol{y}}_1)'X_1 + (\boldsymbol{y}_2 - \hat{\boldsymbol{y}}_2)'X_2 = 0.$$

4. Prove that

$$\hat{\boldsymbol{\beta}}_1'X_1'X_1\hat{\boldsymbol{\beta}} + \hat{\boldsymbol{\beta}}_2'X_2'X_2\hat{\boldsymbol{\beta}} = \hat{\boldsymbol{\beta}}'\tilde{X}'\tilde{X}\hat{\boldsymbol{\beta}}.$$

5. Prove that

$$\hat{\boldsymbol{\beta}}_1'X_1'X_1(\tilde{X}'\tilde{X})^{-1}X_2'X_2\hat{\boldsymbol{\beta}}_1$$
$$= \hat{\boldsymbol{\beta}}_1'X_1'X_1\hat{\boldsymbol{\beta}}_1 - \hat{\boldsymbol{\beta}}_1'X_1'X_1\hat{\boldsymbol{\beta}} + \hat{\boldsymbol{\beta}}_1'X_1'X_1(\tilde{X}'\tilde{X})^{-1}X_2'X_2\hat{\boldsymbol{\beta}}_2.$$

Exercise 3.9: *Show that the numerator in part 1 of Exercise 3.8 is equal to each of the expressions in parts 1, 2 and 3 below.

1. $(\hat{\boldsymbol{\beta}}_1 - \hat{\boldsymbol{\beta}})'(X_1'X_1)(\hat{\boldsymbol{\beta}}_1 - \hat{\boldsymbol{\beta}}) + (\hat{\boldsymbol{\beta}}_2 - \hat{\boldsymbol{\beta}})'(X_2'X_2)(\hat{\boldsymbol{\beta}}_2 - \hat{\boldsymbol{\beta}})$
   [**Hint:** Use the results in parts 2, 4 and 5 of Exercise 3.8.]
2. $\boldsymbol{y}'M\boldsymbol{y} - \boldsymbol{y}_1'M_1\boldsymbol{y}_1 - \boldsymbol{y}_2'M_2\boldsymbol{y}_2$ where $M = I_{n_1+n_2} - \tilde{X}(\tilde{X}'\tilde{X})^{-1}\tilde{X}'$
   [**Hint:** Use the results in part 1 above and part 3 of Exercise 3.8.]
3. $(\boldsymbol{y}_1 - \hat{\boldsymbol{y}}_1)'X_1[(X_1'X_1)^{-1} + (X_2'X_2)^{-1}]X_1'(\boldsymbol{y}_1 - \hat{\boldsymbol{y}}_1)$
   [**Hint:** Use the results in part 1 above and part 3 of Exercise 3.8.]

Exercise 3.10: Suppose a regression model $y_i = \beta_0 + \beta_1 x_{i1} + \beta_2 x_{i2} + \beta_3 x_{i3} + \epsilon_i$, where $1 = 1, \ldots, 10$ and ϵ_i's are independent and identically distributed as $N(0, \sigma^2)$, is to be fitted to the following data:

y	x_1	x_2	x_3	y	x_1	x_2	x_3
60.5	14	25	28	66.3	13	24	20
87.5	10	29	18	39.8	33	24	30
48.2	20	22	16	83.8	15	30	26
40.6	12	17	29	38.2	12	15	14
42.7	13	17	12	15.0	31	13	30

Here

$$(X'X)^{-1} = \begin{pmatrix} 2.8204 & -.0284554 & -.0758535 & -.0264434 \\ -.0284554 & .00256363 & .000971434 & -.00165374 \\ -.0758535 & .000971434 & .00363832 & -.000876241 \\ -.0264434 & -.00165374 & -.000876241 & .00331749 \end{pmatrix}$$

and

$$X'\boldsymbol{y} = \begin{pmatrix} 522.6 \\ 8084 \\ 12354.9 \\ 11413.6 \end{pmatrix}.$$

1. At a 5 per cent level of significance, test the hypothesis that $\beta_1 = \beta_2 = \beta_3 = 0$ against the alternative that $\beta_j \neq 0$ for at least one j.

2. Test the hypothesis that $\beta_2 = \beta_3$ against the alternative that $\beta_2 \neq \beta_3$ at a 5 per cent level.

3. Find a 95 per cent joint confidence region for β_1 and $\beta_2 - \beta_3$.

Exercise 3.11: Exhibit 3.6 provides data on salaries (Y84, Y83) for 1984 and 1983 for chairmen of the 50 largest corporations in the Chicago area. Data are also provided on their age (AGE), the number of shares they hold (SHARES) and the total revenues (REV) and the total income (INC) of the companies they head. Based on the data write a report on the factors that affect the raises given company chairmen.

Exercise 3.12: Using (3.22) explain why significance levels of SI and SP increased so much when we went from the model of part 1 of Exercise 2.18 to that of part 2.

Exercise 3.13: **(a)** For the model given in part 1 of Problem 2.20, test each coefficient to see if it could be zero against the alternative that it is not zero, at a significance level of 5 percent. (All the numbers you will need are routinely provided by most packages.) Write a short note explaining why you think the variables that are significant are so.

Y84	Y83	SHARES	REV	INC	AGE
1481250	1425000	101037	38828	1454.8	65
1239402	1455350	5713459	558.4	29.7	44
1205181	1057707	18367	12595	479.0	60
1012500	862500	134713	4233.6	108.3	66
980000	871056	7896	8321.5	234.1	53
921213	817687	48722	3104	402.6	56
915600	1092570	78513	7000	188.0	47
912500	833000	73120	1245.8	161.6	56
863000	372645	57723	6661.8	490.8	57
882347	736883	38436	3429.8	38.0	55
890000	838462	8247	9758.7	455.8	57
815000	698333	22135	2001.8	109.8	55
748189	647988	50612	28998	2183.0	59
740500	728663	2331816	4907.0	117.7	66
706565	502400	22432	921.6	52.7	63
698923	630000	113906	1468.1	94.2	54
661958	660000	53162	1397.0	89.3	61
654936	350004	23925	875.8	36.7	50
620000	573500	205046	33915.6	206.1	62
607083	483542	224632	4159.7	145.5	64
606977	475176	25369	3344.1	134.8	46
583437	507036	56713	2139.8	36.6	62
567000	498960	379860	1314.8	142.5	53
566446	488543	139200	1794.4	103.0	59
559266	534004	60450	3738.8	141.0	57
551516	454752	232466	2744.6	85.4	49
551154	550000	63220	1041.9	66.4	58
550000	550000	100112	5534.0	387.0	62
545908	518177	25172	1814.5	134.0	50
545000	457375	13200	1374.6	123.4	67
540000	462000	92200	1070.8	41.7	65
499991	469873	180035	3414.8	389.1	52
495915	391035	1036286	1059.3	6.9	57
490000	372500	33002	267.6	16.8	54
489419	434008	1558377	348.6	4.5	65
480000	274820	146407	1365.5	106.2	56
475000	508333	40736	4802.0	55.0	66
473216	300000	2000	1177.2	27.2	47
465000	645000	72786	1800.4	29.1	50
464577	425654	37650	571.6	17.8	64
461250	350250	4826	1716.0	63.6	45
459352	402735	96431	592.3	60.0	54
455998	420422	72226	923.1	26.5	62
451667	371400	107233	590.5	39.7	52
450192	337533	1510150	437.8	13.8	58
450010	267510	2516958	252.2	38.2	47
450000	387500	3346	496.6	31.7	69
434487	313500	307120	220.0	19.0	56
432667	383333	26163	845.1	36.6	57
430000	396000	4022460	1287.7	162.2	47

EXHIBIT 3.6: Data on Corporations and Corporation Chairmen
SOURCE: Reprinted with permission from the May 13, 1985, issue of *Crain's Chicago Business*. © 1985 by Crain's Communications, Inc. The data shown are a portion of the original table.

(b) We know that doctors tend to congregate in urban areas. For the model you constructed in part 3 of Problem 2.20 test the hypothesis that the coefficients of all variables other than UR are all zero against the alternative that they are not all zero.
(c) For the model in part 4 of Problem 2.20, test if any variable in the model you constructed other than divorce rate affects marriage rate.

Exercise 3.14: Moore (1975) reported the results of an experiment to construct a model for total oxygen demand in dairy wastes as a function of five laboratory measurements (Exhibit 3.7). Data were collected on samples kept in suspension in water in a laboratory for 220 days. Although all observations reported here were taken on the same sample over time, assume that they are independent. The measured variables are:

y log(oxygen demand, mg oxygen per minute)
x_1 biological oxygen demand, mg/liter
x_2 total Kjeldahl nitrogen, mg/liter
x_3 total solids, mg/liter
x_4 total volatile solids, a component of x_3, mg/liter
x_5 chemical oxygen demand, mg/liter

1. Fit a multiple regression model using y as the dependent variable and all x_j's as the independent variables.

2. Now fit a regression model with only the independent variables x_3 and x_5. How do the new parameters, the corresponding value of R^2 and the t-values compare with those obtained from the full model?

Exercise 3.15: Using the data of Exhibit 1.19 test if the slopes of least squares lines expressing price in terms of number of pages for paperback and cloth-bound books are different. (We know the intercepts are different.) Examine the residuals from the model you used to carry out the test. Do you feel that any of the conditions required for the test have been violated? If so, which ones?

Most of the data are for books published in 1988. However, two of the cloth-bound books were published in the 1970's, one of the paperbacks in 1989 and another in 1984. Can you guess which ones? Delete these points and repeat the problem described in the last paragraph.

If you were doing this problem for a client, what model(s) would you deliver?

Exercise 3.16: The data in Exhibit 3.8 are for student volunteers who were given a map reading ability test (scores given in column marked sc) and then asked to find routes to given destinations on a transit route map. Their ability in doing this was scored (y). Test if the relationship between the two scores is the same for transit users as it is for non-users against the alternative that the relationships are different.

Day	x_1	x_2	x_3	x_4	x_5	y
0	1125	232	7160	85.9	8905	1.5563
7	920	268	8804	86.5	7388	0.8976
15	835	271	8108	85.2	5348	0.7482
22	1000	237	6370	83.8	8056	0.7160
29	1150	192	6441	82.1	6960	0.3130
37	990	202	5154	79.2	5690	0.3617
44	840	184	5896	81.2	6932	0.1139
58	650	200	5336	80.6	5400	0.1139
65	640	180	5041	78.4	3177	-0.2218
72	583	165	5012	79.3	4461	-0.1549
80	570	151	4825	78.7	3901	0.0000
86	570	171	4391	78.0	5002	0.0000
93	510	243	4320	72.3	4665	-0.0969
100	555	147	3709	74.9	4642	-0.2218
107	460	286	3969	74.4	4840	-0.3979
122	275	198	3558	72.5	4479	-0.1549
129	510	196	4361	57.7	4200	-0.2218
151	165	210	3301	71.8	3410	-0.3979
171	244	327	2964	72.5	3360	-0.5229
220	79	334	2777	71.9	2599	-0.0458

EXHIBIT 3.7: Data on Oxygen Demand in Dairy Wastes
SOURCE: Moore (1975). Reproduced with permission of the author.

Exercise 3.17: Consider the data given in Problem 3.14. Suppose the model is

$$y_i = \beta_0 + \beta_1 x_{i1} + \beta_2 x_{i2} + \beta_3 x_{i3} + \beta_4 x_{i4} + \beta_5 x_{i5} + \epsilon_i,$$

where $i = 1, \dots, n$ and $\boldsymbol{\epsilon} = (\epsilon_1, \dots, \epsilon_n)' \sim N(\mathbf{o}, \sigma^2 I_n)$.

1. Test the hypothesis $\beta_2 = \beta_4 = 0$ at the 5 per cent level of significance.
2. Find a 95 per cent C.I. for β_1.
3. Find a 95 per cent C.I. for $\beta_3 + 2\beta_5$.

Exercise 3.18: It has been conjectured that aminophylline retards blood flow in the brain. But since blood flow depends also on cardiac output (x_1) and carbon dioxide level in the blood (x_2), the following models were postulated:

$$\text{Without aminophylline: } y_i^{(1)} = \beta + \beta_1 x_{i1}^{(1)} + \beta_2 x_{i2}^{(1)} + \epsilon_i^{(1)}$$
$$\text{With aminophylline: } y_i^{(2)} = \beta_0 + \beta_1 x_{i1}^{(2)} + \beta_2 x_{i2}^{(2)} + \epsilon_i^{(2)}$$

Using the data of Exhibit 3.9 test, at a 5% level, the hypothesis $\beta = \beta_0$ against the alternative that $\beta > \beta_0$ assuming the observations are all independent. Actually the observations are not independent: Each row of

Non-users		Users	
y	sc	y	sc
63	2	75	3
62	3	70	3
70	3	99	9
98	9	80	2
85	7	70	4
89	8	70	7
65	4	73	7
71	4	65	4
78	5	60	5
65	3	70	4

EXHIBIT 3.8: Map Reading Test Scores and Route Finding Scores
SOURCE: Prof. Siim Soot, Department of Geography, University of Illinois at Chicago.

No aminophylline			With aminophylline		
x_1	x_2	y	x_1	x_2	y
265	32	9.2	252	35	19.1
348	35	19.3	411	35	20.1
244	43	16.9	229	36	08.1
618	41	22.1	761	29	35.6
434	44	15.6	541	40	22.1
432	28	10.9	313	38	24.7
790	48	16.7	873	52	21.0
245	43	13.0	359	45	14.9
348	36	20.9	433	32	18.3

EXHIBIT 3.9: Blood Velocity Data
SOURCE: Tonse Raju, M.D., Department of Neonatology, University of Illinois at Chicago.

Exhibit 3.9 represents the same subject. Now how would you test the hypothesis?

Exercise 3.19: In rural India, is the relation between IMR and PQL1 different for males and females? How about urban India? (Use the data in Exhibit 1.20.)

Exercise 3.20: For a density of 100 vehicles per mile, what is the predicted speed (Example 1.1, p. 2)? Find a 95 per cent confidence interval for it and also one for the corresponding future observation. Use the model with both density and its square.

Exercise 3.21: Suppose a person has a house to sell in the area from which the data of Exhibit 2.2, p. 32, were gathered. The house has 750 square feet of space, 5 rooms, 2 bedrooms, 1.5 baths, storm windows, a 1-car garage, 1 fireplace and a 25 front-foot lot. What can you tell him about how much he could expect to get for the house?

Exercise 3.22: In a study of infant mortality, a regression model was constructed using birth weight (which is a measure of prematurity, and a good indicator of the baby's likelihood of survival) as a dependent variable and several independent variables, including the age of the mother, whether the birth was out of wedlock, whether the mother smoked or took drugs during pregnancy, the amount of medical attention she had, her income, etc. The R^2 was .11, but each independent variable was significant at a 1 per cent level. An obstetrician has asked you to explain the significance of the study as it relates to his practice. What would you say to him?

Exercise 3.23: For the house price example (Example 2.2, p. 31), obtain a 95 per cent (elliptical) confidence region for the coefficients of RMS and BDR. Obtain the corresponding rectangular region using the Bonferroni Inequality. Display both on the same graph paper and write a short paragraph comparing the two.

CHAPTER 4

Indicator Variables

4.1 Introduction

Indicator or dummy variables are variables that take only two values — 0 and 1. Normally 1 represents the presence of some attribute and 0 its absence. We have already encountered such variables in Example 2.2, p. 31. They have a wide range of uses which will be discussed in this chapter. We shall mainly be concerned with using them as independent variables. In general, their use as dependent variables in least squares analysis is not recommended. Nonetheless, we shall consider the topic further in the final section of this chapter.

4.2 A Simple Application

To see the role dummy variables can play, consider the simplest case where we have a single independent variable, x_{i1}, which is a dummy, i.e., the model

$$y_i = \beta_0 + \beta_1 x_{i1} + \epsilon_i \quad i = 1, \ldots, n \tag{4.1}$$

where

$$x_{i1} = \begin{cases} 0 & \text{when } i = 1, \ldots, n_1 \\ 1 & \text{when } i = n_1 + 1, \ldots, n, \end{cases}$$

and

$$\epsilon_i\text{'s are iid } N(0, \sigma^2), \tag{4.2}$$

i.e., ϵ's are independent and identically distributed as $N(0, \sigma^2)$. Let $\mu_1 = \beta_0$ and $\mu_2 = \beta_0 + \beta_1$. Then (4.1) becomes

$$y_i = \begin{cases} \mu_1 + \epsilon_i, & i = 1, \ldots, n_1 \\ \mu_2 + \epsilon_i, & i = n_1 + 1, \ldots, n. \end{cases} \tag{4.3}$$

This is the model for the familiar two-sample testing problem encountered in a first course in statistics, where we would use the two-sample t test to test $H : \mu_1 = \mu_2$ against, say, $A : \mu_1 \neq \mu_2$. Using model (4.1), the equivalent test would consist of testing $H : \beta_1 = 0$ against $A : \beta_1 \neq 0$ and this would give identical answers (see Problem 4.2).

Notice that we have two means but only one indicator variable, the parameter of which is a difference of means. Suppose, instead, we used a second indicator variable x_{i2} which takes the value 1 for $i = 1, \ldots, n_1$ and zero otherwise. Then the corresponding design matrix X would be

$$\begin{pmatrix} \mathbf{1}_{n_1} & \mathbf{0} & \mathbf{1}_{n_1} \\ \mathbf{1}_{n-n_1} & \mathbf{1}_{n-n_1} & \mathbf{0} \end{pmatrix}$$

where $\mathbf{1}_n$ is a vector of length n consisting only of 1's. Since the columns 2 and 3 would sum up to column 1, we would get a singular matrix. While such a design matrix X can be used, $X'X$ is not non-singular and parameter estimates $\boldsymbol{b}$ must be based on a generalized inverse $(X'X)^-$ of $X'X$. Therefore, $b(= X'X)^- X'\boldsymbol{y})$ is not unique but testing of $H : \mu_1 = \mu_2$ vs. $A : \mu_1 \neq \mu_2$ is still possible.

Year(t)	'62	'63	'64	'65	'66	'67	'68	'69	'70	'71
Deaths (z_t) per 10^8 vehicle-miles	4.9	5.1	5.2	5.1	5.3	5.1	4.9	4.7	4.2	4.2
DFR$_t$=$z_t - z_{t-1}$		.2	.1	-.1	.2	-.2	-.2	-.2	-.5	.0

EXHIBIT 4.1: Traffic Fatality Data for Illinois
SOURCE: Illinois Department of Transportation (1972).

Example 4.1
Exhibit 4.1 presents some data on traffic deaths in Illinois. We wish to test if the annual *increases* (DFR$_t$) in deaths per 100 million vehicle miles before 1966 are different from the rate after 1966, when there was an increase in awareness of traffic safety and many traffic safety regulations went into effect. The dependent variable is clearly DFR$_t$ and the values x_{t1} of the independent variable x_1 are

$$0 \quad 0 \quad 0 \quad 0 \quad 1 \quad 1 \quad 1 \quad 1 \quad 1.$$

On applying a linear least squares package, we get Exhibit 4.2.

Obviously, we wish to test $H : \beta_1 = 0$ vs. $A : \beta_1 \neq 0$ and such a test is routinely carried out by most package programs. In this case, since there is only one independent variable, the t test shown in the upper part of Exhibit 4.2 and the F test shown in the lower part yield identical probabilities. The reader might wish to verify that applying the usual two-sample t test also yields the same t-value. ■

4.3 Polychotomous Variables

Indicator variables, since they take only two values, may be called dichotomous variables. Variables taking a finite number of values — but more than

Variable	b_j	s.e.(b_j)	$t(b_j)$	p-value
Intercept	.10	0.0819	1.220	.2618
x_1	-.32	0.1099	-2.911	.0226

$$R^2 = .5476 \quad R_a^2 = .4830 \quad s = .1639$$

Source	DF	Sum of Squares	Mean Square	F value	p-value
MODEL	1	.2276	.2276	8.473	.0226
ERROR	7	.1880	.0269		
C. TOTAL	8	.4156			

EXHIBIT 4.2: Result of Applying a Least Squares Package Program to Illinois Traffic Data

two — may be called polychotomous variables. An example of a polychotomous variable is the variable 'ownership' which may take the three values, 'public', 'private for-profit' and 'private not-for-profit' (see Example 4.2 below). Another example is a response to a questionnaire with values 'strongly agree', 'agree', 'disagree', 'strongly disagree'. Such polychotomous variables are sometimes called factors and their values are called levels.

While polychotomous variables are usually qualitative (as in the examples just mentioned), they are sometimes useful in the study of numerical variables. For example, in a study of the effect on birth weight of children, the variable 'number of cigarettes consumed by the mother during pregnancy' may be coded into categories (e.g., 0, 1-15, > 15). Often ordinal variables are treated as polychotomous variables although no order is assumed for the levels of such variables.

Consider now a case where we have a single factor with p levels $1, \ldots, p$. For level 1, let there be n_1 observations $y_1, \ldots, y_{n_1}$; for level 2, let there be $n_2 - n_1$ observations; and so on. Then we have the model:

$$y_i = \begin{cases} \mu_1 + \epsilon_i, & i = 1, \ldots, n_1 \\ \mu_2 + \epsilon_i, & i = n_1 + 1, \ldots, n_2 \\ \cdots\cdots\cdots & \cdots\cdots\cdots \\ \mu_p + \epsilon_i, & i = n_{p-1} + 1, \ldots, n_p = n \end{cases} \tag{4.4}$$

and assume that (4.2) still holds. Notice that here we have $N_1 = n_1$ observations with mean μ_1, $N_2 = n_2 - n_1$ observations with mean μ_2, ..., $N_p = n_p - n_{p-1}$ observations with mean μ_p. If we wished to see if our polychotomous variable affected the dependent variable at all, we would test

$$H : \mu_1 = \mu_2 = \cdots = \mu_p \tag{4.5}$$

against $A : \mu_i \neq \mu_j$ for at least one pair i, j with $i \neq j$.

Let

$$\beta_0 = \mu_1 \text{ and } \beta_j = \mu_{j+1} - \mu_1 \text{ for } j = 1, 2, \ldots, p-1. \tag{4.6}$$

Now define $x_{i1}, \ldots, x_{ip-1}$ as follows:

$$x_{ij} = \begin{cases} 1 & \text{if } i = n_j + 1, \ldots, n_{j+1} \\ 0 & \text{otherwise} \end{cases}$$

where $j = 1, \ldots, p-1$. Then (4.4) becomes

$$y_i = \beta_0 + \beta_1 x_{i1} + \beta_2 x_{i2} + \cdots + \beta_{p-1} x_{ip-1} + \epsilon_i \tag{4.7}$$

with $i = 1, \ldots, n$. This is a multiple regression model. Notice that the reparameterization (4.6) to convert (4.4) into a regression model format is far from unique. A slightly different one is illustrated in Example 4.2.

Notice also that here too, as in Section 4.2, and for much the same reasons we have one fewer indicator variable than the number of means. If we had more than one factor, then for each factor we would usually need one less indicator variable than number of levels. Obviously, the hypothesis (4.5) is equivalent to the hypothesis $\beta_j = 0$ for $j = 1, \ldots, p-1$.

Service	Psychometric Scores (QUAL)
Public	61.59, 79.19, 68.89, 72.16, 70.66, 63.17, 53.66, 68.69, 68.75, 60.52, 68.01, 73.06, 55.93, 74.88, 62.55, 69.90, 66.61, 63.80, 45.83, 64.48, 58.11, 73.24, 73.24, 69.94
Private Non-profit	76.77, 68.33, 72.29, 69.48, 59.26, 67.16, 71.83, 64.63, 78.31, 61.48
Private	71.77, 82.92, 72.26, 71.75, 67.95, 71.90

EXHIBIT 4.3: Measures of Quality for Agencies Delivering Transportation for the Elderly and Handicapped
SOURCE: Slightly modified version of data supplied by Ms. Claire McKnight of the Department of Civil Engineering, City University of New York.

Example 4.2

Transportation services for the elderly and handicapped are provided by public, private not-for-profit and private for-profit agencies (although in each case, financial support is mainly through public funding). To see if the quality of the services provided under the three types of ownership was essentially the same, a scale measuring quality was constructed using psychometric methods from results of questionnaires administered to users of such services. Each of several services in the State of Illinois was scored using this scale. Exhibit 4.3 shows the score for each agency.

QUAL	X_1	X_2	QUAL	X_1	X_2
61.59	0	0	58.11	0	0
79.19	0	0	73.23	0	0
68.89	0	0	73.12	0	0
72.16	0	0	69.94	0	0
70.66	0	0	76.77	1	0
63.17	0	0	68.33	1	0
53.70	0	0	72.29	1	0
68.69	0	0	69.48	1	0
68.75	0	0	59.26	1	0
60.52	0	0	67.16	1	0
68.01	0	0	71.89	1	0
73.62	0	0	64.63	1	0
55.93	0	0	78.31	1	0
74.88	0	0	61.48	1	0
62.58	0	0	71.77	1	1
69.90	0	0	82.92	1	1
66.61	0	0	72.26	1	1
63.80	0	0	71.75	1	1
45.83	0	0	67.95	1	1
65.48	0	0	71.90	1	1

EXHIBIT 4.4: Values of x_{i1}'s and x_{i2}'s and Corresponding Values of QUAL

The dependent variable QUAL and the independent variables X_1 and X_2 are shown in Exhibit 4.4. Notice that the definition of the independent variables is slightly different from that given by (4.6), although the latter would have worked about as well. Here it made sense to first distinguish between private and public and then between for-profit and not-for-profit. Portions of the output from a least squares package are shown in Exhibit 4.5. Since we wish to test the hypothesis that coefficients of $X1$ and $X2$ are both zero against the alternative that at least one is not equal to zero, the value of the appropriate statistic is the F-value 2.51, which shows that we can reject the hypothesis at a 10 per cent level but not at a 5 per cent level. We also see that the least squares estimate for the mean level of quality of public services (since this level corresponds to $X1 = 0$ and $X2 = 0$) is about 66.18. For private non-profit systems the estimated mean quality index rises by about 2.78 and the quality index for for-profit organizations rises an additional 4.13. However, neither factor is significant at any reasonable level.

Given the nature of the results obtained, one might be tempted to conjecture that if more privately run services were represented in the data set, stronger results might have been obtained. If this problem had been brought to us by a client, we would then have recommended that they increase the size of the data-set. While on the subject of making recommendations to clients, we would also suggest that the client look into the possibility of finding other independent variables (e.g., was the driver a volunteer? Was the transportation service the main business of the provider? etc.), which by reducing s might help achieve significance. ■

Source	DF	Sum of Squares	Mean Square	F value	p-value
MODEL	2	243.81	121.91	2.511	0.0950
ERROR	37	1796.58	48.56		
C. TOTAL	39	2040.40			

Variable	b_j	s.e.(b_j)	$t(b_j)$	p-value
Intercept	66.18	1.422	46.5	0.0001
x_1	2.78	2.623	1.060	0.2963
x_2	4.13	3.598	1.148	0.2583

$$R^2 = .1195 \quad R_a^2 = .0719 \quad s = 6.968$$

EXHIBIT 4.5: Analysis of Variance Table and Parameter Estimates for Quality Data

4.4 Continuous and Indicator Variables

Mixing continuous and dichotomous or polychotomous independent variables presents no particular problems. In the case of a polychotomous variable, one simply converts it into a set of indicator variables and adds them to the variable list.

Example 4.3

The house-price data of Exhibit 2.2, p. 32, were collected from three neighborhoods or zones; call them A, B and C. For these three levels we need to use two dummy variables. We chose

$$L1 = \begin{cases} 1 & \text{if property is in zone A} \\ 0 & \text{otherwise} \end{cases}$$

$$L2 = \begin{cases} 1 & \text{if property is in zone B} \\ 0 & \text{otherwise.} \end{cases}$$

Obviously, if $L1 = 0$ and $L2 = 0$, the property is in C. Data for $L1$ and $L2$ are also presented in Exhibit 2.2. A portion of the output using these variables is given in Exhibit 4.6. As the output shows, if two identical houses were in zones A and C, the former would cost an estimated \$2700 more and a property in zone B would cost \$5700 more than an identical one in zone C. Notice that simply comparing the means of house values in two

Variable	b_j	s.e.(b_j)	$t(b_j)$	p-value
Intercept	16.964	4.985	3.403	0.0039
FLR	0.017	.0032	5.241	0.0001
RMS	3.140	1.583	1.984	0.0659
BDR	-6.702	1.807	-3.708	0.0021
BTH	2.466	2.462	1.002	0.3323
GAR	2.253	1.451	1.553	0.1412
LOT	0.288	0.127	2.258	0.0393
FP	5.612	3.059	1.835	0.0865
ST	10.017	2.318	4.320	0.0006
L1	2.692	2.867	0.939	0.3626
L2	5.692	2.689	2.117	0.0514

$R^2 = .9258 \quad R_a^2 = .8764 \quad s = 4.442$

EXHIBIT 4.6: Output for House Price Data When L1 and L2 Are Included

areas would give us a comparison of house prices in the areas, not the price difference between identical houses. The two comparisons would be quite different if, say, on the average, houses in one of the two areas were much larger than in the other. For this reason, had we included only $L1$ and $L2$ in the model, and no other variables, the meaning of the coefficients would be quite different.

If we wished to test if location affects property values, we would test the hypothesis that the coefficients of $L1$ and $L2$ are both zero against the alternative that at least one of the coefficients is non-zero. The value of the F test statistic turns out to be 2.343 for which the p-value is .13. ■

4.5 Broken Line Regression

Exhibit 4.9 illustrates a plot of points which would appear to require two lines rather than a single straight line. It is not particularly difficult to fit such a 'broken line' regression. Let us assume the break occurs at the known value x of the independent variable and define

$$\delta_i = \begin{cases} 1 & \text{if } x_{i1} > x \\ 0 & \text{if } x_{i1} \leq x. \end{cases}$$

Then the model

$$y_i = \beta_0 + \beta_1 x_{i1} + \beta_2 (x_{i1} - x)\delta_i + \epsilon_i \tag{4.8}$$

suffices, as can be readily verified. Situations when x is treated as an unknown can be handled using nonlinear regression (see Appendix C, particularly Example C.4, p. 313).

Obs	Country	LIFE	INC	Obs	Country	LIFE	INC
1	AUSTRALIA	71.0	3426	52	CAMEROON	41.0	165
2	AUSTRIA	70.4	3350	53	CONGO	41.0	281
3	BELGIUM	70.6	3346	54	EGYPT	52.7	210
4	CANADA	72.0	4751	55	EL SALVADOR	58.5	319
5	DENMARK	73.3	5029	56	GHANA	37.1	217
6	FINLAND	69.8	3312	57	HONDURAS	49.0	284
7	FRANCE	72.3	3403	58	IVORY COAST	35.0	387
8	WEST GERMANY	70.3	5040	59	JORDAN	52.3	334
9	IRELAND	70.7	2009	60	SOUTH KOREA	61.9	344
10	ITALY	70.6	2298	61	LIBERIA	44.9	197
11	JAPAN	73.2	3292	62	MOROCCO	50.5	279
12	NETHERLANDS	73.8	4103	63	PAPUA	46.8	477
13	NEW ZEALAND	71.1	3723	64	PARAGUAY	59.4	347
14	NORWAY	73.9	4102	65	PHILLIPPINES	51.1	230
15	PORTUGAL	68.1	956	66	SYRIA	52.8	334
16	SWEDEN	74.7	5596	67	THAILAND	56.1	210
17	SWITZERLAND	72.1	2963	68	TURKEY	53.7	435
18	BRITAIN	72.0	2503	69	SOUTH VIETNAM	50.0	130
19	UNITED STATES	71.3	5523	70	AFGHANISTAN	37.5	83
20	ALGERIA	50.7	430	71	BURMA	42.3	73
21	ECUADOR	52.3	360	72	BURUNDI	36.7	68
22	INDONESIA	47.5	110	73	CAMBODIA	43.7	123
23	IRAN	50.0	1280	74	CENTRAL AFRICAN REPUBLIC	34.5	122
24	IRAQ	51.6	560				
25	LIBYA	52.1	3010	75	CHAD	32.0	70
26	NIGERIA	36.9	180	76	DAHOMEY	37.3	81
27	SAUDI ARABIA	42.3	1530	77	ETHIOPIA	38.5	79
28	VENEZUELA	66.4	1240	78	GUINEA	27.0	79
29	ARGENTINA	67.1	1191	79	HAITI	32.6	100
30	BRAZIL	60.7	425	80	INDIA	41.2	93
31	CHILE	63.2	590	81	KENYA	49.0	169
32	COLOMBIA	45.1	426	82	LAOS	47.5	71
33	COSTA RICA	63.3	725	83	MADAGASCAR	36.0	120
34	DOMINICAN REP.	57.9	406	84	MALAWI	38.5	130
35	GREECE	69.1	1760	85	MALI	37.2	50
36	GUATEMALA	49.0	302	86	MAURITANIA	41.0	174
37	ISRAEL	71.4	2526	87	NEPAL	40.6	90
38	JAMAICA	64.6	727	88	NIGER	41.0	70
39	MALAYSIA	56.0	295	89	PAKISTAN	51.2	102
40	MEXICO	61.4	684	90	RWANDA	41.0	61
41	NICARAGUA	49.9	507	91	SIERRA LEONE	41.0	148
42	PANAMA	59.2	754	92	SOMALIA	38.5	85
43	PERU	54.0	334	93	SRI LANKA	65.8	162
44	SINGAPORE	67.5	1268	94	SUDAN	47.6	125
45	SPAIN	69.1	1256	95	TANZANIA	40.5	120
46	TRINIDAD	64.2	732	96	TOGO	35.0	160
47	TUNISIA	51.7	434	97	UGANDA	47.5	134
48	URUGUAY	68.5	799	98	UPPER VOLTA	31.6	62
49	YUGOSLAVIA	67.7	406	99	SOUTH YEMEN	42.3	96
50	ZAMBIA	43.5	310	100	YEMEN	42.3	77
51	BOLIVIA	49.7	193	101	ZAIRE	38.8	118

EXHIBIT 4.7: Data on Per-Capita Income (in Dollars) and Life Expectancy SOURCE: Leinhardt and Wasserman (1979), from the *New York Times* (September, 28, 1975, p. E-3). Reproduced with the permission of the *New York Times*.

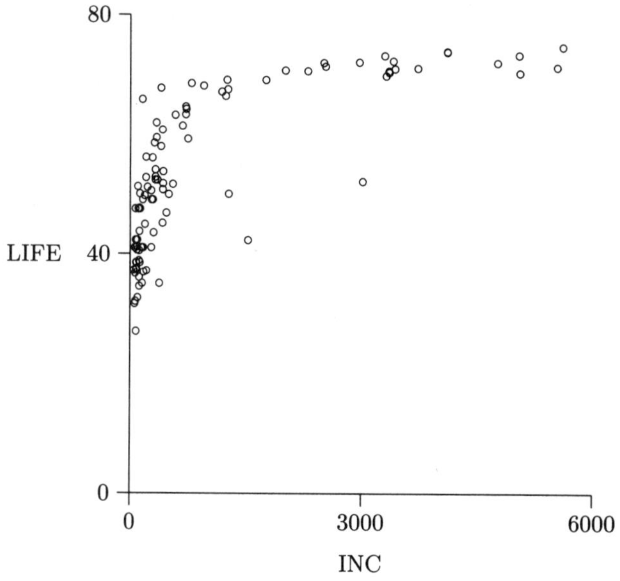

EXHIBIT 4.8: Plot of Life Expectancy Against Per-Capita Income

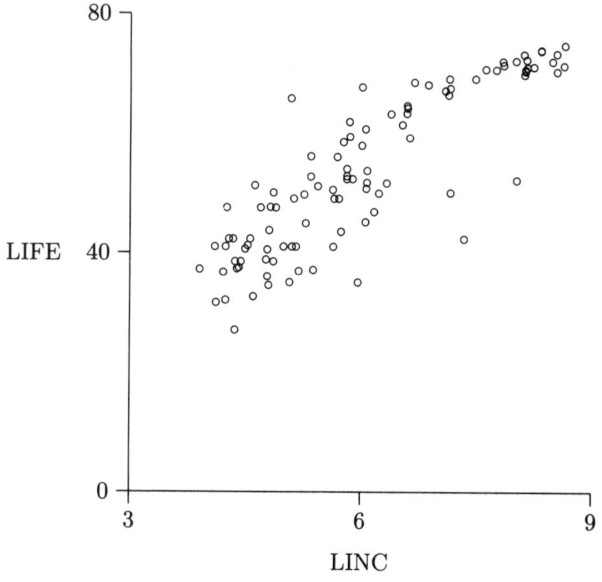

EXHIBIT 4.9: Plot of Life Expectancy Against Log of Per-Capita Income

Example 4.4
Usually poorer countries (i.e., those with lower per capita incomes) have lower life expectancies than richer countries. Exhibit 4.7 gives life expectancies (LIFE) and per capita incomes (INC) in 1974 dollars for 101 countries in the early 70's. Exhibit 4.8 shows a plot which is difficult to read. Taking logarithms of income 'spreads out' the low income points and (see Exhibit 4.9) we discern a pattern that seems to consist of two separate lines: one for the poorer countries, where LIFE increases rapidly with LINC (= log(INC)), and another for the richer countries, where the rate of growth of life expectancy with LINC is much smaller. Therefore, we fitted an equation of the form (4.8) with $\delta_i = 1$ if LINC > 7 and $\delta_i = 0$ otherwise, and obtained

$$\begin{array}{ccccccc} \text{LIFE} & = & -2.40 & + & 9.39\ \text{LINC} & - & 3.36\,[\delta_i(\text{LINC}-7)] \\ & & (4.73) & & (.859) & & (2.42) \end{array} \qquad (4.9)$$

$$(R^2 = .752, s = 6.65)$$

where, as before, the parenthetic quantities are standard errors. The 7 was found by inspecting Exhibit 4.9. We shall return to this example in future chapters. ■

4.6 Indicators as Dependent Variables

While it is not desirable to use dichotomous dependent variables in a linear least squares analysis (typically logit, probit or contingency table analysis is used for this purpose), if we are willing to aggregate our data, least squares analysis may still be used. The example below illustrates such a case. Another case is illustrated in Chapter 9.

Example 4.5
An interesting problem for political scientists is to determine how a particular group of people might have voted for a particular candidate. Typically such assessments are made using exit polls. However, with adequate data, regression procedures might be used to obtain estimates.

Consider the data of Exhibit 4.10 in which the columns Garcia, Martinez and Yanez give the total votes for each of those candidates. (Note that votes for the three candidates may not add to the total turnout because of write-in votes, spoilt ballots, etc.) Let p_L be the probability that a Latino casts a valid vote for (say) Garcia and p_N the probability that a non-Latino casts a valid vote for him. If LATV_i and NONLV_i are, respectively, the total Latino and non-Latino votes cast in each precinct i, the expected number of votes for Garcia is

$$p_L\,\text{LATV}_i + p_N\,\text{NONLV}_i.$$

Since we have the total vote count for Garcia, p_L and p_N can be readily estimated by least squares and we obtain

$$\begin{array}{rcccc} \text{GARCIA} & = & .37\ \text{LATV} & + & .64\ \text{NONLV} \\ & & (.043) & & (.052) \end{array}$$

$$(R^2 = .979, s = 18.9).$$

Therefore, we estimate that roughly 37 per cent of the Latinos voted for Garcia and about 64 per cent of the others voted for him. ■

Variables such as all those in Exhibit 4.10 will be called *counted* variables since they are obtained by counting. We might prefer to use as dependent variable the proportion of all voters who voted for Garcia. Such a variable will be called a *proportion of counts*. Both counted variables and proportions of counts usually require special care, as we shall see in Chapters 6 and 9.

Pr.	LATV	NONLV	TURNOUT	GARCIA	MARTINEZ	YANEZ
1	114	78	192	95	59	15
2	143	100	243	120	74	41
3	105	91	196	120	58	18
4	176	97	273	138	71	26
5	169	141	310	143	85	48
6	190	110	300	158	97	29
7	1	305	306	206	15	11
8	190	132	322	128	125	43
9	120	62	182	79	70	27
10	186	224	410	169	158	49
11	152	85	237	105	81	24
12	164	89	253	124	60	29
13	168	64	232	111	89	13
14	75	157	232	143	27	25
15	177	60	237	98	87	21
16	140	121	261	128	92	40
17	178	115	293	150	66	52
18	157	85	242	108	78	31
19	76	124	200	124	24	14
20	120	59	179	73	70	11
21	84	65	149	52	65	12
22	119	92	211	123	55	15
23	172	144	316	136	127	30
24	87	59	146	118	21	7
25	134	59	193	114	55	20
26	137	60	197	83	67	39
27	167	131	298	147	112	42

EXHIBIT 4.10: Votes from Chicago's Twenty-Second Ward by Precinct (Pr.)
SOURCE: Ray Flores, The Latino Institute, Chicago.

Problems

Exercise 4.1: The number of buses y assigned to an urban bus route is usually based on the total number (x) of passengers carried by all buses during the rush hour at a point on the route called the peak load point. Providers of bus service claim that, for a predetermined level (L) of x, a certain fixed number (μ) of buses is assigned as long as $x < L$, and when $x > L$, y is a linear function $y = a + bx$ of x, i.e.,

$$y = \begin{cases} \mu & \text{when } x < L \\ a + bx & \text{when } x \geq L \end{cases} \tag{4.10}$$

where $\mu = a + bL$. If L is known, write (4.10) in a form so that the other parameters can be estimated using linear least squares. The function (4.10) is called a transit supply function (see Sen and Johnson, 1977).

Exercise 4.2: Let $\bar{y}_1$ and s_1 be the sample mean and standard deviation of the first n_1 of the y_i's in the model (4.1) and let $\bar{y}_2$ and s_2 be those of the last $n_2 = n - n_1$ of the y_i's. Show that

$$\frac{b_1}{\text{s.e.}(b_1)} = \frac{\bar{y}_2 - \bar{y}_1}{\sqrt{[n_1 s_1^2 + n_2 s_2^2][n_1^{-1} + n_2^{-1}]/[n-2]}}.$$

[**Hint:** The design matrix X corresponding to (4.1) is

$$\begin{pmatrix} \mathbf{1}_{n_1} & \mathbf{0} \\ \mathbf{1}_{n_2} & \mathbf{1}_{n_2} \end{pmatrix}.$$

Hence

$$(X'X)^{-1} = \begin{pmatrix} n_1^{-1} & -n_1^{-1} \\ -n_1^{-1} & n/(n_1 n_2) \end{pmatrix}, \quad \boldsymbol{b} = (X'X)^{-1}X'\boldsymbol{y} = \begin{pmatrix} \bar{y}_1 \\ \bar{y}_2 - \bar{y}_1 \end{pmatrix}$$

and it follows that

$$X(X'X)^{-1}X' = \begin{pmatrix} n_1^{-1}\mathbf{1}_{n_1}\mathbf{1}'_{n_1} & 0 \\ 0 & n_2^{-1}\mathbf{1}_{n_2}\mathbf{1}'_{n_2} \end{pmatrix}.$$

Therefore

$$\begin{aligned} \boldsymbol{e}'\boldsymbol{e} &= \boldsymbol{y}'[I - X(X'X)^{-1}X']\boldsymbol{y} \\ &= \boldsymbol{y}_1'[I_{n_1} - n_1^{-1}\mathbf{1}_{n_1}\mathbf{1}'_{n_1}]\boldsymbol{y}_1 + \boldsymbol{y}_2'[I_{n_2} - n_2^{-1}\mathbf{1}_{n_2}\mathbf{1}'_{n_2}]\boldsymbol{y}_2 \\ &= n_1 s_1^2 + n_2 s_2^2, \end{aligned}$$

where $\boldsymbol{y}' = (\boldsymbol{y}_1', \boldsymbol{y}_2')$.]

Exercise 4.3: Consider the model $y_i = \beta_0 + \beta_1 x_{i1} + \cdots + \beta_{ik} x_{ik} + \delta z_i + \epsilon_i$ where $i = 1, \ldots, n$ and

$$z_j = \begin{cases} 1 & \text{if } i = 1 \\ 0 & \text{otherwise,} \end{cases}$$

and ϵ_i's are independently and identically normally distributed with mean 0 and variance σ^2.

1. Obtain explicitly the least squares estimate of δ.
2. Obtain a relationship between the residual sum of squares of the model given above and that of the same model with $\delta = 0$.
3. Obtain a test of hypothesis for $\delta = 0$ against the alternative $\delta \neq 0$.

Exercise 4.4: An experiment was conducted to examine the effects of air pollution on interpersonal attraction. Twenty-four subjects were each placed with a stranger for a 15 minute period in a room which was either odor free or contaminated with ammonium sulfide. The stranger came from a culture which was similar or dissimilar to that of the subject. Thus, there were four possible environments for each subject:

1. Culturally similar stranger, odor-free room;
2. Culturally dissimilar stranger, odor-free room;
3. Culturally similar stranger, room contaminated with ammonium sulfide;
4. Culturally dissimilar stranger, room contaminated with ammonium sulfide.

At the end of the encounter, each subject was asked to assess his degree of attraction towards the stranger on a Likert scale of 1-10 with 10 indicating strong attraction. The full data set is given in Srivastava and Carter (1983). A portion of the data set is reproduced below with the permission of the authors (the numbers are values of the Likert Index).

	Culturally Similar Stranger	Culturally Dissimilar Stranger
Odor-free Room	9, 10, 4, 7, 8, 9	2, 2, 1, 6, 2, 2
Contaminated Room	6, 9, 7, 8, 7, 3	1, 3, 3, 2, 2, 3

Set up a regression model to estimate the effects of cultural similarity and room odor on interpersonal attraction. Conduct appropriate tests and report your findings.

Exercise 4.5: Using Exhibit 4.10, estimate the support for Yanez and Martinez among Latinos and others.

Exercise 4.6: Data on the cost of repairing starters, ring gears or both starters and ring gears are presented in Exhibit 4.11. Define the variables

$$x_{i1} = \begin{cases} 1 & \text{if starter repaired} \\ 0 & \text{otherwise,} \end{cases}$$

Part	Repair Cost in $'s
Starter	37 127 118 75 66 59 499 420 526 141 126 142 137 471 172 149 315 506 575 81 67 36 130 110 126 189 27 88 67 13 432 148 94 432 108 648 81 108 150 79 420 34 236 27 67 42 161 506 468 97 189 551 79 40 420 220 126 261 192 202 101 180 58 61 72 49 189 73 236 306 64
Ring gear	425 474 276 229 256 431 252 1069 190 765 621 310 604 540 81 641 432 252 431 310 256 236 276 609 472 603 431 304 414 241 741
Both	499 420 526 229 471 315 506 575 67 431 190 765 621 432 540 432 648 81 420 310 236 276 506 468 609 472 603 431 551 304 414

EXHIBIT 4.11: Data on Cost of Repairing Starters, Ring Gears or Both Starters and Ring Gears in Diesel Engines
SOURCE: M.R. Khavanin, Department of Mechanical Engineering, University of Illinois at Chicago.

and

$$x_{i2} = \begin{cases} 1 & \text{if ring gear repaired} \\ 0 & \text{otherwise.} \end{cases}$$

Obtain a regression model, with no intercept term, expressing cost in terms of x_{i1}, x_{i2} and the product $x_{i1}x_{i2}$. Give a physical interpretation of the parameter estimates.

Exercise 4.7: Using the data set in Exhibit 1.20, construct a single model for infant mortality rate (IMR), using suitably defined indicator variables for rural-urban and male-female distinctions.

Exercise 4.8: Exhibit 4.12 gives data on numbers of patient contacts for April 19–25, 1987, on screening (SC), diet class (DC), meal rounds (MR) and team rounds (TR) for 11 professional dietitians and 13 dietitian interns. The sum of the times taken for all of these activities is also given. Use a suitable model to estimate the average time taken for each activity by professional dietitians. Make a similar estimate for interns. Test if these average times are the same for the two groups.

Note that dietitians perform several other activities which are not given in the tables. The data were made available to one of the authors by a student.

Exercise 4.9: Exhibit 4.13 presents data on the sex, the attending physician (MD), severity of illness (Svty), total hospital charges (Chrg) and age

Professional Dietitians					Dietitian Interns				
Time	SC	DC	MR	TR	Time	SC	DC	MR	TR
219	3	3	77	29	316	3	4	163	30
264	3	3	95	27	251	2	1	141	16
226	6	2	68	24	216	4	2	135	16
242	6	5	80	25	303	13	3	135	16
220	7	1	70	19	280	4	3	138	18
229	3	1	66	30	285	6	2	141	22
253	7	2	81	24	268	2	4	139	25
233	3	2	86	27	269	2	3	152	18
260	4	6	85	25	307	5	3	143	16
235	8	2	72	21	204	5	0	135	17
247	8	0	82	26	283	4	4	151	16
					233	3	2	126	25
					266	4	1	148	20

EXHIBIT 4.12: Time Taken by Professional Dietitians and Interns for Four Patient Contact Activities

for 49 patients, all of whom had an identical diagnosis. Estimate a model expressing the logarithm of charges against age and the other variables expressed as suitable indicator variables. Test the hypothesis that the attending physician has no effect on the logarithm of hospital charges against the alternative that this factor has an effect.

Sex	MD	Svty	Chrg	Age	Sex	MD	Svty	Chrg	Age
M	730	2	8254	57	F	499	1	6042	61
M	730	4	24655	93	F	499	1	11908	89
M	730	1	1487	17	F	499	3	24121	86
M	730	1	5420	61	M	499	3	15600	72
M	730	2	18823	61	F	499	3	25561	92
M	730	3	20280	61	F	499	1	2499	39
F	730	1	4360	44	M	499	3	12423	69
M	730	3	22382	90	M	499	3	21311	92
M	730	3	22165	90	M	499	3	15969	60
M	730	4	22632	70	F	499	3	16574	72
F	730	4	22642	77	F	1021	3	24214	89
F	730	2	14111	85	F	1021	4	28297	79
M	730	2	9763	62	F	1021	4	64465	71
F	730	2	13343	65	F	1021	3	17506	71
M	730	1	4886	54	F	1021	4	27150	76
F	730	3	22712	87	F	1021	2	13652	82
M	730	2	7194	50	M	1021	1	4402	41
F	730	3	24809	73	F	1021	1	4427	40
M	730	1	9405	62	F	1021	3	22734	66
M	499	1	9990	63	M	1021	2	8759	56
M	499	3	24042	67	F	1021	4	29636	80
F	499	4	17591	68	F	1021	1	15466	67
F	499	2	10864	85	M	1021	2	18016	45
M	499	2	3535	20	M	1021	1	7510	57
					F	1021	2	12953	65

EXHIBIT 4.13: Data on Hospital Charges
SOURCE: Dr. Joseph Feinglass, Northwestern Memorial Hospital, Chicago.

CHAPTER 5

The Normality Assumption

5.1 Introduction

In much of the work presented in the last four chapters we have assumed that the Gauss-Markov conditions were true. We have also sometimes made the additional assumption that the errors and, therefore, the dependent variables were normally distributed. In practice, these assumptions do not always hold; in fact, quite often, at least one of them will be violated. In this and the next four chapters we shall examine how to check whether each of the assumptions actually holds and what, if anything, we can do if it does not. This chapter is devoted to the normality assumption.

In Section 5.2 we present three methods to check whether the errors ϵ_i are approximately normal. One of the methods is graphical and the other two are formal tests. Graphical methods give a greater 'feel' for the data and, if the problem lies with a few of the observations, such methods help identify them. Formal tests require less 'judgment' on the part of the analyst, but even they cannot be considered totally objective in this context.

When we accept a hypothesis, all we can say is that there was not enough evidence to reject it; we cannot say that the ϵ's are, in fact, normal. But we can possibly say that they are close enough to being normal so that our inferences under that assumption are valid. This introduces a subjective element. In general, sample sizes need to be fairly large for us to make that claim. On the other hand, with large enough sample sizes, normality could get rejected even when the distribution is nearly normal. Then we need to be somewhat loose in our interpretations. Similar discussions also apply to like tests in the following chapters.

The methods we use to check for normality assume that no Gauss-Markov violations exist. In fact, in Chapter 9 (see Section 9.4.3, p. 204), we use methods, which were originally developed to change non-normal data to nearly normal data, in order to achieve compliance with Gauss-Markov conditions. In the next few chapters which are devoted to the Gauss-Markov conditions, when we check for violations of one of these conditions, we assume that the others hold. Thus outputs from the methods given in this and the following chapters need to be considered simultaneously rather than sequentially. Regression is an art and its practice is fraught with misdiagnoses and dead ends. Fortunately, with experience, diagnoses get better and the number of dead ends declines.

In the experience of the authors, failure of G-M conditions affects tests for normality more than failure of normality affects the diagnoses of violations of G-M conditions. In that sense, this chapter is inappropriately located.

Notice that in the last two chapters it was really the normality of b_j's rather than the y_i's that was required. Since b_j's are linear combinations of y_i's, under certain conditions, we are able to invoke the central limit theorem to show that b_j's are approximately normal even when y_i's are not. These conditions are presented in Section 5.3. If even the b_j's are not normal, we can get an approximate idea of their distribution and use it to make inferences. A method of doing this, called *bootstrapping*, is discussed in Section 5.4. Another method, called the transform-both-sides method (Carroll and Ruppert, 1988, Ch. 4), which applies Box-Cox transformations (see Section 9.4.3, p. 204) to both sides of a regression model, is not treated in this book.

5.2 Checking for Normality

For all methods presented below, the underlying assumptions are that the errors are identically distributed and that if the vector of errors $\boldsymbol{\epsilon}$ is non-normal, the vector of residuals $\boldsymbol{e} = M\boldsymbol{\epsilon} = (I - H)\boldsymbol{\epsilon}$ will be too (by (2.3) and (2.13)).

If $H = (\boldsymbol{h}_1, \ldots, \boldsymbol{h}_n)'$ and $H = (h_{ij})$, then

$$e_i = \epsilon_i - \boldsymbol{h}_i'\boldsymbol{\epsilon}. \tag{5.1}$$

Since H is an idempotent matrix, $HH' = H$, and $\boldsymbol{h}_i'\boldsymbol{h}_i = h_{ii}$. Hence,

$$\text{var}\,(\boldsymbol{h}_i'\boldsymbol{\epsilon}) = \boldsymbol{h}_i'\boldsymbol{h}_i\sigma^2 = h_{ii}\sigma^2 \to 0 \text{ as } h_{ii} \to 0. \tag{5.2}$$

Therefore, from Chebyshev's inequality (p. 284), the second term of (5.1) goes to zero in probability. Thus, when h_{ii}'s are small, e_i's may be used in lieu of ϵ_i's. However, as we shall see in the next section, if h_{ii}'s are small, we might not need the normality of ϵ_i's.

5.2.1 Probability Plots

Let $Z_{(1)} < \ldots < Z_{(n)}$ be the ordered values of n independent and identically distributed N(0,1) random variables $Z_1, \ldots, Z_n$; here due regard is given to the sign of an observation. Then the mean value of $Z_{(i)}$ can be approximated by

$$\text{E}(Z_{(i)}) \approx \gamma_i = \Phi^{-1}[(i - 3/8)/(n + 1/4)] \tag{5.3}$$

(see, e.g., Blom, 1958), where Φ^{-1} is the inverse function ($\Phi(x) = a \Rightarrow x = \Phi^{-1}(a)$) of the standard normal cdf

$$\Phi(x) = (2\pi)^{-\frac{1}{2}} \int_{-\infty}^{x} e^{-\frac{1}{2}t^2}\,dt.$$

If $u_{(1)} < \ldots < u_{(n)}$ are ordered values of n independent and identically distributed (i.e., iid) $N(\mu, \sigma^2)$ random variables, then $E([u_{(i)} - \mu]/\sigma) \approx \gamma_i$. Consequently,

$$E[u_{(i)}] \approx \mu + \sigma\gamma_i,$$

and a plot of $u_{(i)}$'s against γ_i would be taken to be approximately a straight line. We shall call such a plot a *rankit plot*, although it would be more precise to call it an approximate rankit plot, given the approximation in (5.3).

In order to check if the residuals are approximately normally distributed, we would plot $e_{(i)}$ against $\gamma_i = \Phi^{-1}[(i - 3/8)/(n + 1/4)]$ where $e_{(i)}$'s are the ordered values of e_i. Alternatively, and preferably, ordered values of the standardized residuals

$$e_i^{(s)} = \frac{e_i}{s\sqrt{1 - h_{ii}}}, \tag{5.4}$$

i.e., the residuals divided by their standard errors (see (2.27)), could be plotted against γ_i's. Rankit plots of the Studentized residuals[1], to be discussed in Section 8.3, p. 156, could also be used. If the plot is approximately a straight line, the residuals would be taken to be approximately normally distributed, and, by the assumption discussed above, we would also consider the ϵ's to be approximately normal. (These plots have been discussed in detail in Madansky, 1988, Ch. 1)

Rankit plots are available from several statistical packages, although, frequently, they are not available from the regression portion of the package (e.g., in SAS, 1985a, it is an option within PROC UNIVARIATE). Programs can be readily written to draw rankit plots and this is often preferable when plots provided by a package are too small and cluttered to be easily read. In the Linear Least Squares Curve Fitting Program that is a companion to Daniel and Wood (1980) such plots are routinely given.

Example 5.1

Exhibit 5.1 shows the rankit plot of the standardized residuals from a regression of the log of charges against sex, age, severity of illness and attending physician, this last factor being expressed as two indicator variables. The data are those given in Exhibit 4.13, p. 99. The plot is fairly straight but not entirely so, indicating that the residuals are close to normal but are probably not normal.

'Reading' such plots requires some experience. Daniel and Wood (1980, see especially pp. 33–43) have given a large number of rankit plots of random normal data which the reader might wish to examine in order to develop a feel for them. ■

[1]There is some confusion in the literature about names. What we have called standardized residuals is sometimes called Studentized; then what we have called Studentized residuals (and also RSTUDENT) is called R-STUDENT or RSTUDENT.

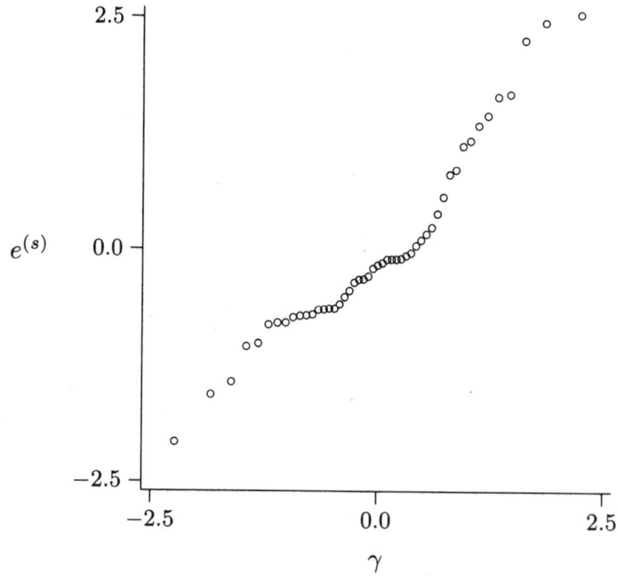

EXHIBIT 5.1: Rankit Plot of Standardized Residuals from Medical Charge Data

Example 5.2 (Continuation of Example 4.4, Page 92)
Exhibit 5.2 shows a rankit plot for the standardized residuals from the model (4.9). The plot does not show a straight line. We then removed the cases corresponding to the four smallest and the two largest residuals, reran the regression and constructed the plot shown in Exhibit 5.3. This one is much straighter, indicating that much of the apparent non-normality indicated in Exhibit 5.2 was due to the presence of outliers.

However, given the very large number of data points, Exhibit 5.3 still indicates the possibility of slight non-normality. ■

While the computations for the rankit plot can be carried out on a computer, they can also be done by hand quite easily on a special graph paper called normal probability paper. On such paper the axes have been scaled in such a way that plotting $e_{(i)}$ against $100\ [i - 3/8]/(n + 1/4)]$ or some approximation of it ($100i/n$, $100(i - \frac{1}{2})/n$ and $100i/(n+1)$ are commonly used) results in approximately a straight line if e_i's are normal. Although, properly, such plots are called *normal plots,* the expression normal plots is sometimes used for what we have called rankit plots.

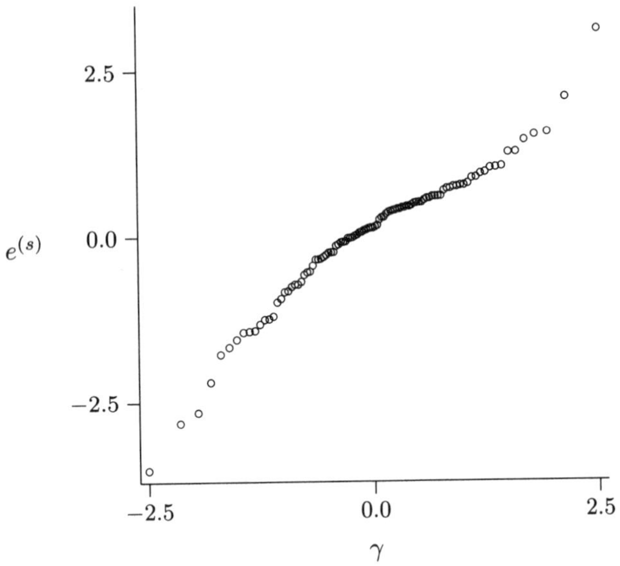

EXHIBIT 5.2: Rankit Plot of Standardized Residuals from Life Expectancy Data

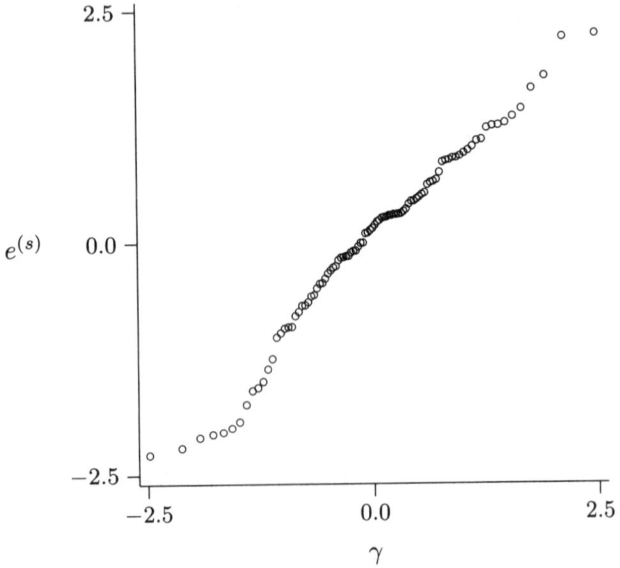

EXHIBIT 5.3: Rankit Plot of Standardized Residuals from Life Expectancy Data after Deletion of Outliers

5.2.2 Tests for Normality

Among several tests for normality, the Shapiro-Wilk (1965) test has become somewhat standard for small sample sizes (e.g., < 50) and is given in many statistical packages. The test can be described as follows.

Let $u_1, \ldots, u_n$ be independently and identically distributed and assume that $u_{(1)} < \ldots < u_{(n)}$ are their ordered values. Set $s^2 = (n-1)^{-1} \sum (u_i - \bar{u})^2$, where $\bar{u} = n^{-1} \sum u_i$. Then the Shapiro-Wilk test statistic is given by

$$W = \sum a_i u_{(i)}/s, \tag{5.5}$$

where $a_1, \ldots, a_n$ depend on the expected values of the order statistics from a standard normal distribution and are tabulated in Shapiro and Wilk (1965), Madansky (1988) and elsewhere. The null hypothesis of normality is rejected if $W \leq W_\alpha$, where W_α is a tabulated critical point. Tables of W_α are also given in Shapiro and Wilk (1965) and in Madansky (1988), although computer packages which provide this statistic provide the a_i's as well as the p-values. The test statistic W takes values between 0 and 1, with values close to 1 indicating near-normality. The residuals e_i replace the u_i's in usual applications of the Shapiro-Wilk test to regression. An alternative to the Shapiro-Wilk statistic is the square of the correlation coefficient between $u_{(i)}$'s and γ_i's (see Shapiro and Francia, 1972), which was originally suggested as an approximation to the Shapiro-Wilk statistic.

Another alternative to the Shapiro-Wilk test is Kolmogorov's test, the latter being used most frequently when n is large, since then (5.5) is difficult to compute. Kolmogorov's test is actually quite general and may be used to test if a set of observations $u_1, \ldots, u_n$ come from any specified distribution function $F_H(x)$. Let $F(x)$be the empirical distribution of the u_i's, i.e., $F(x) = u_x/n$, where u_x is the number of u_i's that are not greater than x. Then Kolmogorov's statistic is

$$D = \sup_x |F(x) - F_H(x)|, \tag{5.6}$$

and the hypothesis that the u_i's have the distribution given by F_H is rejected for large values of D. In our case, of course, $F_H(x)$ is the distribution function of the normal distribution. The mean is the same as that of the residuals (which, of course, is zero if there is a constant term) and the variance is s^2. This test is also widely available in packages which also provide corresponding p-values.

These tests are rather fully discussed in Madansky (1988, Ch. 1).

Example 5.3 (Continuation of Example 5.1, Page 102)

The value of W for the standardized residuals from the hospital charge data is .926 and the probability of getting a smaller value under the hypothesis of normality is less than .01. Therefore, it would appear that the residuals are not normal. However, they are fairly close to being normal since W is so near 1. ■

Example 5.4 (Continuation of Example 5.2, Page 103)
The value of D corresponding to Exhibit 5.2 is .105, which is significant at the 1 per cent level. For Exhibit 5.3 it is .077, which is not significant at a .15 level. ■

5.3 Invoking Large Sample Theory

As mentioned in Section 5.1, if the number of cases is large, we might be able to test hypotheses or obtain confidence regions using the methods discussed in Chapter 3, even if y_i's are not normal. The exact condition is given in the following theorem, due to Srivastava (1971), which is proved in Section 5.5.

Theorem 5.1 *Let $\boldsymbol{b}$ be the least squares estimate of $\boldsymbol{\beta}$ in the usual multiple regression model and assume that Gauss-Markov conditions hold. If, in addition, the observations are independent and*

$$\max_{1\leq i\leq n} h_{ii} \to 0, \tag{5.7}$$

where h_{ii} are the diagonal elements of the matrix $X(X'X)^{-1}X'$, then

$$s^{-2}(C\boldsymbol{b} - C\boldsymbol{\beta})'[C(X'X)^{-1}C']^{-1}(C\boldsymbol{b} - C\boldsymbol{\beta}) \to \chi^2_r \text{ as } n \to \infty, \tag{5.8}$$

where C is an $r \times (k+1)$ matrix of rank $r \leq (k+1)$.

However, we suggest using

$$\frac{1}{rs^2}(C\boldsymbol{b} - C\boldsymbol{\beta})'[C(X'X)^{-1}C']^{-1}(C\boldsymbol{b} - C\boldsymbol{\beta}) \sim F_{r,n-k-1} \tag{5.9}$$

in place of (5.8) since it appears to yield a better approximation. Notice that the statistic given in (5.9) is the one on which all our tests and confidence regions are based (see Chapter 3). Thus, all these procedures may be used without change if the h_{ii}'s are small.

By (A.9) of Appendix A (Page 278),

$$\sum_{i=1}^{n} h_{ii} = \text{tr}H = k+1 \tag{5.10}$$

if X is of dimension $(k+1) \times n$. Therefore, $\max h_{ii}$ is small if n/k is large and a few h_{ii}'s are not much larger than the rest of them. In Section 8.2, we shall see that y_i's corresponding to relatively large values of h_{ii} tend to have larger weights in the computation of b_j's. Therefore, the condition (5.7) is eminently reasonable in the context of the central limit theorem that is used to prove Theorem 5.1. Because h_{ii}'s often identify unduly influential points, they are frequently available from packages. The matrix H is sometimes called the *hat matrix* and h_{ii} the *leverage*.

It is difficult to specify a number such that if $\max h_{ii}$ falls below it we can safely carry out our tests. Apart from the fact that such a number would depend on how critical we wish to be, it would also depend on how non-normal the distribution of ϵ_i's is. However, as a *very rough* rule of thumb, $\max h_{ii} < .2$ may be taken as small enough for most applications if the original distribution of ϵ's is not too 'horrible' (e.g., excessively long-tailed, or J-shaped).

Example 5.5 (Continuation of Example 5.3, Page 105)
For the regression model described in Example 5.1, $\max h_{ii} = .266$. Since, as we have already seen, the residuals, and therefore the errors, are near normal already, we feel that assuming normality would not lead to conclusions which are too far wrong. ■

Example 5.6 (Continuation of Example 5.4, Page 106)
Before deleting the outliers, the independent variables in the model (4.9) yielded $\max h_{ii} = .0985$. In our judgment this is small enough that we could have carried out the usual F tests had normality been the only concern we had. After deleting the outliers, we get $\max h_{ii} = .10$. ■

5.4 *Bootstrapping

When h_{ii}'s are small and the sample size large, (5.8) and (5.9) would give reasonably good approximations. However, when the sample size is not too large (and h_{ii}'s are small), an improvement in the above approximation can be made by a method called *bootstrapping*. This method, which is due to Efron (1979), is based on resampling from the observed data which are considered fixed. In our case, since ϵ_i's are unknown, we resample from the e_i's, which are considered fixed or population values. The exact steps, for cases when the model contains an intercept term, are described below. A theoretical justification is given in Freedman (1981).

Step (i) Draw a random sample of size n *with replacement* from $e_1, \ldots, e_n$. Denote its members as $e^*_{(q)1}, \ldots, e^*_{(q)n}$ and let $\mathbf{e}^* = (e^*_{(q)1}, \ldots, e^*_{(q)n})'$.

Step (ii) Based on this sample calculate

$$(\mathbf{e}^*_{(q)})'X(X'X)^{-1}C'[C(X'X)^{-1}C']^{-1}C(X'X)^{-1}X'\mathbf{e}^*_{(q)}/s^{*2}_{(q)} \quad (5.11)$$

where

$$s^{*2}_{(q)} = n\frac{(n-1)^{-1}}{n-k-1}\sum_{i=1}^{n}(e^*_{(q)i} - \bar{e}^*_{(q)})^2,$$

and $\bar{e}^*_{(q)} = n^{-1}\sum_{i=1}^{n} e^*_{(q)i}$.

Step (iii) Repeat steps (i) and (ii) Q times, i.e., for $q = 1, 2, \ldots, Q$. In most studies, the value of Q is set between 200 and 1000. From these iterations obtain an empirical distribution of (5.11). Obtain the necessary significance points from this empirical distribution.

For testing the hypothesis $H : C\boldsymbol{\beta} - \mathbf{d} = \mathbf{o}$, calculate the statistic

$$(C\boldsymbol{b} - \mathbf{d})'[C(X'X)^{-1}C']^{-1}(C\boldsymbol{b} - \mathbf{d})/s^2$$

and use the significance points obtained in Step (iii).

If there is no intercept term β_0 in the regression model, $\sum e_i$ may not be zero. Hence, we center these e_i's and define

$$\tilde{e}_i = e_i - \bar{e}, \quad \text{for all } i = 1, \ldots, n.$$

Then steps (i) to (iii) are carried out with $\tilde{e}_1, \ldots, \tilde{e}_n$ replacing $e_1, \ldots, e_n$.

5.5 *Asymptotic Theory

To prove Theorem 5.1, we need the central limit theorem. The following version, proved in Gnedenko and Kolmogorov (1954), is convenient.

Theorem 5.2 *Let $Z_1, \ldots, Z_n$ be independently and identically distributed with mean zero and variance σ^2. For $i = 1, \ldots, n$ let $\{a_{n_i}\}$ be a sequence of constants such that*

$$\max_{1\le i\le n} |a_{n_i}| \to 0 \text{ and } \sum_{i=1}^{n} a^2_{n_i} \to 1, \text{ as } n \to \infty.$$

Then, as $n \to \infty$, $\sum_{i=1}^{n} a_{n_i} Z_i \to N(0, \sigma^2)$, i.e., $\sum_{i=1}^{n} a_{n_i} Z_i$ converges to a random variable which has a normal distribution with mean 0 and variance σ^2.

Proof of Theorem 5.1: For the sake of convenience in presentation, we set $C = I$. The proof for the general case is similar.

We first derive the asymptotic distribution of

$$\begin{aligned}
&\sigma^{-2}(\boldsymbol{b} - \boldsymbol{\beta})'(X'X)(\boldsymbol{b} - \boldsymbol{\beta})\\
&= \sigma^{-2}[(X'X)^{-1}X'\boldsymbol{y} - \boldsymbol{\beta}]'[X'X][(X'X)^{-1}X'\boldsymbol{y} - \boldsymbol{\beta}]\\
&= \sigma^{-2}[(X'X)^{-1}(X'\boldsymbol{y} - X'X\boldsymbol{\beta})]'[X'X][(X'X)^{-1}(X'\boldsymbol{y} - X'X\boldsymbol{\beta})]\\
&= \sigma^{-2}\boldsymbol{\epsilon}'X(X'X)^{-1}X'\boldsymbol{\epsilon} = \sigma^{-2}\boldsymbol{\epsilon}'H\boldsymbol{\epsilon}.
\end{aligned}$$

Let $L = (\ell_{ijn}) = (\boldsymbol{\ell}_n^{(1)}, \ldots, \boldsymbol{\ell}_n^{(n)}) = (X'X)^{-\frac{1}{2}}X'$. Then $H = L'L$. and $LL' = I_{k+1}$.

Consider the linear combination $\boldsymbol{a}'L\boldsymbol{\epsilon}$, where $\boldsymbol{a}'\boldsymbol{a} = 1$. Let $b_n^{(i)} = \boldsymbol{a}'\boldsymbol{\ell}_n^{(i)}$. Then $\boldsymbol{a}'L\boldsymbol{\epsilon} = \sum_{i=1}^n b_n^{(i)}\epsilon_i$ and

$$|b_n^{(i)}| = |\boldsymbol{a}'\boldsymbol{\ell}_n^{(i)}| \le (\boldsymbol{a}'\boldsymbol{a})^{\frac{1}{2}}(\boldsymbol{\ell}_n^{(i)\prime}\boldsymbol{\ell}_n^{(i)})^{\frac{1}{2}} = (\boldsymbol{\ell}_n^{(i)\prime}\boldsymbol{\ell}_n^{(i)})^{\frac{1}{2}} \to 0$$

when $\max_{1\le i\le n} h_{ii} = \max_{1\le i\le n} \boldsymbol{\ell}_n^{(i)\prime}\boldsymbol{\ell}_n^{(i)} \to 0$. Moreover, $\sum_{i=1}^n (b_n^{(i)})^2 = \boldsymbol{a}'LL'\boldsymbol{a} = 1$. Therefore, it follows from Theorem 5.2 that $\boldsymbol{a}'L\boldsymbol{\epsilon} \to N(0, \sigma^2)$. Since $\boldsymbol{a}'L\boldsymbol{\epsilon} \to N(0, \sigma^2)$ for every vector $\boldsymbol{a}$, it follows that (see Appendix B, just above Lemma B.1, p. 289) $L\boldsymbol{\epsilon} \to N(\mathbf{o}, \sigma^2 I)$. Hence, if $\max_{1\le i\le n} h_{ii} \to 0$, as $n \to \infty$, then as $n \to \infty$

$$\sigma^{-2}(\boldsymbol{b} - \boldsymbol{\beta})'(X'X)(\boldsymbol{b} - \boldsymbol{\beta}) \to \chi^2_{k+1}.$$

Equivalently, since $s^2 \to \sigma^2$ in probability,

$$s^{-2}(\boldsymbol{b} - \boldsymbol{\beta})'(X'X)(\boldsymbol{b} - \boldsymbol{\beta}) \to \chi^2_{k+1}.$$

□

Problems

Exercise 5.1: Consider the regression model

$$y_t = \beta_0 + \beta_1 t + \epsilon_t \text{ for } t = 1, \ldots, n.$$

(This model is called a linear trend model.) The design matrix X for this model is given by

$$X' = \begin{pmatrix} 1 & 1 & \ldots & 1 \\ 1 & 2 & \ldots & n \end{pmatrix}.$$

Show that $n^{-1}(X'X) \to \infty$ (i.e., at least one element of $X'X \to \infty$) as $n \to \infty$ but that $h_{ii} \to 0$ as $n \to \infty$.

Exercise 5.2: Would it be appropriate, as far as normality is concerned, to use the F test given in Chapter 3 to test hypotheses about $\boldsymbol{\beta}$ in the problem considered in Example 4.3, p. 88.

Exercise 5.3: Examine the residuals and h_{ii}'s from the model of Exercise 4.4, p. 96. Discuss the appropriateness of the conclusions you reached when you did Exercise 4.4.

Exercise 5.4: From the point of view of normality, comment on each of the tests you ran in Exercise 3.13, p. 77.

Exercise 5.5: For the two regressions in Exercise 3.14, p. 79, can one assume normality of the errors?

Exercise 5.6: Test for normality of observations in the data set used for Exercise 2.15, p. 53.

Exercise 5.7: *Using a bootstrap sample size of 200, obtain the 90 per cent point of the distribution of the test statistic in Exercise 1.11, p. 24 (after removing the outlier). Compare it with the one you obtained using a t distribution.

Exercise 5.8: *Using bootstrapping, obtain the 90 per cent points of the distributions of the parameter estimates you obtained in part 1 of Exercise 3.14, p. 79. Use a sample size of 500.

CHAPTER 6

Unequal Variances

6.1 Introduction

One of the great values of the Gauss-Markov theorem is that it provides conditions which, if they hold, assure us that least squares is a good procedure. These conditions can be checked and if we find that one or more of them are seriously violated, we can take action that will cause at least approximate compliance. This and the next few chapters will deal with various ways in which these G-M conditions can be violated and what we would then need to do.

This chapter is devoted to the second G-M condition, which states that $\text{var}(\epsilon_i) = \text{var}(y_i)$ is a constant, σ^2. Violation of this condition is often called *heteroscedasticity*, while compliance is referred to as *homoscedasticity*. Recall that heteroscedasticity does not bias the least squares estimates of β_j's, but it causes variances of parameter estimates to be large and can affect R^2, s^2 and tests substantially. The test of the general linear hypothesis (Chapter 2) is affected also because under heteroscedasticity, $s^2(X'X)^{-1}$ need no longer be an unbiased estimate of the covariance matrix of $\boldsymbol{\beta}$.

6.2 Detecting Heteroscedasticity

Very frequently, we can determine if heteroscedasticity is likely to be present from an understanding of the underlying situation and also (as we shall see in later sections) determine what corrective measures might be taken. For example, if the dependent variable is a counted variable, it is likely to have approximately a Poisson distribution (as in the case of telephone mains in Example 1.2, p. 10); then the variance σ_i^2 of the ith observation is approximately $\text{E}(y_i)$. If $y_i = m_i/n_i$ is a proportion of counts m_i and n_i, its variance would probably be close to $\text{E}(y_i)(1 - \text{E}(y_i))/n_i$. When y_i is the mean $\sum_{\alpha=1}^{n_i} z_\alpha/n_i$ of homoscedastic variables $z_1, \ldots, z_{n_i}$, then $\sigma_i^2 \propto n_i^{-1}$.

Even where the distribution cannot be guessed, some idea of the variance can be. Consider house prices for an entire metropolitan area. It would appear less likely that a house worth \$50,000 would sell for \$100,000 than that a million dollar one would sell for \$1,050,000. To continue this intuition-based discussion, it appears to be more likely that the less expensive house would sell for \$60,000 than the more expensive one for \$1.2 million. Thus, the standard deviation of the selling price is not constant, nor does it vary in proportion to the intrinsic value. Rather, it is something in between.

At this stage the reader might wish to recall some of the random variables he/she might have encountered and see if their variances (or quantities proportional to them) can be guessed (the reader might be surprised at how often this is possible!).

Another way of checking to see if heteroscedasticity is present is through plots. If $\sigma_i^2 = \text{var}(\epsilon_i)$ varies with $E(y_i)$, a plot of the residuals (which are estimates of ϵ_i's) against the $\hat{y}_i$'s (which are estimates of $E(y_i)$'s) *might* show the residuals e_i to be more spread out for some values of $\hat{y}_i$ than for others. Standardized or Studentized residuals (see Section 8.3, p. 156) could also be used and might even be preferable (see Cook and Weisberg, 1982).

d.	sp.	d.	sp.	d.	sp.	d.	sp.
4	4	14	10	29	18	57	27
2	5	17	10	34	18	78	27
4	5	11	12	47	18	64	28
8	5	19	12	30	19	84	28
8	5	21	12	48	20	54	29
7	7	15	13	39	21	68	29
7	7	18	13	42	21	60	30
8	8	27	13	55	21	67	30
9	8	14	14	56	24	101	30
11	8	16	14	33	25	77	31
13	8	16	15	48	25	85	35
5	9	14	16	56	25	107	35
5	9	19	16	59	25	79	36
13	9	34	16	39	26	138	39
8	10	22	17	41	26	110	40
		29	17			134	40

EXHIBIT 6.1: Data on Automobile Speed (sp.) and Distance Covered to Come to a Standstill After Braking (d.)
SOURCE: Ezekiel and Fox (1959). Reproduced, with permission, from Ezekiel, M. and F.A. Fox, *Methods of Correlation and Regression Analysis.* © 1959 John Wiley & Sons, Inc.

Example 6.1

Exhibit 6.2 illustrates a plot of e_i's against $\hat{y}_i$ after fitting an ordinary least squares model

$$\text{distance} = \beta_1 \text{speed} + \beta_2 \text{speed}^2 \qquad (6.1)$$

to the data in Exhibit 6.1. The plot here would seem to indicate the existence of heteroscedasticity. (By contrast, Exhibit 6.6 seems to indicate virtually no heteroscedasticity.)

If there are enough data points, plots like Exhibit 6.2 can also give us an idea of how the variance of the y_i's varies with the $\mathrm{E}(y_i)$'s. Divide the range of the $\hat{y}_i$'s into three portions, making a reasonable compromise between getting portions of roughly equal widths and getting roughly equal numbers of points in each portion. In the case of Exhibit 6.2, suitable break points might be 25 and 72. Let the medians of the $\hat{y}_i$'s within each such partition be $y^{(1)}$, $y^{(2)}$ and $y^{(3)}$ and let the corresponding inter-quartile range of the e_i's be $Q^{(1)}$, $Q^{(2)}$ and $Q^{(3)}$. A plot of the $Q^{(k)}$'s against the $y^{(k)}$'s can help identify a relationship between $\mathrm{var}\,(y_i)$ and $\mathrm{E}(y_i)$. In the case of Exhibit 6.2 such a plot is approximately a straight line, suggesting that the standard deviations of y_i's are roughly proportional to $\mathrm{E}(y_i)$'s and, therefore, $\mathrm{var}\,(y_i) \propto [\mathrm{E}(y_i)]^2$. The reader is requested to carry out these steps in Exercise 6.9. ■

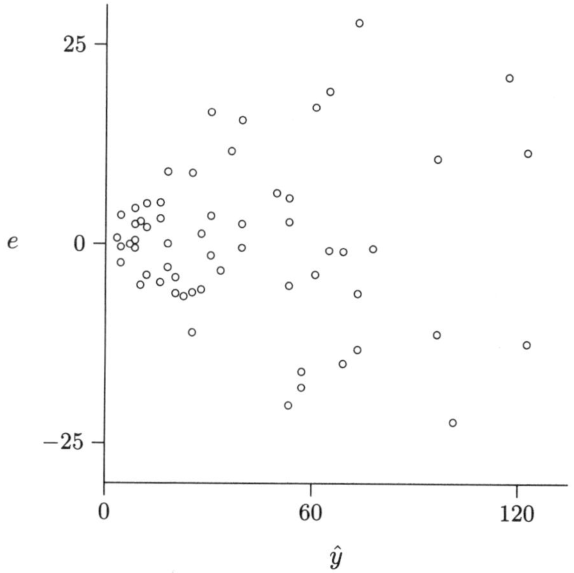

EXHIBIT 6.2: Plot of Residuals against Predicted for Speed-Braking Distance Data

Sometimes σ_i varies with one or more x_{ij}'s. For example, if we were regressing total hospital charges against severity of illness, the attending physician and the sex and age of the patient (as in Exercise 4.9, p. 97), it is not unlikely that the variance of charges might vary with one or more of the independent variables. Some physicians might order essentially the same set of preliminary laboratory and other medical tests for all patients, while others might tailor the order to the individual case. As people get older they become more susceptible to a wider range of diseases. Consequently, whether certain medical tests are more likely to be ordered for older patients and whether they are ordered or not affect variability of charges. However, it should be emphasized that it is the variation in the variance of the *dependent* variable that is a violation of the second G-M condition.

It is also possible that the variance of the y_i's could vary with changes in variables not included in the model. For example, it is known that *some* respondents inflate their income when they are attracted to the interviewer. Then the variance of income would vary with the interviewer. Similarly, different laboratory equipment, different machines, etc., can affect variance.

In such cases, it may be useful to examine plots of residuals against each independent variable and each variable that we expect affects the variance. Many careful analysts routinely obtain plots of e_i against all the x_{ij}'s and against $\hat{y}_i$. However, none of these plots are entirely safe in that heteroscedasticity can be present and not be apparent from them.

A number of other plots have also been suggested in the literature, including plotting the absolute values, squares or logarithms of the absolute values of residuals or the Standardized residuals against predicted values, other variables and even $(1 - h_{ii})$s. One advantage cited is that some of these plots make identification of the nature of the heteroscedasticity (e.g., the relationship between $\text{var}(y_i)$ and $E(y_i)$) easier. For example, since the log of the absolute values of residuals may be considered to be a proxy for the log of standard deviations of y_i's, and log of the predicteds that for the log of the expectation of the y_i's, the slope of a line fitting their plot would yield α when heteroscedasticity is described by $\text{var}(y_i) \propto [E(y_i)]^{2\alpha}$. For further discussion of these and other methods, see Carroll and Ruppert (1988, p.29 *et seq.*) and Cook and Weisberg (1982).

6.2.1 Formal Tests

A number of formal tests have also been proposed. A large number of them essentially test whether the variances σ^2 of individual ϵ_i's are related to some other variable(s), e.g., the independent variables or functions of them. A fair number of these approaches attempt to relate $|e_i|$'s, e_i^2's or the rank of $|e_i|$'s to other variables. A review of several procedures is given in Judge *et al.* (1985, see especially pp. 446–454) and Madansky (1988, p. 75 *et seq.*). One such test consists of testing for significance the correlation between the ranks of the absolute values of the residuals with the ranks of $\hat{y}_i$'s or those of individual independent variable values (such a correlation between ranks is called the Spearman correlation).

Another such test is that given by White (1980). It may be shown that under homoscedasticity, if each $h_{ii} \to 0$, $S_1 = n^{-1}s^2X'X$ and $S_2 = n^{-1}\sum_{i=1}^{n} e_i^2 \boldsymbol{x}_i \boldsymbol{x}_i'$ are asymptotically equivalent, while the presence of heteroscedasticity can cause them to be quite different. Therefore, one can base a test on the comparison of S_1 and S_2. Such a test is available, for example, in SAS. A relatively simple test statistic based on this principle (and also, incidentally, on the principle mentioned in the last paragraph) is $nR^2_{(h)}$, where $R^2_{(h)}$ is the usual R^2 from a regression of the e_i^2's against the independent variables x_{ij}'s and all their square and product terms (including a constant term even if one is not present in the original model and with any redundant variables eliminated). Under the hypothesis of no heteroscedasticity and provided the fourth moment of all the observations are the same, $nR^2_{(h)}$ has asymptotically a chi-square distribution with degrees of freedom equal to one less than the number of independent variables in the above mentioned regression.

Like many other tests for violations of specific Gauss-Markov conditions, White's test is also sensitive to other violations. Therefore, one needs to examine plots or in other ways assure oneself that it is indeed heteroscedasticity that is causing $nR^2_{(h)}$ to be high.

It might be noted in passing that S_2, which is provided by SAS, can be used to estimate the covariance matrix of $\boldsymbol{b}$ when heteroscedasticity is present — see White (1980) and SAS (1985b).

6.3 Variance Stabilizing Transformations

When heteroscedasticity occurs we can take one of two types of actions to make the σ_i's approximately equal. One consists of transforming y_i appropriately when the variance of y_i depends on its mean; the other involves weighting the regression. We consider the former in this section; the latter will be examined in the next section.

For any function $f(y)$ of y with continuous first derivative $f'(y)$ and finite second derivative $f''(y)$, we know from elementary calculus that

$$f(y_i) - f(\eta_i) = (y_i - \eta_i)f'(\eta_i) + \tfrac{1}{2}(y_i - \eta_i)^2 f''(\theta), \tag{6.2}$$

where θ lies between y_i and η_i, and $\eta_i = \mathrm{E}(y_i)$. Thus, when $(y_i - \eta_i)^2$ is small, we have

$$f(y_i) - f(\eta_i) \approx f'(\eta_i)(y_i - \eta_i). \tag{6.3}$$

Squaring and taking expectations of both sides of (6.3), we get approximately,

$$\mathrm{var}\,(f(y_i)) \approx (f'(\eta_i))^2 \sigma_i^2(\eta_i), \tag{6.4}$$

where $\sigma_i^2(\eta_i)$ is the variance of the random variable y_i with mean η_i. Thus, in order to find a suitable transformation f of y_i which would make $\mathrm{var}\,(f(y_i))$ approximately a constant, we need to solve the equation

$$f'(\eta_i) = c/\sigma_i(\eta_i), \tag{6.5}$$

where c is any constant. Such a transformation f is called a variance stabilizing transformation.

As an example, consider the case where y_i is a counted variable. Then $\sigma_i^2(\eta_i) \propto \eta_i$ and we need an f such that

$$f'(\eta_i) = c/\eta_i^{1/2}. \tag{6.6}$$

Clearly, if we choose $c = 1/2$, then $f(\eta_i) = \eta_i^{1/2}$ solves (6.6); and therefore in this case $y_i^{1/2}$ is a variance stabilizing transformation. Suppose now we have $y_i = m_i/n_i$ which is a proportion of counts and hence has a binomial distribution. Then $\text{var}\,(y_i) = n_i^{-1}\eta_i(1-\eta_i)$, where $\eta_i = \text{E}(y_i)$. Thus, we need to solve

$$f'(\eta_i) = cn_i^{1/2}/(\eta_i^{1/2}(1-\eta_i)^{1/2}). \tag{6.7}$$

Integrating both sides with respect to η_i, we get

$$f(\eta_i) = \int f'(\eta_i)d\eta_i = cn_i^{1/2}\int \frac{d\eta_i}{\eta_i^{1/2}(1-\eta_i)^{1/2}} = 2cn_i^{1/2}\sin^{-1}\sqrt{\eta_i}$$

yielding $n_i^{1/2}\sin^{-1}\sqrt{y_i}$ as an appropriate transformation. As another example, if $\sigma_i = \eta_i$, then (6.5) yields $f(\eta_i) = \log(\eta_i)$.

The Box-Cox transformations, which will be described in Section 9.4.3, p. 204, are often useful for alleviating heteroscedasticity when the distribution of the dependent variable is not known.

Example 6.2 (Continuation of Example 1.1, Page 2)
In Example 1.1 we took the square root of the dependent variable. That was an attempt at a variance stabilizing transformation, since we expected that σ_i and η_i were related although it was not immediately obvious what the relationship was. Exhibit 6.3 shows a plot of e_i's against y_i's when such a transformation was not made. The right hand mass of points is much more spread out (in terms of size of residuals) than the mass on the left. Exhibit 6.4, which is the same as Exhibit 1.5, shows the plot corresponding to the square root transformation. Now the two masses have more nearly equal spreads. Therefore, Exhibit 6.3 illustrates greater heteroscedasticity than does Exhibit 6.4. Exhibit 6.5 gives a similar plot, when log (speed) is the dependent variable. Now we seem to have slightly overdone it. The appropriate choice would appear to be between the square root transformation and the log transformation, or perhaps something in between.

Note, however, that the comparisons made above are not entirely fair, since we are changing more than just the variance of the y_i's. As mentioned below, this is one reason why variance stabilizing transformations are sometimes not very useful. ∎

While variance stabilizing transformations are sometimes very useful — particularly in simple regression — quite often they are not. As an example, consider house prices again. Since they are usually based on lot prices plus improvements, one would expect that y_i is a linear combination of at least

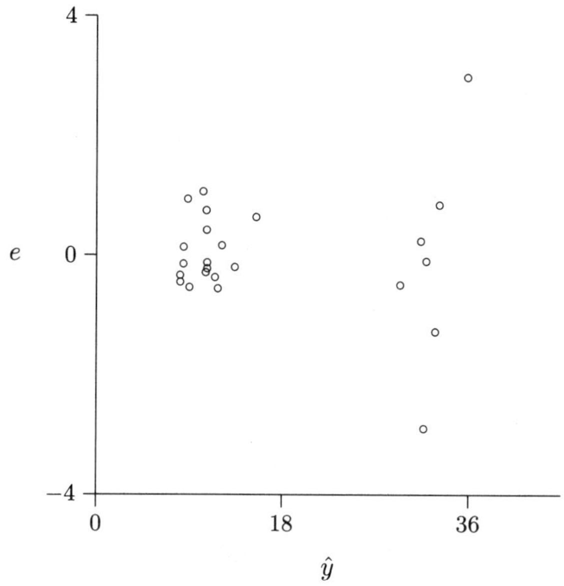

EXHIBIT 6.3: Residual vs. Predicted Plot for the Regression of Speed Against Density and Density^2

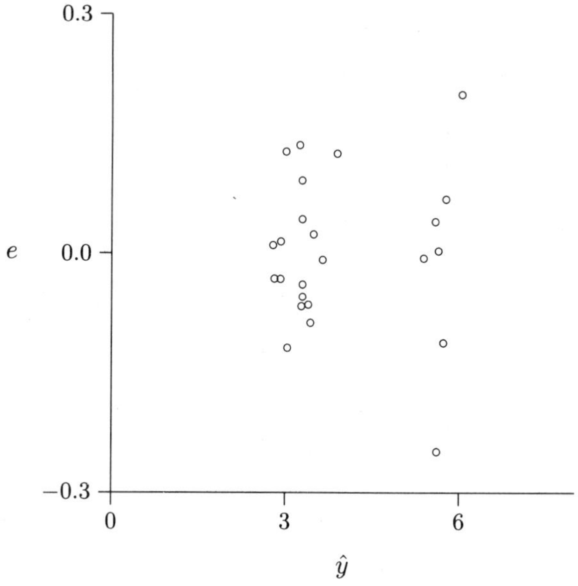

EXHIBIT 6.4: Residual vs. Predicted Plot for the Regression of Square Root of Speed Against Density and Density^2

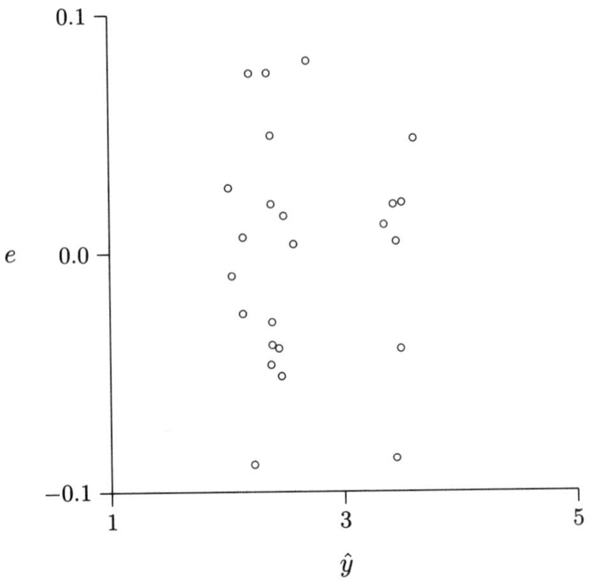

EXHIBIT 6.5: Residual vs. Predicted Plot for the Regression of Logarithm of Speed Against Density and Density^2

some of the x_{ij}'s. This simple relationship would be lost if we replaced y_i by, say, $\sqrt{y_i}$. In other cases, a transformation of y which is not variance stabilizing may be desirable for other reasons (see Chapter 9). In some situations we may expect that σ_i is a function of one of the independent variables and it is this relationship we wish to exploit. Fortunately, in such cases, homoscedasticity may be achieved in another way, as we shall see in the next section.

6.4 Weighting

Suppose $\mathrm{var}(\epsilon_i) = \sigma_i^2 = c_i^2\sigma^2$ where c_i^2 are known constants. Then constancy of variance can also be achieved by dividing both sides of each of the equations of the regression model,

$$y_i = \beta_0 + \beta_1 x_{i1} + \cdots + \beta_k x_{ik} + \epsilon_i, \quad i = 1, \ldots, n,$$

by c_i, i.e., by considering

$$y_i/c_i = \beta_0/c_i + \cdots + \beta_k x_{ik}/c_i + \epsilon_i/c_i, \quad i = 1, \ldots, n. \tag{6.8}$$

Model (6.8) is clearly homoscedastic. Each $w_i = (c_i)^{-2}$ is called a weight, the nomenclature coming from the fact that now we are minimizing a weighted sum of squares,

$$\sum w_i(y_i - \beta_0 - \beta_1 x_{i1} \cdots - \beta_k x_{ik})^2. \tag{6.9}$$

Obviously, when the σ_i's (or a quantity proportional to them) are known, weights are not difficult to compute.

The estimate of $\boldsymbol{\beta}$ obtained from the model (6.8), i.e., by minimizing (6.9), is called a *weighted least squares* (WLS) estimate of β and will be denoted by $\boldsymbol{b}_{WLS}$. When $c_i = 1$, i.e., when least squares is not 'weighted', we call it *ordinary least squares (OLS)* — which is what we have been doing until this point in the book. Nowadays, WLS estimates can be obtained from just about all statistical packages.

Example 6.3

Suppose for each value $\boldsymbol{x}_i$ of the independent variables, w_i observations $y_{i\ell_i}$ are taken. Assume that the model is $y_{i\ell_i} = \boldsymbol{x}_i'\boldsymbol{\beta} + \epsilon_{i\ell_i}$ where $\ell_i = 1, \ldots, w_i$, $i = 1, \ldots, n$ and $\epsilon_{i\ell_i}$'s meet the Gauss-Markov conditions. In particular, let $\text{var}(\epsilon_{i\ell_i}) = \sigma^2$. Write $\bar{y}_i = \sum_{\ell_i=1}^{w_i} y_{i\ell_i}/w_i$. Suppose that only these averages have been recorded, i.e., the individual observations $y_{i\ell_i}$ are not available.

Then, since $\text{E}[\bar{y}_i] = \boldsymbol{x}_i'\boldsymbol{\beta}$, one might be tempted to use OLS in order to obtain an estimate of $\boldsymbol{\beta}$, i.e., to implicitly minimize $\sum_{i=1}^n (\bar{y}_i - \boldsymbol{x}_i'\boldsymbol{\beta})^2$. But

$$\text{var}(\bar{y}_i) = w_i^{-2} \sum_{\ell_i=1}^{w_i} \text{var}(y_{i\ell_i}) = w_i^{-2} \sum_{\ell_i=1}^{w_i} \text{var}(\epsilon_{i\ell_i}) = \sigma^2/w_i.$$

Therefore, this approach would violate Gauss-Markov conditions and could lead to inferior estimates. Intuitively speaking also, the approach violates the principle of 'one observation – one vote.'

Obviously, it would be preferable to minimize $\sum_{i=1}^n \sum_{\ell_i=1}^{w_i} (y_{i\ell_i} - \boldsymbol{x}_i'\boldsymbol{\beta})^2$. But since $\sum_{\ell_i=1}^{w_i} [(y_{i\ell_i} - \bar{y}_i)] = 0$, this equals

$$\sum_{i=1}^{n} \sum_{\ell_i=1}^{w_i} (y_{i\ell_i} - \bar{y}_i + \bar{y}_i - \boldsymbol{x}_i'\boldsymbol{\beta})^2 = \sum_{i=1}^{n} \sum_{\ell_i=1}^{w_i} [(y_{i\ell_i} - \bar{y}_i)^2 + (\bar{y}_i - \boldsymbol{x}_i'\boldsymbol{\beta})^2]$$

$$= \sum_{i=1}^{n} \sum_{\ell_i=1}^{w_i} (y_{i\ell_i} - \bar{y}_i)^2 + \sum_{i=1}^{n} w_i (\bar{y}_i - \boldsymbol{x}_i'\boldsymbol{\beta})^2.$$

Since the first term in the last expression does not include $\boldsymbol{\beta}$, minimizing it is equivalent to minimizing

$$\sum_{i=1}^{n} w_i (\bar{y}_i - \boldsymbol{x}_i'\boldsymbol{\beta})^2.$$

Since $\text{var}(\bar{y}_i) = \sigma^2/w_i$, we see from (6.9) that this yields the appropriate WLS estimate.

It should be pointed out that while the OLS estimate of $\boldsymbol{\beta}$ using all the observations is the same as the WLS estimate using the means $\bar{y}_i$'s, the estimates of the error variances from the two models could be different. If all the observations were used in an OLS model, then an unbiased estimate of σ^2 would be

$$\left(\sum_{i=1}^{n} w_i - k - 1\right)^{-1} \sum_{i=1}^{n} \sum_{\ell_i=1}^{w_i} [(y_{i\ell_i} - \bar{y}_i)^2 + (\bar{y}_i - \boldsymbol{x}_i'\boldsymbol{b}_{WLS})^2], \tag{6.10}$$

when $\boldsymbol{\beta}$ is a $(k+1)$-vector. On the other hand, if the averages were used in a WLS procedure, an unbiased estimator of the error variance would be

$$(n-k-1)^{-1} \sum_{i=1}^{n} w_i (\bar{y}_i - \boldsymbol{x}_i'\boldsymbol{b}_{WLS})^2,$$

as we shall see shortly. ■

In order to obtain expressions for various estimates, let us now describe weighting in matrix notation. Let Ω be a diagonal matrix with diagonal elements $c_1^2, \ldots, c_n^2$. When we do weighted regression, the original model $\boldsymbol{y} = X\boldsymbol{\beta} + \boldsymbol{\epsilon}$, with $\mathrm{E}(\boldsymbol{\epsilon}) = \mathbf{o}$ and $\mathrm{cov}(\boldsymbol{\epsilon}) = \sigma^2\Omega$, is transformed to the model $\boldsymbol{y}^{(\Omega)} = X^{(\Omega)}\boldsymbol{\beta} + \boldsymbol{\epsilon}^{(\Omega)}$, where $\boldsymbol{y}^{(\Omega)} = C\boldsymbol{y}$, $X^{(\Omega)} = CX$ and $\boldsymbol{\epsilon}^{(\Omega)} = C\boldsymbol{\epsilon}$ and C is a diagonal matrix with non-zero elements $c_1^{-1}, \ldots, c_n^{-1}$. Since $C\Omega C' = I$, it follows that $\mathrm{cov}(\boldsymbol{\epsilon}^{(\Omega)}) = \sigma^2 I$. Hence, the variables with superscript (Ω) satisfy the Gauss-Markov conditions, and least squares analysis can be carried out using them.

However, if we prefer to work with the original variables, we may write the estimate of $\boldsymbol{\beta}$ as (since $CC' = C'C = \Omega^{-1}$)

$$\boldsymbol{b}_{WLS} = (X^{(\Omega)\prime}X^{(\Omega)})^{-1}X^{(\Omega)\prime}\boldsymbol{y}^{(\Omega)} = (X'\Omega^{-1}X)^{-1}X'\Omega^{-1}\boldsymbol{y}. \tag{6.11}$$

Therefore,

$$\begin{aligned}\mathrm{cov}(\boldsymbol{b}_{WLS}) &= (X'\Omega^{-1}X)^{-1}X'\Omega^{-1}(\sigma^2\Omega)\Omega^{-1}X(X'\Omega^{-1}X)^{-1} \\ &= \sigma^2(X'\Omega^{-1}X)^{-1}.\end{aligned}$$

The residual vector is

$$\begin{aligned}\boldsymbol{e}^{(\Omega)} = \boldsymbol{y}^{(\Omega)} - \hat{\boldsymbol{y}}^{(\Omega)} &= \boldsymbol{y}^{(\Omega)} - X^{(\Omega)}\boldsymbol{b}_{WLS} = C\boldsymbol{y} - CX\boldsymbol{b}_{WLS} \\ &= C(\boldsymbol{y} - X\boldsymbol{b}_{WLS}) = C(\boldsymbol{y} - \hat{\boldsymbol{y}}_{WLS}),\end{aligned}$$

where $\hat{\boldsymbol{y}}_{WLS} = X\boldsymbol{b}_{WLS}$. Hence,

$$e_i^{(\Omega)} = (y_i - \hat{y}_{i,WLS})/c_i = \sqrt{w_i}(y_i - \hat{y}_{i,WLS}),$$

where $\hat{y}_{i,WLS}$ is the ith component of $\hat{\boldsymbol{y}}_{WLS}$. Thus, an unbiased estimate of σ^2 is given by

$$(n-k-1)^{-1}\sum_{i=1}^{n}(e_i^{(\Omega)})^2 = (n-k-1)^{-1}\sum_{i=1}^{n} w_i(y_i - \hat{y}_{i,WLS})^2.$$

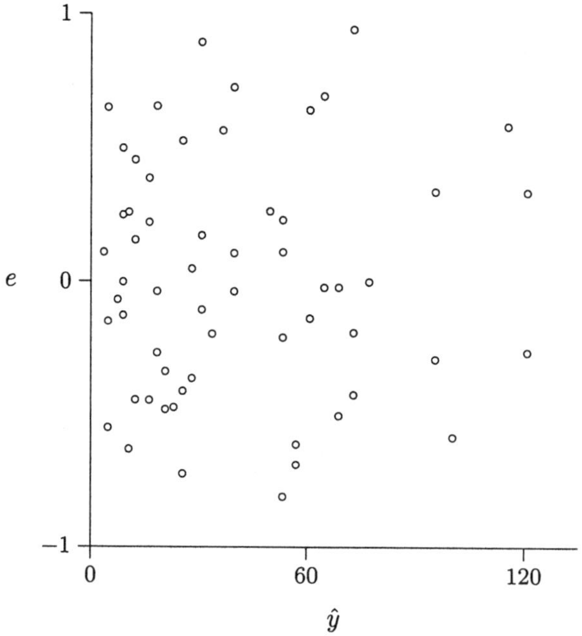

EXHIBIT 6.6: Residual vs. Predicted Plot for the Weighted Regression of Braking Distance Against Speed

Had we ignored the presence of heteroscedasticity, and obtained the OLS estimator $\boldsymbol{b}_{OLS} = (X'X)^{-1}X'\boldsymbol{y}$, then

$$\text{cov}(\boldsymbol{b}_{OLS}) = (X'X)^{-1}X'(\sigma^2\Omega)X(X'X)^{-1} = \sigma^2(X'X)^{-1}X'\Omega X(X'X)^{-1}.$$

From the Gauss-Markov theorem, it follows that for a non-null vector $\boldsymbol{a}$,

$$\text{var}\,(\boldsymbol{a}'\boldsymbol{b}_{WLS}) \leq \text{var}\,(\boldsymbol{a}'\boldsymbol{b}_{OLS}),$$

where, of course, $\boldsymbol{a}'\boldsymbol{b}_{WLS}$ and $\boldsymbol{a}'\boldsymbol{b}_{OLS}$ are estimates of $\boldsymbol{a}'\boldsymbol{\beta}$. Therefore, under heteroscedasticity, appropriate weighting yields preferable estimates. Weighted least squares is a special case of *generalized least squares* considered in Chapter 7.

Example 6.4 (Continuation of Example 6.1, Page 112)
Exhibit 6.6 shows the residual versus predicted plot resulting from running a regression on the data of Exhibit 6.1 using speed^{-2} as weight. (This weight has also been suggested in Hald, 1960; it is equivalent to using

$$\text{distance/speed} = \beta_1 + \beta_2\,\text{speed}$$

as the model.) The reader is invited to compare Exhibit 6.6 with Exhibit 6.2.

The 'residuals' in Exhibit 6.6 are $\sqrt{w_i}[y_i - \hat{y}_{i,WLS}]$ which, apart from being theoretically appropriate, are also the ones to plot if one wishes to check if homoscedasticity has been approximately achieved. ∎

When running WLS, the user of regression packages needs to bear in mind the fact that usually packages will give $y_i - \hat{y}_{i,WLS}$ as residuals instead of $\sqrt{w_i}(y_i - \hat{y}_{i,WLS})$, which we have seen are the appropriate ones. Moreover, some packages respond to a command to weight with integer-valued w_i's by making each data point $(y_i, \boldsymbol{x}_i)$ into w_i copies of it. (If w_i's are not integers, some packages will truncate the w_i's to their integer values.) When w_i's are integers this form of weighting yields the same estimates we would get had we done the weighting the usual way (by minimizing (6.9)). The estimate of $\boldsymbol{b}_{WLS}$ is the same and the covariance matrix of $\boldsymbol{b}_{WLS}$ is $\sigma^2(X'WX)^{-1}$ where $W = \text{diag}(w_1, \ldots, w_n)$. However, we need to be careful regarding the estimate of σ^2. As we have already seen, an unbiased estimate of σ^2 is $\sum_{i=1}^n w_i(y_i - \hat{y}_{i,WLS})^2/(n-k-1)$, but if an OLS package program is used unaltered, it would compute the estimate of σ^2 to be $\sum_{i=1}^n w_i(y_i - \hat{y}_{i,WLS})^2/(\sum_{i=1}^n w_i - k - 1)$, which would be wrong and will frequently be extremely small (See also Example 6.3 and Exercise 6.6).

Example 6.5 (Continuation of Example 6.3, Page 119)
The reason why 'brute force' application of OLS, after making copies of the data points, yields a poor estimate of σ^2 can be seen from the discussion of Example 6.3. The 'brute force' application is the same as the problem considered there, if we set $y_{i\ell_i} = \bar{y}_i$ for $\ell_i = 1, \ldots, w_i$ and each i. But this makes the first term in (6.10) equal to zero and yields a (frequently severe) underestimate of σ^2. The fact is that we would be treating each set of w_i $\bar{y}_i$'s as independent observations, when they are not! ∎

In Example 6.4, we weighted with a function of the independent variable; we may also weight using the dependent variable. For the reader's convenience, Exhibit 6.7 presents a table of weights for various types of dependent variables and various transformations of them (weights for the transformations were computed from variances obtained by using (6.4)). *However, we should point out that in practice the theoretical distribution of the dependent variable is not the only cause of heteroscedasticity.* The error term can be affected by variables left out. For example, consider the dependent variable average household size by state and assume that the data were obtained from the census. Although this is a mean of n_i counted variables, the n_i's are so large that the appropriate formula in Exhibit 6.7 would give variances which would be nearly zeros. Therefore, if we encounter a non-zero s, it would imply that the reason for the variance is not just the theoretical distribution of average sample size.

Weighting with functions $w[\mathrm{E}(y_i)]$ of $\mathrm{E}(y_i)$ presents a problem since $\mathrm{E}(y_i)$ is not known. On occasion one might be able to use y_i as an estimate

Type of Variable	Untransformed y_i	$\sqrt{y}_i$	$\log(y_i)$
Counts (Poisson)	z_i^{-1}	1 (i.e., apply ordinary least squares)	z_i
Proportion of counts (of form $y_i = m_i/n_i$)	$n_i z_i^{-1}(1-z_i)^{-1}$	$n_i(1-z_i)^{-1}$	$n_i z_i(1-z_1)^{-1}$
Homoscedastic variable	1 (obviously)	z_i	z_i^2
Means of n_i homoscedastic variables	n_i	$n_i z_i$	$n_i z_i^2$
Mean of n_i counted variables	$n_i z_i^{-1}$	n_i	$n_i z_i$

LEGEND AND NOTES: $z_i = \mathrm{E}(y_i)$. Columns for $\sqrt{y_i}$ and $\log(y_i)$ were computed using the approximate formula

$$\mathrm{var}\,(f(y)) = (\mathrm{var}\,(y))(f'(z))^2;$$

where f' represents the derivative, and z represents the mean of y. The entries in the table are the reciprocal of the variances.

EXHIBIT 6.7: Suggested Weights

of $\mathrm{E}(y_i)$ and weight with $w[y_i]$'s but this usually leads to bias. Frequently, a better approach is to obtain ordinary least squares estimates, compute $\hat{y}_i$'s, and *then* run a least squares procedure using as weights $w[\hat{y}_i]$'s.

Of course, this would usually give different, and presumably better, estimates of $\boldsymbol{\beta}$ and $\mathrm{E}(y_i)$'s. One could then use the most recently obtained estimates of $\mathrm{E}(y_i)$'s to compute weights and run a weighted least squares procedure again. These iterations can be continued until some convergence criterion is satisfied. This procedure is often called *iteratively reweighted least squares.* Computer programs for it are also available, although most common linear least squares packages do not include it. Since nonlinear least squares involves iterative procedures anyway, they can often be used to advantage to carry out such iterations (see Appendix C, especially Section C.2.4, p. 305; also see SAS, 1985b, especially pp. 597–598; and Wilkinson, 1987, especially p. NONLIN-25). As discussed in Section C.2.4, when some nonlinear least squares programs are applied to weighted linear least squares, the iterative steps are exactly the same as repeated applications of OLS as described earlier in this paragraph. One difficulty with all these procedures is that, on occasion, weights might turn out to be negative, in which case it is not always clear what needs to be done.

Example 6.6

The data shown in the left part of Exhibit 6.8 were collected in 1976 by Louise Stanten-Maston, a student of one of the authors, from 54 dial-a-ride systems in the U.S. and Canada. The variables were the number of riders (RDR) using the system, the number of vehicles (VH) in operation, hours of operation (HR), the fare (F) and the population (POP) and area (AR) of the place where the service was provided. In addition, there is a subjective rating called IND. Dial-a-ride services can vary in several ways. They can provide service from several points to several other points (many-many service) or connect several points to a few or one point (many-one); they can provide door to door service or to designated stops; they may require advance registration. Some of these features increase ridership while others may provide better quality. IND is a composite measure which is 1 when several ridership enhancing features were present and is 0 otherwise.

The reason for collecting the data was to construct a travel demand model, i.e., a model that expresses number of riders in terms of other variables. Such models are used to forecast ridership when new systems are planned. For reasons that will be discussed in Chapter 9, it seems desirable to take logarithms of all variables except, obviously, IND. Set

$$\text{LRIDERS} = \log[\text{RDR} + \tfrac{1}{2}] \quad \text{LPOP} = \log[\text{POP}] \quad \text{LAREA} = \log[\text{AR}]$$

$$\text{LVEHS} = \log[\text{VH}] \quad \text{LFARES} = \log[\text{F}] \quad \text{LHOURS} = \log[\text{HR}]$$

The reason behind the use of $\frac{1}{2}$ in the definition of LRIDERS also will be discussed in Ch. 9 (see p. 185). Because logarithms were being taken, three of the fares were set at one cent; the services were actually free in those cases.

Since the number of riders is a counted variable, its logarithm, which is the dependent variable, would call for weighting by E(RDR) (Exhibit 6.7). Exhibit 6.9 shows the parameter estimates obtained for the first few iterations (It.# 0 through It.# 3) and after 14 iterations (It.# ∞) when some rather stringent convergence criteria were met. Computations for the first few iterations were made using a least squares program, simply using the predicteds from each step to compute weights for the next. In the multiple iterations case we used PROC NLIN in SAS, but the computations are essentially identical to those of repeating a linear least squares procedure several times (See Appendix C, Section C.2.4, p. 305). It.# 0 represents ordinary least squares. Weights used for each iteration are shown in Exhibit 6.8, with the column labeled w giving weights for the fourteenth iteration.

Obs	POP	AR	RDR	HR	VH	F	IND	$w^{(1)}$	$w^{(2)}$	$w^{(3)}$	w
1	100000	13.6	2718	18.5	22	.25	1	2050	2545	2624	2652
2	8872	2.3	250	12.0	3	.35	0	149	133	132	132
3	17338	4.3	350	12.0	2	.60	1	225	220	217	215
4	26170	4.6	186	12.0	4	.50	0	216	201	198	196
5	60000	17.0	600	12.0	14	.50	0	543	550	546	546
6	40000	7.0	420	12.0	5	.50	1	493	529	531	531
7	30850	3.9	249	12.0	2	.50	1	253	253	250	248
8	25000	6.5	350	13.0	8	.25	0	335	323	322	322
9	44000	10.9	925	24.0	19	.30	0	1019	1091	1100	1103
10	24300	6.4	514	24.0	12	.60	0	750	783	791	794
11	21455	2.6	117	8.0	4	.50	0	177	167	167	168
12	47000	7.0	450	12.0	7	.50	0	351	345	342	342
13	45000	7.0	275	12.0	5	.50	0	268	254	249	247
14	23000	6.0	360	12.0	6	.25	1	478	510	515	518
15	20476	3.8	307	11.0	4	.50	1	379	401	406	408
16	20504	5.1	227	12.0	5	.50	0	240	225	222	222
17	71901	15.8	208	12.0	4	.60	0	219	199	190	185
18	70000	12.0	700	16.3	13	1.00	0	743	784	784	784
19	30000	10.0	440	12.0	7	1.00	0	326	313	309	308
20	26689	3.5	275	15.0	4	.01	1	310	309	308	305
21	9790	4.6	201	15.5	3	.50	0	165	145	141	139
22	19805	10.4	314	14.0	5	.50	1	436	450	447	444
23	5321	4.7	95	15.0	2	.50	0	103	85	82	80
24	56828	51.6	679	12.0	15	.60	1	896	982	973	969
25	11995	5.1	224	12.0	4	.50	0	180	162	159	158
26	10490	6.1	277	12.0	4	.50	1	319	322	321	323
27	7883	4.1	67	10.0	2	.50	0	89	74	71	70
28	3025	2.4	83	8.0	2	.50	0	69	57	56	56
29	17074	7.5	245	12.0	4	.50	1	341	347	345	344
30	7728	4.3	148	12.0	3	.50	0	135	117	114	114
31	27137	14.2	266	12.0	6	.50	0	244	223	216	214
32	12287	4.1	270	12.0	4	.50	1	355	368	371	374
33	24090	568.0	56	10.5	3	.50	0	66	46	40	37
34	9521	4.3	236	12.0	4	.50	1	333	341	345	347
35	18404	408.0	251	12.0	5	.50	1	202	170	154	147
36	7253	4.6	150	12.0	3	.50	0	131	113	111	110
37	28500	10.0	370	10.0	5	.25	1	353	360	358	356
38	35176	24.9	464	16.8	10	.50	0	430	410	398	392
39	12988	5.2	260	12.0	5	.50	0	217	200	198	198
40	9892	251.0	63	12.0	3	.50	0	68	48	43	40
41	15136	15.7	341	14.5	6	.50	1	451	460	455	452
42	26321	17.8	222	12.0	6	.50	0	233	210	203	200
43	18000	10.0	200	10.0	3	.50	1	233	226	221	219
44	9500	5.5	228	11.5	3	.25	1	228	220	218	218
45	27600	9.0	900	12.0	4	.10	1	302	298	293	290
46	24127	7.2	199	12.0	6	.60	0	275	260	257	256
47	14000	4.0	600	20.0	6	.25	1	639	696	709	713
48	53860	3.0	300	17.0	10	.01	0	434	433	436	436
49	18000	28.0	310	14.5	9	.25	0	282	253	244	240
50	29103	2.5	369	15.2	4	.20	1	484	525	534	536
51	102711	9.5	400	16.0	11	.01	0	420	406	398	393
52	25000	5.0	140	5.0	2	.35	1	127	121	119	118
53	32000	5.0	3400	18.7	12	.35	1	1256	1498	1555	1580
54	35000	7.0	200	4.0	4	.35	1	192	192	193	194

EXHIBIT 6.8: Dial-a-Ride Data and Weights

One purpose for showing results from several iterations is to demonstrate that obtaining weights from a single OLS run is usually not good enough, but if a suitable iterative program is not available, running a few iterations of a weighted least squares program may not be too bad. The t values given show that running OLS when weighting is required could lead to different conclusions (this example was not specially selected to show this; in extreme examples very different conclusions can come from weighting). Notice that in neither case is the distribution of $t(b_j)$ a t distribution. In the OLS case it is not a t distribution because the standard errors, computed as they are as the diagonal elements of $s^2(X'X)^{-1}$, are incorrect because of heteroscedasticity. In the case of weighted least squares, weighting by $\hat{y}_{i,WLS}$'s which are functions of the observations gives rise to a more complicated distribution than the t. However, use of t tables in the latter case is usually not too far wrong for moderate to large samples (see, for example, Carroll and Ruppert, 1988, p. 23 *et seq.*).

As a practical matter, after suitable weights have been found, they can simply be appended to the data and used in future computations, e.g., variable search computations of Chapter 11. ■

	It.#0		It.#1	It.#2	It.#3	It.#∞	
Variable	b_j	$t(b_j)$	b_j	b_j	b_j	b_j	$t(b_j)$
Intercept	0.911	0.920	0.358	0.364	0.396	0.415	0.278
LPOP	0.213	2.320	0.248	0.246	0.244	0.243	1.845
LAREA	-0.179	-4.419	-0.232	-0.257	-0.266	-0.270	-3.356
LVEHS	0.782	6.128	0.882	0.916	0.929	0.936	5.170
LFARES	0.120	2.388	0.141	0.145	0.147	0.148	2.419
LHOURS	0.633	3.324	0.666	0.660	0.651	0.647	2.343
IND	0.636	7.136	0.765	0.787	0.793	0.796	7.631

EXHIBIT 6.9: Coefficients and t-values for Selected Weighting Iterations for Dial-a-Ride Data

We have only considered the case where c_i's were known functions of either the x_{ij}'s or of $E(y_i)$'s. Weights can be constructed under less restrictive conditions. While a complete discussion of this subject is beyond the scope of this book, we discuss one particular case of common occurrence below and another in Section 7.3.1, p. 134.

Suppose the variance of y_i's is a step function of some other known variables. Then we could divide our observations into several subsets and the variance of y_i's would be a constant for each subset. If each of these subsets contains enough observations, then variances can be estimated for each subset separately and used to compute weights. This estimation would be based on residuals from an OLS procedure. Either the sum of squares of residuals in each subset divided by the number of cases in the subset could be used or the interquartile range for residuals in each subset, as a quantity proportional to the standard deviation, could be used.

Alternatively, if the ratio of the number of observations to the number of independent variables is large enough, we could run least squares on each subset separately and use the estimates of the error variance from each to obtain weights. For a review on the performance of this procedure, see Judge *et al.* (1985, p. 428 *et seq.*)

Problems

Exercise 6.1: An analyst is conducting a study of factors affecting waiting times. He assumes that his dependent variable waiting time (y_i) has a negative exponential distribution for which $\text{E}(y_i) = \lambda_i$ and $\text{var}\,(y_i) = \lambda_i^2$. Obtain a variance stabilizing transformation for y_i.

Exercise 6.2: Show that if instead of weighting by $c_1^{-2}, \dots, c_n^{-2}$, we had weighted by $\alpha c_1^{-2}, \dots, \alpha c_n^{-2}$ where α is any positive number, then $\boldsymbol{b}_{WLS}$ and its covariance matrix would be unaffected.

Exercise 6.3: Discuss how you would estimate the variance of a future observation (Section 3.8.2, p. 71), in the case of weighted regression.

Exercise 6.4: Consider the model $y_i = \beta x_i + \epsilon_i$, where ϵ_i's are independent and distributed as $N(0, \sigma^2 x_i^2)$. Find the weighted least squares estimator for β and its variance. Give reasons why you would not wish to use ordinary least squares in this case.

Exercise 6.5: Suppose $y_1, \dots, y_n$ are independently distributed and $y_i = \beta x_i + \epsilon_i$ with $x_i > 0$, $\text{E}(\epsilon_i) = 0$, $\text{var}\,(\epsilon_i) = \sigma^2 x_i$ and $i = 1, \dots, n$. Find the best unbiased estimator of β and its variance.

Exercise 6.6: Consider the model $y_i = \boldsymbol{x}_i'\boldsymbol{\beta} + \epsilon_i$, where $i = 1, \dots, n$, $\boldsymbol{x}_i$'s are $k+1$-vectors, the ϵ_i's are independently distributed with means zero and variances $w_i^{-1}\sigma^2$ and the w_i's are known positive integers. Show that the weighted least squares estimate of $\boldsymbol{\beta}$ can be obtained using ordinary least squares in the following way: Construct a data set in which each of the cases $(y_i, \boldsymbol{x}_i)$ is repeated w_i times. Show that the ordinary least squares estimate of $\boldsymbol{\beta}$ obtained from this data set is $(X'WX)^{-1}X'W\boldsymbol{y}$, and an unbiased estimate of σ^2 is $\sum_{i=1}^n w_i(y_i - \hat{y}_i)^2/(n-k-1)$, where $X' = (\boldsymbol{x}_1, \dots, \boldsymbol{x}_n)$ and $W = \text{diag}\,(w_1, \dots, w_n)$. (These estimates are therefore the same as the corresponding weighted least squares estimates using the w_i's as weights.) [**Hint:** Let $\mathbf{1}_i = (1, \dots, 1)'$ be a vector of 1's of dimension w_i. To obtain the OLS estimator, we are using the model $D\boldsymbol{y} = DX\boldsymbol{\beta} + D\boldsymbol{\epsilon}$, where

$$D = \begin{pmatrix} \mathbf{1}_1 & \mathbf{o} & \dots & \mathbf{o} \\ \mathbf{o} & \mathbf{1}_2 & \dots & \mathbf{o} \\ \dots & \dots & \dots & \dots \\ \dots & \dots & \dots & \dots \\ \mathbf{o} & \mathbf{o} & \dots & \mathbf{1}_n \end{pmatrix} : \sum_{i=1}^n w_i \times n.]$$

Exercise 6.7: Consider Exercise 4.8. Assuming that the time taken for each contact of each type is independent of the others, would you expect heteroscedasticity to be present in the model you constructed to carry out the test? Draw various residual plots to check if your conjecture is true. Write a paragraph discussing your findings.

Exercise 6.8: Each case in Exhibit 6.10 represents a pair of zones in the city of Chicago. The variable x gives travel times which were computed from bus timetables augmented by walk times from zone centroids to bus-stops (assuming a walking speed of 3 m.p.h) and expected waiting times for the bus (which were set at half the headway, i.e., the time between successive buses). The variable y was the average of travel times as reported to the U.S. Census Bureau by n travelers. The data were selected by one of the authors from a larger data set compiled by Cæsar Singh from Census tapes, timetables and maps.

Plot y against x. What do you notice? In order to obtain a linear expression for perceived travel time in terms of computed travel times, what weights would you use? Carry out the appropriate regression exercise and plot suitable residuals. Ignoring the outlier(s), do you think you have adequately taken care of heteroscedasticity?

Obs #	n	x	y	Obs #	n	x	y
1	1	26	35.0	17	5	25	34.0
2	1	40	57.0	18	4	29	32.5
3	7	32	34.3	19	3	24	28.3
4	3	36	38.3	20	6	34	40.8
5	2	27	37.5	21	7	28	29.3
6	4	39	36.3	22	5	21	26.0
7	4	29	31.3	23	4	40	47.0
8	3	22	35.0	24	2	35	40.0
9	1	34	30.0	25	3	24	30.0
10	1	25	30.0	26	2	35	45.0
11	10	37	40.5	27	2	31	37.5
12	2	36	47.5	28	9	21	25.0
13	1	20	30.0	29	17	36	51.1
14	2	26	40.0	30	2	36	42.5
15	2	31	30.0	31	1	35	45.0
16	3	22	26.7	32	3	24	25.0

EXHIBIT 6.10: Data on Perceived (y) and Computed (x) Travel Times by Bus

Exercise 6.9: Carry out the steps, discussed in Example 6.1, to graphically determine a relationship between $\text{var}(y_i)$ and $\text{E}(y_i)$.

Exercise 6.10: For the OLS model used in Example 6.1, plot

1. the absolute values of the residuals against predicted values, and
2. the logarithm of the absolute values of the residuals against the logarithm of predicted values.

In each case, devise and demonstrate the efficacy of a method for determining what the correct weights should be.

Exercise 6.11: As noted in Chapter 1, a classical problem in regression is that of relating heights of sons to heights of fathers. Using the data of

Exhibit 6.11, obtain an appropriate relationship. Obviously, one needs to weight. If the number of sons for each height category were available, that variable would have provided the appropriate weights (why?). What would you do with the data given and why?

Height of Father (nearest inch)	62	63	64	65	66	67	68	69	70	71	72	73
Average Height of Sons (ins.)	65.5	66.5	66.8	66.8	67.6	67.8	68.6	69.1	69.5	70.6	70.3	72.0
# of Fathers	2	6	12	19	27	26	26	26	20	15	8	5

EXHIBIT 6.11: Heights of Fathers and Sons

SOURCE: Dacey (1983, Ch. 1) from McNemar (1969, p. 130). Reproduced, with permission, from McNemar, Q., *Psychological Statistics.* © 1969 John Wiley & Sons, Inc.

Exercise 6.12: Would you guess that heteroscedasticity would be a problem in the model of Exercise 4.6, p. 96? Examine the residuals and also carry out a test to see if your conjecture is correct. What action, if any, is required?

Exercise 6.13: Do you think the least squares exercise of Example 4.5, p. 92, should have been weighted? If so, carry out the appropriate least squares exercise. In this case are the estimates very different? Comment.

Exercise 6.14: Plot residuals against predicted values and each independent variable for Model 2 of Exercise 2.15, p. 53. Would weighting help? If so, what weights would you apply?

Now, instead of DAO, consider log[DAO] as the dependent variable (and the same independent variables as in Model 2 of Exercise 2.15). Does it now appear that weighting is called for?

Compute the variance of a future observation for DAO, corresponding to GNP=1500, CP=3500 and OP=700, for each of the models: Model 1 of Exercise 2.15, Model 2 of Exercise 2.15 and the model with the logged dependent variable that you just constructed. Which has the best variance?

Exercise 6.15: Construct appropriate residual plots and check if house prices (Example 4.3; the data are on p. 32) seem to be heteroscedastic. When such data are collected from a small area, as these data are, variance is usually small for houses of the predominant price range (because it is easy to get selling price information for 'comparables') and is larger for other houses. Check if this is the case here. If so, find appropriate weights and obtain weighted least squares estimates comparable to those in Exhibit 4.6, p. 89. Also run the tests we ran in Chapter 3. Compare the results and state any conclusions you come to.

Exercise 6.16: Do you feel the variances of the raises given company chairmen (Exhibit 3.6, p. 78) are equal? If you feel they are not equal, what transformation could be used to improve matters? Make appropriate plots and give your conclusions.

Exercise 6.17: For the models in Exercise 2.20, p. 55, which ones call for weighting? Construct appropriate plots to verify your conjectures. When weighting is indicated carry out the appropriate weighted procedures. Verify, using residual plots, if heteroscedasticity has been rendered negligible. If not, choose better weights and repeat the exercises.

CHAPTER 7

*Correlated Errors

7.1 Introduction

Continuing with our examination of violations of Gauss-Markov conditions, in this chapter we examine the case where

$$\mathrm{E}(\boldsymbol{\epsilon}\boldsymbol{\epsilon}') = \sigma^2\Omega \tag{7.1}$$

could be non-diagonal; i.e., some $\mathrm{E}(\epsilon_j\epsilon_j)$'s may be non-zero even when $i \neq j$. Cases of this kind do occur with some frequency. For example, observations of the same phenomena (e.g., per capita income) taken over time are often correlated (serial correlation), observations (e.g., of median rent) from points or zones in space that are close together are often more alike than observations taken from points further apart (spatial correlation), and observations from the same production run or using the same laboratory equipment often have more semblance than those from distinct runs.

While a non-diagonal Ω does not bias the estimates (as we saw in Section 2.6, p. 35), even fairly small non-diagonal elements can cause the variance of estimates to increase substantially. Consider the estimate $\boldsymbol{\ell}'\boldsymbol{b}$ of $\boldsymbol{\ell}'\beta$ (Section 2.9, p. 41). If $\boldsymbol{y}$ has covariance matrix $\sigma^2\Omega$, then the variance of $\boldsymbol{\ell}'\boldsymbol{b}$ is of the form $\sigma^2\boldsymbol{c}'\Omega\boldsymbol{c}$ with $\boldsymbol{c}' = \boldsymbol{\ell}'(X'X)^{-1}X'$. This contains $n(n-1)$ terms involving non-diagonal elements of Ω. Therefore, even if each such element is small, their combined effect *can be* considerable. Even worse is the fact that, when we use ordinary least squares, computer packages typically compute estimates of variance under the assumption that G-M conditions hold, i.e., $\Omega = I$. Therefore, unaccounted for non-diagonal elements can substantially affect any inferences we reach.

Unfortunately, since there are $n(n+1)/2$ distinct elements in Ω, it would be impossible to reliably estimate all of them on the basis of n observations. Consequently, general methods of handling the model (7.1) are not available. However, if there are known relationships involving very few parameters among the elements of Ω, then estimation procedures become available. But each type of relationship usually requires a distinct approach and this leads to a vast number of cases. In this chapter we consider a few such cases which we think are of relatively common occurrence. A larger number of them are examined in more advanced (typically econometrics) texts, e.g., Judge *et al.* (1985). A fairly general treatment of the subject is Rao's theory of MINQUE (minimum norm quadratic unbiased estimators; see Rao, 1970), which is beyond the scope of this book.

Even when methods are available they are not always foolproof. For example, sometimes we get negative estimates of variances. While for the cases we consider we have sometimes been able to suggest methods of circumventing this problem, in general, the ingenuity of the user is the only recourse. Therefore, it is usually preferable to collect data in such a way that errors can be considered uncorrelated.

In the next two sections we examine fairly general Ω's. In subsequent sections the above mentioned special cases are considered.

7.2 Generalized Least Squares: Case When Ω Is Known

Consider the usual regression model

$$y = X\beta + \epsilon, \tag{7.2}$$

where y is the response vector of n observations, X is an $n \times (k+1)$ matrix of known constants and β is the $(k+1)$-vector of unknown regression parameters. But now assume that, while $\mathrm{E}(\epsilon) = \mathbf{o}$, $\mathrm{E}(\epsilon\epsilon')$ is given by (7.1) with Ω a known symmetric, positive definite matrix of order n. In this section we show that under these conditions, a preferred estimate of β is the *generalized least squares* estimator

$$b_{GLS} = (X'\Omega^{-1}X)^{-1}X'\Omega^{-1}y. \tag{7.3}$$

From Section A.13, p. 279, we know that we can write Ω as $\Omega = \Gamma D\Gamma' = \Xi\Xi'$ where Γ is orthogonal, D is a diagonal matrix of positive diagonal elements and $\Xi = \Gamma D^{1/2}$. Pre-multiplying both sides of equation (7.2) by Ξ^{-1}, we get

$$\Xi^{-1}y = \Xi^{-1}X\beta + \Xi^{-1}\epsilon.$$

Therefore, if we let $y^{(\Omega)} = \Xi^{-1}y$, $X^{(\Omega)} = \Xi^{-1}X$ and $\epsilon^{(\Omega)} = \Xi^{-1}\epsilon$, we get the model

$$y^{(\Omega)} = X^{(\Omega)}\beta + \epsilon^{(\Omega)}, \tag{7.4}$$

where $\mathrm{E}(\epsilon^{(\Omega)}) = \mathrm{E}(\Xi^{-1}\epsilon) = \mathbf{o}$, and

$$\begin{aligned} \mathrm{cov}(\epsilon^{(\Omega)}) &= \mathrm{cov}(\Xi^{-1}\epsilon) = \Xi^{-1}\,\mathrm{cov}(\epsilon)(\Xi')^{-1} \\ &= \sigma^2\Xi^{-1}\Omega(\Xi^{-1})' = \sigma^2\Xi^{-1}(\Xi\Xi')(\Xi')^{-1} = \sigma^2 I. \end{aligned}$$

Thus, we see that the G-M conditions hold for model (7.4). Consequently, the BLUE estimator for β is the ordinary least squares estimator which is

$$\begin{aligned} b_{GLS} &= (X^{(\Omega)\prime}X^{(\Omega)})^{-1}X^{(\Omega)\prime}y^{(\Omega)} \\ &= (X'(\Xi')^{-1}\Xi^{-1}X)^{-1}X'(\Xi')^{-1}\Xi^{-1}y = (X'\Omega^{-1}X)^{-1}X'\Omega^{-1}y. \end{aligned}$$

It is easy to verify that $\text{cov}(\boldsymbol{b}_{GLS}) = \sigma^2(X'\Omega^{-1}X)^{-1}$.

The vector of residuals may be written as

$$\begin{aligned}\boldsymbol{e}^{(\Omega)} &= \boldsymbol{y}^{(\Omega)} - \hat{\boldsymbol{y}}^{(\Omega)} = \boldsymbol{y}^{(\Omega)} - X^{(\Omega)}(X^{(\Omega)\prime}X^{(\Omega)})^{-1}X^{(\Omega)\prime}\boldsymbol{y}^{(\Omega)} \\ &= [I - X^{(\Omega)}(X^{(\Omega)\prime}X^{(\Omega)})^{-1}X^{(\Omega)\prime}]\boldsymbol{y}^{(\Omega)}\end{aligned}$$

and an estimate of σ^2 is given by $s^2 = (n-k-1)^{-1}\boldsymbol{e}^{(\Omega)\prime}\boldsymbol{e}^{(\Omega)}$. Using the new variables $\boldsymbol{y}^{(\Omega)}$, $X^{(\Omega)}$, and $\boldsymbol{e}^{(\Omega)}$, all procedures on testing and confidence intervals can be carried out without any change. For example, if we wish to test the hypothesis

$$H : C\boldsymbol{\beta} = \mathbf{o} \text{ vs } A : C\boldsymbol{\beta} \neq \mathbf{o},$$

where C is an $r \times (k+1)$ matrix with $r \leq k+1$, then H is rejected if

$$(\boldsymbol{b}_{GLS}'C')[C(X'\Omega^{-1}X)^{-1}C']^{-1}(C\boldsymbol{b}_{GLS}) \geq s^2 rF_{r,n-k-1,\alpha},$$

where

$$s^2 = (n-k-1)^{-1}\boldsymbol{e}^{(\Omega)\prime}\boldsymbol{e}^{(\Omega)} = (n-k-1)[\boldsymbol{y}'\Omega^{-1}\boldsymbol{y} - \boldsymbol{b}_{GLS}'X'\Omega^{-1}\boldsymbol{y}].$$

7.3 Estimated Generalized Least Squares

Unfortunately, the matrix Ω is not usually known and needs to be estimated. If $\hat{\Omega}$ is an estimate of Ω, then $\boldsymbol{b}_{EGLS} = (X'\hat{\Omega}^{-1}X)^{-1}X'\hat{\Omega}^{-1}\boldsymbol{y}$ has been called an *estimated generalized least squares estimate* (or an EGLS estimate) of $\boldsymbol{\beta}$. In general, small sample properties of such estimates are hard to come by, except by model-specific Monte Carlo methods. However, fairly general asymptotic properties are available. For example, under certain conditions, given in Theil (1971, Ch. 8) and in Judge *et al.* (1985, p.176), $\boldsymbol{b}_{GLS}$ and $\boldsymbol{b}_{EGLS}$ are both consistent and have the same asymptotic distribution. Moreover, both estimates are asymptotically normal with mean $\boldsymbol{\beta}$ and covariance matrix $n^{-1}\sigma^2\Psi^{-1}$, and $\sqrt{n}(\boldsymbol{b}_{EGLS} - \boldsymbol{b}_{GLS}) \xrightarrow{P} 0$. Under some further conditions,

$$\hat{\sigma}^2 = (\boldsymbol{y} - X\boldsymbol{b}_{EGLS})'\hat{\Omega}^{-1}(\boldsymbol{y} - X\boldsymbol{b}_{EGLS})/(n-k)$$

is a consistent estimator of σ^2.

7.3.1 A Special Case: Error Variances Unequal and Unknown

A special case where we need to consider empirical estimation of Ω occurs when the non-diagonal elements of Ω are zero and the diagonal elements are unknown. That is, $\sigma^2\Omega = \text{diag}(\sigma_1^2, \ldots, \sigma_n^2)$ and the σ^2 are unknown. This is a case of heteroscedasticity with unknown variances.

There is now one unknown variance corresponding to each observation. Such situations do not lend themselves well to reliable estimation procedures, but let us examine this case anyway. Suppose we apply ordinary least squares and obtain the vector $\boldsymbol{e}$ of residuals e_i. It is reasonable to base any estimation of σ_i^2 on these residuals. A standard method consists of considering $\boldsymbol{e}^{(2)} = (e_1^2, \ldots, e_n^2)'$. Since $\boldsymbol{e} = M\boldsymbol{\epsilon}$, where, as in (2.13), $M = (m_{ij}) = I - X(X'X)^{-1}X'$, it follows that $e_i = \sum_{j=1}^n m_{ij}\epsilon_j$, and hence, since for $i \neq j$, $\text{E}[\epsilon_i\epsilon_j] = 0$,

$$\text{E}[e_i^2] = \sum_{j=1}^n m_{ij}^2\sigma_j^2 \text{ for } i = 1, \ldots, n. \tag{7.5}$$

Let $M^{(2)} = (m_{ij}^2)$ and $\boldsymbol{\sigma}^{(2)} = (\sigma_1^2, \ldots, \sigma_n^2)'$. Then we may write the system of equations (7.5) as

$$\text{E}(\boldsymbol{e}^{(2)}) = M^{(2)}\boldsymbol{\sigma}^{(2)}. \tag{7.6}$$

Replacing $\text{E}(\boldsymbol{e}^{(2)})$ by its estimate $\boldsymbol{e}^{(2)}$ and $\boldsymbol{\sigma}^{(2)}$ by its estimate $\hat{\boldsymbol{\sigma}}^{(2)} = (\hat{\sigma}_1^2, \ldots, \hat{\sigma}_n^2)'$ we get

$$\boldsymbol{e}^{(2)} = M^{(2)}\hat{\boldsymbol{\sigma}}^{(2)}. \tag{7.7}$$

We can solve this set of equations to get the required estimates $\hat{\sigma}_i^2$. These estimates are also known to be MINQUE.

A major difficulty with these estimates is that some $\hat{\sigma}_i^2$'s can turn out to be negative. Some alternative estimates of σ_i^2 have also been proposed (see Judge *et al.*, 1985).

Although, as mentioned earlier, it is not desirable to estimate individual σ_i^2's in this way, the method can be useful if there is a relationship among the σ_i^2's. For example, assume that

$$\sigma_i^2 = \boldsymbol{z}_i'\boldsymbol{\alpha}$$

for $i = 1, \ldots, n$ where $\boldsymbol{z}_i$ is an m-dimensional known vector, with $m < n$, and $\boldsymbol{\alpha}$ is a vector of parameters. Let $Z' = (\boldsymbol{z}_1, \ldots, \boldsymbol{z}_n)$. Then from (7.6),

$$\text{E}(\boldsymbol{e}^{(2)}) = M^{(2)}Z\boldsymbol{\alpha}, \tag{7.8}$$

which prompts the estimation of $\boldsymbol{\alpha}$ as

$$\hat{\boldsymbol{\alpha}} = (Z'M^{(2)}M^{(2)}Z)^{-1}Z'M^{(2)}\boldsymbol{e}^{(2)}, \tag{7.9}$$

which in turn can be used to estimate σ_i^2's. Since $(M^{(2)})^{-1}\,\text{E}(\boldsymbol{e}^{(2)}) = Z\boldsymbol{\alpha}$, another estimate of $\boldsymbol{\alpha}$ is

$$\hat{\boldsymbol{\alpha}} = (Z'Z)^{-1}Z'(M^{(2)})^{-1}\boldsymbol{e}^{(2)}. \tag{7.10}$$

The difference between the two estimates stems obviously from different error distributions for the model (7.8). Yet another alternative (which, incidentally, is a MINQUE — see Froehlich, 1973) is

$$\hat{\boldsymbol{\alpha}} = (Z'M^{(2)}Z)^{-1}Z'\mathbf{e}^{(2)}. \tag{7.11}$$

For none of these estimators is there any guarantee of positivity of estimates of the σ_i^2's. Judge *et al.* (1985) have given other methods of estimation in this case as well as for situations where σ^2 has other algebraic forms.

7.4 Nested Errors

Suppose that out of the total of $n = mM$ observations $y_1, y_2, \ldots$, each of the M sets of m observations, $y_1, \ldots, y_m$; $y_{m+1}, \ldots, y_{2m}$; ...... was obtained in some common way. For example, each set could have been obtained from a common production run, or result from experiments using common equipment, or be obtained by the same survey-taker. Then these sets could contain mutually correlated observations.

A simple case of this occurs when all the correlations within each group are the same and we can write $\mathrm{E}[\epsilon\epsilon'] = \sigma^2(1-\rho)I + \sigma^2\Phi$ where

$$\Phi = \begin{pmatrix} \Sigma & 0 & \cdots & 0 \\ 0 & \Sigma & \cdots & 0 \\ \vdots & \vdots & \ddots & \vdots \\ 0 & 0 & \cdots & \Sigma \end{pmatrix},$$

and

$$\Sigma = \rho\mathbf{1}\mathbf{1}' = \rho \begin{pmatrix} 1 & \cdots & 1 \\ \vdots & \ddots & \vdots \\ 1 & \cdots & 1 \end{pmatrix}.$$

Since $m^{-1}\mathbf{1}\mathbf{1}'$ is clearly idempotent with trace 1, its eigenvalues consist of 1 one and $m-1$ zeros. The eigenvalues of Σ, therefore, consist of $m-1$ zeros and one ρm. Let G be an orthogonal matrix that diagonalizes Σ, i.e.,

$$G\Sigma G' = D \text{ and } GG' = I$$

where D is a diagonal matrix $D = \mathrm{diag}\,(d_1, \ldots, d_m)$ of the eigenvalues of Σ, which are $d_1 = \rho m$ and $d_\ell = 0$ when $\ell \neq 1$. An example of such a matrix G is the matrix with first row $m^{-1/2}(1, \ldots, 1)' = m^{-1/2}\mathbf{1}'$ and the remaining

$(m-1)$ rows

$$\begin{array}{ccccc} \frac{1}{\sqrt{2}} & -\frac{1}{\sqrt{2}} & 0 & \cdots & 0, \\ \frac{1}{\sqrt{6}} & \frac{1}{\sqrt{6}} & \frac{-2}{\sqrt{6}} & \cdots & 0, \\ \multicolumn{5}{c}{\dots\dots\dots\dots\dots\dots\dots\dots\dots\dots,} \\ \multicolumn{5}{c}{\dots\dots\dots\dots\dots\dots\dots\dots\dots\dots,} \\ \frac{1}{\sqrt{m(m-1)}} & \frac{1}{\sqrt{m(m-1)}} & \frac{1}{\sqrt{m(m-1)}} & \cdots & -\frac{m-1}{\sqrt{m(m-1)}}, \end{array}$$

although one would typically use a matrix program like SAS PROC MATRIX (SAS, 1982b), or MINITAB[1] (MINITAB, 1988) to obtain a suitable G. Let Γ be the n dimensional matrix with G's on its diagonal and 0's elsewhere, i.e.,

$$\Gamma = \begin{pmatrix} G & 0 & \cdots & 0 \\ 0 & G & \cdots & 0 \\ \vdots & \vdots & \ddots & \vdots \\ 0 & 0 & \cdots & G \end{pmatrix}.$$

If we multiply both sides of our original model $\boldsymbol{y} = X\boldsymbol{\beta} + \boldsymbol{\epsilon}$ by Γ on the left, we get

$$\boldsymbol{w} = Z\boldsymbol{\beta} + \boldsymbol{\eta}$$

where $\boldsymbol{w} = \Gamma\boldsymbol{y}$, $Z = \Gamma X$ and $\boldsymbol{\eta} = \Gamma\boldsymbol{\epsilon}$. It is straightforward to verify that $\mathrm{E}[\boldsymbol{\eta}\boldsymbol{\eta}']$ is a diagonal matrix with diagonal elements $\delta_1, \ldots, \delta_n$, where M of the δ_i's (one corresponding to each block of m observations) are $\tau^2 + \rho m$ and the remainder are τ^2 where $\tau^2 = \sigma^2(1-\rho)$. In the case of the example of G given above,

$$\delta_i = \begin{cases} \tau^2 + \rho m & \text{when } i = tm + 1 \text{ where } t = 0, \ldots, M-1 \\ \tau^2 & \text{otherwise.} \end{cases}$$

Thus we have removed all non-diagonal terms in the error covariance matrix and are left with a situation with two subsets of observations within each of which the variances are the same. If a computer program is used to obtain G, an examination of the eigenvalues will show which observation belongs to which subset.

Then we can apply any of the methods mentioned at the end of Section 6.4, p. 118. Alternatively, if M is not too large, we could lose a few observations by simply deleting M cases, w_{tm+1}, where $t = 0, \ldots, M-1$, and thereby reducing the model to a purely homoscedastic model. The above discussion can be extended to the case when the M subsets are of unequal size (Srivastava and Ng, 1988).

[1]MINITAB is a registered trademark of MINITAB, Inc., State College, PA

7.5 The Growth Curve Model

When nothing is known of the form of Ω, it is still possible to estimate it, if the experiment can be replicated an adequate number of times, i.e., if we have the model

$$\boldsymbol{y}_t = X\boldsymbol{\beta} + \boldsymbol{\epsilon}_t$$

where X is an $m \times p$ matrix, $t = 1, 2, \ldots, M$, $\mathrm{E}(\boldsymbol{\epsilon}_t) = \mathbf{o}$ and $\mathrm{cov}(\boldsymbol{\epsilon}_t) = \Omega$. Here $\boldsymbol{y}_1, \ldots, \boldsymbol{y}_M$ are independent m dimensional vectors and σ^2 has been absorbed in Ω.

Let $\bar{\boldsymbol{y}} = M^{-1}\sum_{t=1}^{M}\boldsymbol{y}_t$ and $\bar{\boldsymbol{\epsilon}} = M^{-1}\sum_{t=1}^{M}\boldsymbol{\epsilon}_t$. Then

$$\bar{\boldsymbol{y}} = X\boldsymbol{\beta} + \bar{\boldsymbol{\epsilon}}$$

where $\mathrm{E}(\bar{\boldsymbol{\epsilon}}) = \mathbf{o}$ and $\mathrm{cov}(\bar{\boldsymbol{y}}) = \mathrm{cov}(\bar{\boldsymbol{\epsilon}}) = M^{-1}\Omega$. Hence, the generalized least squares estimate of $\boldsymbol{\beta}$ is given by

$$\boldsymbol{b}_{GLS} = (X'\Omega^{-1}X)^{-1}X'\Omega^{-1}\bar{\boldsymbol{y}}.$$

However, Ω is not known. But since $\mathrm{E}(\boldsymbol{y}_t - \bar{\boldsymbol{y}}) = 0$, for all $t = 1, \ldots, M$, an unbiased estimator of Ω is given by

$$\hat{\Omega} = (M-1)^{-1}\sum_{t=1}^{M}(\boldsymbol{y}_t - \bar{\boldsymbol{y}})(\boldsymbol{y}_t - \bar{\boldsymbol{y}})'.$$

Therefore, an estimated generalized least squares estimate of $\boldsymbol{\beta}$ is

$$\boldsymbol{b}_{EGLS} = (X'\hat{\Omega}^{-1}X)^{-1}X'\hat{\Omega}^{-1}\bar{\boldsymbol{y}}.$$

Under the assumption that $\boldsymbol{\epsilon}_t$ is multivariate normal, it can be shown that this $\boldsymbol{b}_{EGLS}$ is an unbiased estimator (see Exercise 7.3).

Under normality, the hypothesis $H : C\boldsymbol{\beta} = 0$ against $A : C\boldsymbol{\beta} \neq 0$ (where C is $r \times p$ dimensional with $r \leq p$) is rejected if

$$\frac{M-r-m+p}{(M-1)r}\,\frac{\boldsymbol{b}'_{EGLS}C'(CEC')^{-1}C\boldsymbol{b}_{EGLS}}{1+(M-1)^{-1}T^2} > F_{r,M-r-m+p,\alpha}$$

where $E = (X'\hat{\Omega}^{-1}X)^{-1}$, $T^2 = M\bar{\boldsymbol{y}}'G'(G\hat{\Omega}G')^{-1}G\bar{\boldsymbol{y}}$ and $G : (m-p)\times m$ is such that $GX = 0$. Alternatively, if it is inconvenient to find a suitable G, one could use $T^2 = M\bar{\boldsymbol{y}}'[\hat{\Omega}^{-1} - \hat{\Omega}^{-1}X(X'\hat{\Omega}^{-1}X)^{-1}X'\hat{\Omega}^{-1}]\bar{\boldsymbol{y}}$. (This result is obtained with the help of Lemma A.1, p. 279, of Appendix A).

Example 7.1

Exhibit 7.1 shows dental measurements for girls from 8 to 14 years old. Each measurement is the distance, in millimeters, from the center of the pituitary to the ptery-maxillary fissure. Suppose we wish to relate these measurements to age and write our model as

$$y_{ts} = \beta_0 + \beta_1 x_{t1} + \epsilon_{ts},$$

where $x_{t1} = \text{age} - 11$. Then

$$X = \begin{pmatrix} 1 & 1 & 1 & 1 \\ -3 & -1 & 1 & 3 \end{pmatrix}'.$$

Clearly, for the same subject s the $m = 4$ measurements y_t are not independent and

$$\text{cov}(\boldsymbol{\epsilon}_s) = \Omega, \text{ where } \boldsymbol{\epsilon}_s = (\epsilon_{1s}, \ldots, \epsilon_{4s})'.$$

However, we have $M = 11$ replications of the experiment.

Age	Subjects										
in Years	1	2	3	4	5	6	7	8	9	10	11
8	21.0	21.0	20.5	23.5	21.5	20.0	21.5	23.0	20.0	16.5	24.5
10	20.0	21.5	24.0	24.5	23.0	21.0	22.5	23.0	21.0	19.0	25.0
12	21.5	24.0	24.5	25.0	22.5	21.0	23.0	23.5	22.0	19.0	28.0
14	23.0	25.5	26.0	26.5	23.5	22.5	25.0	24.0	21.5	19.5	28.0

EXHIBIT 7.1: Data on Dental Measurements
SOURCE: Pothoff and Roy (1964). Reproduced from *Biometrika* with the permission of Biometrika Trustees.

Since $\bar{\boldsymbol{y}} = (21.2, 22.2, 23.1, 24.1)'$, an estimate of Ω is given by $\hat{\Omega} = \sum_{i=1}^{11} (\boldsymbol{y}_i - \bar{\boldsymbol{y}})(\boldsymbol{y}_i - \bar{\boldsymbol{y}})'/10$

$$= \begin{pmatrix} 4.51 & 3.36 & 4.43 & 4.36 \\ 3.36 & 3.62 & 4.02 & 4.08 \\ 4.33 & 4.03 & 5.59 & 5.47 \\ 4.36 & 4.08 & 5.47 & 5.94 \end{pmatrix}.$$

Hence

$$\boldsymbol{b}_{EGLS} = (X'\hat{\Omega}^{-1}X)^{-1}X'\hat{\Omega}^{-1}\bar{\boldsymbol{y}} = \begin{pmatrix} 22.70 \\ 0.482 \end{pmatrix}.$$

Suppose we wish to test the hypothesis that the linear term is zero. That is, $H : \beta_1 = 0$ against $A : \beta_1 \neq 0$. In this case, $C = (0, 1)$, $r = 1$, $p = 2$ and

$$E = (X'\hat{\Omega}^{-1}X)^{-1} = \begin{pmatrix} 3.807 & 0.160 \\ -0.160 & 0.160 \end{pmatrix}$$

and $C\,\text{E}C' = 0.045$. The matrix

$$G = \begin{pmatrix} 1 & -1 & -1 & 1 \\ -1 & 3 & -3 & 1 \end{pmatrix}$$

is such that $GX = 0$. Therefore, $T^2 = 11\bar{\boldsymbol{y}}'G'(G\hat{\Omega}G')^{-1}G\bar{\boldsymbol{y}} = 0.11$ and

$$F = \frac{M - r - m + p}{(M-1)r} \frac{\boldsymbol{b}'_{EGLS}C'(CEC')^{-1}C\boldsymbol{b}_{EGLS}}{1 + T^2/10} = 45.94.$$

Therefore, at 1 and 8 degrees of freedom, we reject the hypothesis at a 5 per cent level. ■

7.6 Serial Correlation

Frequently when the observations y_i are taken over successive time intervals, the ϵ_i's are correlated. This type of correlation is called serial correlation. We consider the particular case where the ϵ_i's follow a first order autoregressive (often called AR(1)) process:

$$\epsilon_t = \rho\epsilon_{t-1} + u_t, \tag{7.12}$$

where $|\rho| < 1$ and for all $t = 1, \ldots, n$, u_t's are independent and identically distributed with mean 0 and variance σ_u^2. The model (7.12) is called autoregressive since ϵ_t is linearly related to lagged values of itself. It is said to be of first order because the maximum lag is one.

The AR(1) is only one possible model of serial correlation. Other models include higher order autoregressive, AR(r), processes given by

$$\epsilon_t = \sum_{s=1}^{r} \rho_s \epsilon_{t-s} + u_t,$$

moving average, MA(m), processes

$$\epsilon_t = u_t + \sum_{s=1}^{m-1} \alpha_s u_{t-s},$$

and mixtures, ARMA(r,m), of autoregressive and moving average processes. Typically, analysis consists of first identifying the kind of process, its order (r, m) and then the relevant parameters ρ, ρ_s or/and α_s. The literature on such time series processes is huge, much of it stemming from the seminal work of Box and Jenkins (1970), and a reasonably complete treatment of it is well beyond the scope of this book. Consequently, we shall briefly discuss the AR(1) process, which is by far the most popular of these processes. We refer the interested reader to Judge *et al.* (1985) and to Anderson (1971) for a fuller treatment of the AR(1) and other models.

If the AR(1) process has been in operation since the indefinite past, then by repeated application of (7.12) we have, since $|\rho| < 1$,

$$\epsilon_t = \lim_{n\to\infty} (\rho^{n+1}\epsilon_{t-n-1} + \sum_{s=0}^{n} \rho^s u_{t-s}) = \sum_{s=0}^{\infty} \rho^s u_{t-s}.$$

Hence

$$\mathrm{E}(\epsilon_t) = 0, \ \mathrm{var}\,(\epsilon_t) = \sigma_u^2 \sum_{s=0}^{\infty} (\rho^2)^s = \sigma_u^2/(1-\rho^2)$$

and

$$\mathrm{cov}(\epsilon_t, \epsilon_{t+r}) = \sum_{s=0}^{\infty} \rho^{2s+r}\sigma_u^2 = \rho^r\sigma_u^2/(1-\rho^2).$$

Therefore, $\text{cov}(\boldsymbol{\epsilon}) = \sigma_u^2 \Omega$ is

$$\sigma_u^2(1-\rho^2)^{-1} \begin{pmatrix} 1 & \rho & \rho^2 & \cdots & \rho^{n-1} \\ \rho & 1 & \rho & \cdots & \rho^{n-2} \\ \rho^2 & \rho & 1 & \cdots & \rho^{n-3} \\ \vdots & \vdots & \vdots & \ddots & \vdots \\ \rho^{n-1} & \rho^{n-2} & \rho^{n-3} & \cdots & 1 \end{pmatrix}. \tag{7.13}$$

Let $e_1, \ldots, e_n$ be the usual ordinary least squares residuals from the model

$$\boldsymbol{y} = X\boldsymbol{\beta} + \boldsymbol{\epsilon}.$$

Then

$$\hat{\rho} = \sum_{i=2}^{n} e_i e_{i-1} \Big/ \sum_{1}^{n} e_i^2 \tag{7.14}$$

is an estimate of ρ which may be shown to be consistent if for all i, $h_{ii} \to 0$ where, as in Section 5.2, p. 101 h_{ii}'s are the diagonal elements of $H = X(X'X)^{-1}X'$. Let $\hat{\Omega}$ be the result of replacing ρ by $\hat{\rho}$ in the matrix Ω in (7.13). Then an estimated generalized least squares estimator of $\boldsymbol{\beta}$ in the model is given by

$$\boldsymbol{b}_{EGLS} = (X'\hat{\Omega}^{-1}X)^{-1}X'\hat{\Omega}^{-1}\boldsymbol{y}. \tag{7.15}$$

This-two step method of first estimating ρ using (7.14) after running ordinary least squares and then using the resulting residuals to obtain an estimated generalized least squares estimate is often called the Prais-Winsten (1954) procedure.

Since $\hat{\Omega}$ is a matrix of typically fairly large dimensions, it is computationally preferable to obtain the estimate (7.15) in a different way. Let

$$\hat{\Psi} = \begin{pmatrix} \sqrt{1-\hat{\rho}^2} & 0 & 0 & \cdots & 0 & 0 \\ -\hat{\rho} & 1 & 0 & \cdots & 0 & 0 \\ 0 & -\hat{\rho} & 1 & \cdots & 0 & 0 \\ \vdots & \vdots & \vdots & \ddots & \vdots & \vdots \\ 0 & 0 & 0 & \cdots & 1 & 0 \\ 0 & 0 & 0 & \cdots & -\hat{\rho} & 1 \end{pmatrix}.$$

Then it may be verified that $\hat{\Psi}'\hat{\Psi} = (1-\hat{\rho}^2)\hat{\Omega}^{-1}$. Since least squares estimation is unaffected by scalar multiplication, we may apply the transformation $X^{(\hat{\Omega})} = \hat{\Psi}X$, $\boldsymbol{y}^{(\hat{\Omega})} = \hat{\Psi}\boldsymbol{y}$ and $\boldsymbol{\epsilon}^{(\hat{\Omega})} = \hat{\Psi}\boldsymbol{\epsilon}$ in (7.15) and get

$$\boldsymbol{b}_{EGLS} = (X^{(\hat{\Omega})\prime}X^{(\hat{\Omega})})^{-1}X^{(\hat{\Omega})\prime}\boldsymbol{y}^{(\hat{\Omega})} \tag{7.16}$$

which is an OLS estimator. This transformed model may be written as

$$y_t - \hat{\rho}y_{t-1} = \sum_{j=0}^{k}(x_{tj} - \hat{\rho}x_{t-1,j})\beta_j + u_t \tag{7.17}$$

for $t = 2, \ldots, n$ where, as before, n is the total number of observations, k is the number of independent variables and $x_{t0} = 1$. When $t = 1$,

$$\sqrt{1-\hat{\rho}^2}y_1 = \sqrt{1-\hat{\rho}^2}\sum_{j=0}^{k}\beta_j x_{1j} + \sqrt{1-\hat{\rho}^2}\epsilon_1. \tag{7.18}$$

Therefore, all one need do is to apply ordinary least squares to the model consisting of (7.17) and (7.18), which is roughly equivalent to what most computer packages (e.g., SAS PROC AUTOREG — SAS, 1982a) do for the AR(1) model. Notice that the models (7.17) and (7.18) could also have been directly obtained from (7.12). An alternative to this approach is to apply maximum likelihood estimation directly, as illustrated on p. 144 for the case of spatial correlation.

7.6.1 The Durbin-Watson Test

A test of the hypothesis $\rho = 0$ against the alternative $\rho \neq 0$ is based on the Durbin-Watson statistic,

$$d = \sum_{t=2}^{n}(e_t - e_{t-1})^2 / \sum_{t=1}^{n} e_t^2.$$

While this test is primarily to detect the existence of the AR(1) process, it is frequently used, in practice, to detect the presence of any kind of serial correlation, under the assumption that most serially correlated data would exhibit, at least partially, the behavior of an AR(1) process. Notice that $d = 0$ when $e_t = e_{t-1}$, $d = 4$ when $e_t = -e_{t-1}$, while a value of d close to 2 indicates a low or zero valued ρ.

Unfortunately, the percentage points of d cannot be given in tables since its distribution depends on X. However, for any chosen level of significance, numbers d_L and d_U independent of X have been tabulated (and are given on p. 326 for the 5 per cent level) such that, when ϵ_t are normal,

1. if $d_U < d < 4 - d_U$, the hypothesis H is accepted,
2. if $d < d_L$ or $d > 4 - d_L$, it is rejected, and
3. if $d_L < d < d_U$ or $4 - d_U < d < 4 - d_L$, the test is inconclusive.

However, given X, the distribution of d under H can be computed. For example, methods have been given by L'Esperance *et al.* (1976), Koerts and Abrahamse (1969), White (1978) and Srivastava and Yau (1989), and one of these can be used to give appropriate probability values. Also, d is asymptotically normal (see Srivastava, 1987; also see Section 7.7.1), which can also be used to obtain approximate critical values.

7.7 Spatial Correlation

Just as when data are taken over time we frequently have serial correlation, when data are taken over contiguous geographical areas (e.g., census tracks, counties or states in a country) we frequently encounter spatial correlation. This is because nearby areas are often much alike; e.g., mean household income for some city block will usually not be too different from that for a neighboring block.

7.7.1 Testing for Spatial Correlation

One can use an obvious generalization of the Durbin-Watson statistic to test for spatial correlation. This test, known to geographers as Geary's (1954) test, is based on the statistic

$$c \sum_{i,j=1}^{n} w_{ij}(e_i - e_j)^2/s^2 \tag{7.19}$$

where c is a constant and w_{ij} is a monotonically declining function of the distance between the ith and jth regions. Both c and w_{ij}'s are chosen by the user so that $w_{ij} = w_{ji}$ and $w_{ii} = 0$. The most common choice (see Haggett *et al.*, 1977, and Bartels, 1979) of w_{ij} is to set $w_{ij} = a_{ij}$ where

$$a_{ij} = \begin{cases} 1 & \text{if } i \text{ and } j \text{ are contiguous} \\ 0 & \text{otherwise.} \end{cases}$$

An alternative to (7.19) is the statistic

$$c \sum_{i,j} w_{ij} e_i e_j / s^2 \tag{7.20}$$

which is often called Moran's (1950) statistic. Both (7.19) and (7.20) can be written in the form

$$c\boldsymbol{e}'V\boldsymbol{e}/s^2 \tag{7.21}$$

where V is a suitable $n \times n$ matrix.

When $\boldsymbol{\epsilon} \sim N(0, \sigma^2 I)$, an exact distribution of (7.21) can be obtained (using methods in Sen, 1990, or procedures similar to those in, say, Koerts and Abrahamse, 1969), but it is unreasonable to expect to find tabulated values since we would need such values for each of many possible V's, which would vary from application to application. Consequently, most users of these statistics invoke the fact that (7.21) is asymptotically normal (Sen, 1976, 1990, Ripley, 1981, p. 100 *et seq.*) with mean $c\,\text{tr}B$ and variance $2c^2(n-k+1)^{-1}[(n-k-1)\,\text{tr}B^2 - (\,\text{tr}B)^2]$ where $B = M'VM$ and $M = I - X(X'X)^{-1}X'$.

If ϵ is not normal, but the Gauss-Markov conditions continue to hold, (7.21) is still asymptotically normal (under some mild conditions — see Sen, 1976) with mean $c\,\mathrm{tr}B$ and variance $2c^2\,\mathrm{tr}B^2$.

For more on the properties of these and other tests of spatial correlation see Cliff and Ord (1981).

7.7.2 Estimation of Parameters

Unlike serial correlation, the literature on estimation in the presence of spatial correlation is relatively sparse although spatial correlation frequently has more serious effects than does serial correlation (owing to the larger number of non-zero elements in the error covariance matrix — see also Krämer and Donninger, 1987).

The first order spatial autoregressive model for the error vector ϵ, a generalization of the serial correlation model, is

$$\epsilon = \rho A\epsilon + u \tag{7.22}$$

where $\mathrm{E}[u] = 0$, the components u_i of u are uncorrelated, i.e., $\mathrm{E}[uu'] = \sigma^2 I$, and $A = (a_{ij})$ with $a_{ij} = 1$ if i and j represent contiguous zones and $a_{ij} = 0$ otherwise. The model (7.22) implies that each ϵ_i is ρ times the sum of ϵ_ℓ's of contiguous zones plus an independent disturbance. Assuming that $I - \rho A$ is nonsingular, (7.22) can be written as $\epsilon = (I - \rho A)^{-1}u$ and hence

$$\mathrm{E}[\epsilon\epsilon'] = \sigma^2(I - \rho A)^{-2}. \tag{7.23}$$

One method of estimating the model $y = X\beta$ in this case is to use generalized least squares. The estimate is $b_\rho = [X'(I - \rho A)^2 X]^{-1}X'(I - \rho A)^2 y$. The catch is that we do not know ρ. If ϵ is normally distributed, what may be done now is to run distinct generalized least squares for several values of ρ (say between 0 and .25 in increments of .01) and select that which yields the smallest value of

$$(y - Xb_\rho)'(I - \rho A)^2(y - Xb_\rho) - 2\log(\det[(I - \rho A)]). \tag{7.24}$$

This criterion is, of course, based on maximum likelihood (see also Ord, 1975; Warnes and Ripley, 1987).

Alternatively, we could proceed as follows. Since A is symmetric, there exists an orthogonal matrix Γ such that

$$\Gamma A\Gamma' = \mathrm{diag}\,(\lambda_1, \ldots, \lambda_n)$$

where λ_i's are the eigenvalues of A and Γ is a matrix of the corresponding eigenvectors. Then $\Gamma(I - \rho A)\Gamma'$ is also diagonal and has diagonal elements $1 - \rho\lambda_i$. Furthermore, $\Gamma(I - \rho A)^{-1}\Gamma'$ is diagonal, as is $\Gamma(I - \rho A)^{-2}\Gamma'$, and this last matrix has diagonal elements $(1 - \rho\lambda_i)^{-2}$. Thus if $X^{(A)} = \Gamma X$,

$\boldsymbol{y}^{(A)} = \Gamma \boldsymbol{y}^{(A)}$ and $\boldsymbol{\epsilon}^{(A)} = \Gamma \boldsymbol{\epsilon}$, we can write our regression model $\boldsymbol{y} = X\boldsymbol{\beta} + \boldsymbol{\epsilon}$ as

$$\boldsymbol{y}^{(A)} = X^{(A)}\boldsymbol{\beta} + \boldsymbol{\epsilon}^{(A)}. \tag{7.25}$$

But now, because of (7.23),

$$\begin{aligned} \mathrm{E}(\boldsymbol{\epsilon}^{(A)}\boldsymbol{\epsilon}^{(A)\prime}) &= \Gamma\,\mathrm{E}(\boldsymbol{\epsilon}\boldsymbol{\epsilon}')\Gamma' = \Gamma(I - \rho A)^{-2}\Gamma' \\ &= \mathrm{diag}\left(\frac{1}{(1-\rho\lambda_1)^2}, \ldots, \frac{1}{(1-\rho\lambda_n)^2}\right) \end{aligned}$$

is diagonal. Therefore, we can perform a weighted regression on the transformed model (7.25), using as weights $1 - \rho\lambda_i$. Again, an estimate of ρ could be obtained by comparing, for several ρ's, the values of (7.24) — which, in this case, is $\boldsymbol{e}^{(A)\prime}\boldsymbol{e}^{(A)} - 2\log(\det[(I - \rho A)])$, where $\boldsymbol{e}^{(A)}$ is the vector of residuals. Macros for spatial correlation for use with MINITAB have been written by Griffith, 1989. A spatial equivalent of the Prais-Winsten procedure (p. 141) for serially correlated AR(1) models is also possible as an alternative.

The first order spatial moving average model is written as $\boldsymbol{\epsilon} = (I + \alpha A)\boldsymbol{u}$ where A is as in (7.22). It can be easily verified that for this model $\mathrm{E}[\boldsymbol{\epsilon}\boldsymbol{\epsilon}'] = \sigma^2(I + 2\alpha A + \alpha^2 A^2)$. Since it may be shown that if $A^r = (a_{ij}^r)$, then a_{ij}^r is the number of sequences of exactly $r-1$ zones one must traverse in getting from i to j, a higher order spatial moving average model is $\boldsymbol{\epsilon} = [I + \sum_{\ell=1}^{L} \alpha_\ell A^\ell]\boldsymbol{u}$. For this model, the covariance of $\boldsymbol{\epsilon}$ takes the form

$$\mathrm{E}(\boldsymbol{\epsilon}\boldsymbol{\epsilon}') = \sigma^2(I + \rho_1 A + \rho_2 A^2 + \cdots + \rho_q A^q), \tag{7.26}$$

where the ρ_k's are unknown parameters.

It can readily be verified that

$$\Gamma A^2 \Gamma' = \Gamma A \Gamma' \Gamma A \Gamma' = \mathrm{diag}\,(\lambda_1^2, \ldots, \lambda_n^2),$$

and in general

$$\Gamma A^k \Gamma' = \mathrm{diag}\,(\lambda_1^k, \ldots, \lambda_n^k).$$

Consequently,

$$\Gamma\,\mathrm{E}(\boldsymbol{\epsilon}'\boldsymbol{\epsilon})\Gamma' = \sigma^2 \mathrm{diag}\,(1 + \sum_{k=1}^{q} \rho_k \lambda_1^k, \ldots, 1 + \sum_{k=1}^{q} \rho_k \lambda_n^k). \tag{7.27}$$

Thus, again we have reduced the spatially correlated case to one that only requires weighted regression, but here, because of the large number of parameters, it might be difficult to estimate weights by minimizing (7.24). But weights may be obtainable using (7.9), (7.10) or (7.11).

For more on these methods, as well as other methods of estimation under spatially autoregressive errors, see Anselin (1988), Griffith (1988), Mardia and Marshall (1984), Cook and Pocock (1983), Haining (1987) and Vecchia (1987). Mardia and Marshall have established consistency and asymptotic normality for a wide range of procedures.

Problems

The time series computations can be done using just about any of the major statistical packages, as well as one of the many special purpose time series packages. All of the exercises can be done quite easily using a matrix program like SAS PROC MATRIX (SAS, 1982b) or PROC IML (SAS, 1985c) or MINITAB (MINITAB, 1988).

Exercise 7.1: For the model (7.2), show that:

1. The estimators $\boldsymbol{b}_{OLS}$ and $\boldsymbol{b}_{GLS}$ are unbiased estimators, where $\boldsymbol{b}_{OLS}$ is the OLS estimator.
2. $\text{cov}(\boldsymbol{b}_{GLS}) = \sigma^2(X'\Omega^{-1}X)^{-1}$.
3. $\text{cov}(\boldsymbol{b}_{OLS}) = \sigma^2(X'X)^{-1}(X'\Omega X)(X'X)^{-1}$.

Using the Gauss-Markov theorem, state which of the two following numbers is bigger:

$$\boldsymbol{\ell}'(X'X)^{-1}(X'\Omega X)(X'X)^{-1}\boldsymbol{\ell} \text{ or } \boldsymbol{\ell}'(X'\Omega^{-1}X)^{-1}\boldsymbol{\ell},$$

where $\boldsymbol{\ell}$ is a non-null p-dimensional vector.

Exercise 7.2: Show that for the model (2.42) on page 43, OLS and GLS estimates are identical.
[**Hint:** Let $A = I - n^{-1}\mathbf{1}\mathbf{1}'$. Then $A^{-} = A$ and $AZ = Z$ where Z is as in (2.40).]

Exercise 7.3: Show that $\boldsymbol{b}_{EGLS}$ in Section 7.5 is an unbiased estimator of $\boldsymbol{\beta}$ under the assumption of multivariate normality.
[**Hint:** Under the assumption of normality, $\hat{\Omega}$ and $\bar{\boldsymbol{y}}$ are independently distributed. Take the expectation given $\hat{\Omega}$.]

Exercise 7.4: The data of Exercise 3.14, p. 79, were collected over time. At a 5 per cent level, test the hypothesis that there is no serial correlation against the alternative that there is serial correlation.

Exercise 7.5: Apply the Durbin-Watson test to each of the models considered in Exercise 2.19, p. 54. When can you accept the hypothesis of no serial correlation?

Exercise 7.6: In Exercise 2.15, p. 53, we suggested taking first differences of the dependent variable as an attempt to combat serial correlation. For each of the models described, compute the Durbin-Watson statistic and explain if we succeeded. In the cases where we did not succeed, can you do better using a first order autoregressive model?

Exercise 7.7: U.S. population (y_t) in thousands for the years 1790, 1800, ..., 1970 is:

Year	Oil	Gas	Bit.	Anth.	Year	Oil	Gas	Bit.	Anth.
1950	80.7	11.7	34.5	69.9	1966	64.7	18.5	23.3	43.9
1951	76.4	11.6	32.9	70.4	1967	63.6	18.3	23.3	43.9
1952	75.3	12.1	32.3	67.2	1968	61.4	17.8	22.5	45.6
1953	78.5	13.9	32.0	68.3	1969	61.4	17.4	23.0	48.7
1954	80.4	15.3	29.1	59.8	1970	59.9	16.9	27.9	51.5
1955	78.6	15.3	28.4	53.6	1971	60.8	17.2	30.4	53.5
1956	76.6	15.4	29.6	54.5	1972	58.4	16.9	31.9	52.9
1957	82.1	15.7	30.2	57.9	1973	63.5	18.7	33.6	55.7
1958	78.6	16.2	28.3	56.5	1974	102.9	24.1	57.7	85.5
1959	74.0	17.2	27.5	51.9	1975	105.1	32.3	65.9	109.6
1960	72.3	18.3	26.6	48.0	1976	106.7	40.1	63.4	112.6
1961	71.8	19.6	25.8	48.8	1977	105.5	51.6	62.3	107.4
1962	70.8	19.8	24.8	46.5	1978	103.2	55.3	64.6	99.7
1963	69.5	20.0	24.0	49.8	1979	133.3	66.0	64.1	106.5
1964	68.3	19.2	24.0	50.8	1980	204.5	81.7	59.4	105.5
1965	66.3	19.1	23.5	47.5	1981	273.6	96.0	57.4	104.0

EXHIBIT 7.2: Prices of Crude Oil, Natural Gas, Bituminous Coal and Lignite, and Anthracite by Year

SOURCE: Darrell Sala, Institute of Gas Technology, Chicago.

3929, 5308, 7239, 9638, 12866, 17069, 23191, 31443, 39818, 50155, 62947, 75994, 91972, 105710, 122775, 131669, 151325, 179323, 203211.
Fit a model of the form

$$\sqrt{y_t} = \beta_0 + \beta_1 t + \beta_2 t^2 + \beta_3 t^3$$

to these data, assuming the errors to be first order autoregressive. Why did we take the square root of the dependent variable?

Exercise 7.8: In Exercise 7.7, $y_{t+1} - y_t$ is approximately equal to the number of births less deaths plus net immigration during the period t to $t+1$. Since births and deaths are approximately proportional to y_t, it is reasonable to propose a model of the form

$$z_t = y_{t+1} - y_t = \beta_0 + \beta_1 y_t + \epsilon_t.$$

Assuming ϵ_t to be first order autoregressive, estimate β_0 and β_1. How would you forecast the 1980 population?

In this exercise we have ignored the fact that y_t is possibly heteroscedastic. Can you suggest a way to, at least crudely, compensate for it?

Exercise 7.9: It is usually conjectured that, over the last several years, prices of other fuels have been determined by the price of oil. Exhibit 7.2 gives prices for crude oil, natural gas, bituminous coal and lignite, and anthracite in 1972 cents per 1000 BTU. Assuming first order autoregressive errors, estimate a model expressing bituminous coal and lignite prices in terms of oil prices. Since it would appear that after the 'energy crisis' of 1973–74, the price of bituminous coal and lignite responded over about two years, introduce suitable lagged variables to take this into account. Test the hypothesis that the price of this energy source is determined by oil prices.

Do the same for anthracite and natural gas prices. Can the use of broken line regression be helpful in the latter case?

Exercise 7.10: The data set in Exhibit 7.3 gives ratios u_t of fluid intake to urine output over five consecutive 8-hour periods ($t = 1, \ldots, 5$) for 19 babies divided into two groups (G). The babies in group 1 received a surfactant treatment. The seven babies in group 2 were given a placebo and constitute a control group.

1. Estimate a model expressing u_t as a linear function of t for the control group.

2. Assuming that the covariance matrix is the same for both groups, test if the same linear function suffices for both groups against the alternative that the functions are different.

3. An examination of a plot of the means of u_t for each time period over all surfactant subjects will reveal that the plot is not quite a straight line. Can you find a more appropriate function?

Exercise 7.11: Human immuno-deficiency virus (HIV) infection causing acquired immuno-deficiency syndrome (AIDS) is known to affect the functioning of a variety of organ systems, including the central nervous system. In order to test its effect on the developing brain, the following study was carried out.

Five baby chimpanzees were injected with a heavy dose of HIV intravenously at 1 hour of age, subsequent to which the babies were allowed to be nursed by the mother. After six months, under general anesthesia, the radio-active microspheres technique was used to measure brain blood flow. Since there is a federal restriction on the use of this primate model, only a limited number of studies could be done. Therefore, the investigators obtained biopsies from five regions of the brain and measured the radioactive counts; the results (y) were expressed as cerebral blood flow in ml/100 grams of brain tissue. The partial pressure (x) of carbon dioxide in millimeters of mercury (also called PCO_2) was also obtained. After the biopsy, the animals were returned to their cages for future studies.

Assume that all pairs of observations for the same chimpanzee have the same correlation, while observations for different chimpanzees are uncorrelated. With y as the dependent variable and x as the independent variable, test the hypothesis that the slope is the same for all regions of the brain against the alternative that region 2 and 4 observations have different slopes than those from other regions.

Exercise 7.12: Trucks can be weighed by two methods. In one, a truck needs to go into a weighing station and each axle is weighed by conventional

G	u_1	u_2	u_3	u_4	u_5
1	0.26	0.23	0.38	0.38	0.35
1	0.00	0.28	0.14	0.47	1.17
1	0.72	1.00	0.27	0.56	0.64
1	1.81	0.72	1.02	0.68	1.00
1	0.43	0.80	0.87	1.99	0.68
1	0.16	0.24	0.40	0.39	1.02
1	0.03	0.36	0.52	0.93	1.25
1	0.09	0.19	0.37	0.57	0.78
1	0.53	0.58	0.44	1.64	0.82
1	0.19	0.30	0.53	0.69	0.93
1	0.32	0.32	0.20	1.08	1.17
1	0.60	0.49	0.80	1.20	1.11
2	0.46	0.46	1.20	0.63	0.40
2	0.00	0.58	0.89	0.50	0.96
2	0.29	0.41	0.79	1.68	2.70
2	0.00	0.33	0.34	0.28	0.73
2	0.29	0.62	0.40	0.67	0.41
2	0.80	0.60	0.62	0.85	1.38
2	0.62	0.17	0.46	0.63	0.79

EXHIBIT 7.3: Data on Intake/Output Ratio
SOURCE: Rama Bhat, M.D., Department of Pediatrics, University of Illinois at Chicago. This data is a part of a larger data set. A full discussion of how the data were gathered is given in Bhat et al. (1989).

means. The other is a newer and a somewhat experimental method where a thin pad is placed on the highway and axles are weighed as trucks pass over it. Former weights are called static weights and are given in Exhibit 7.5 as sw, while the latter are called weight in motion or wim. The parenthetic superscripts in the exhibit are the axles: 1 represents axle 1, 23 the combination of axles 2 and 3 and 45 the combination of 4 and 5. Obviously, one would expect the axle weights for the same truck to be correlated. Assume that the off-diagonal terms of the correlation matrix are the same and the matrix is the same for all trucks.

1. Assuming weight in motion to have the same variance for all axles or axle pairs (i.e., $wim^{(1)}$, $wim^{(23)}$ and $wim^{(45)}$ have the same variance), estimate a linear model of the form $wim - sw = \beta_0 + \beta_1 sw$. Test the hypothesis $\beta_1 = 0$ against the alternative that it is not zero.

2. If you cannot assume that the variances are the same, but assume instead that the variance of each wim is proportional to a function of sw, how would you proceed?

3. Examine suitable residuals from the results of part 1 above to verify

Chimp. #	Frontal (Region 1)		Parietal (Region 2)		Occipital (Region 3)		Temporal (Region 4)		Cerebellum (Region 5)	
	x	y	x	y	x	y	x	y	x	y
1	30.3	64.3	30.3	99.6	30.3	71.7	29.3	86.5	30.3	61.8
2	35.1	56.8	34.0	62.3	34.4	40.1	35.2	97.0	35.0	47.1
3	36.1	62.4	36.8	92.9	34.9	51.6	37.4	83.0	37.2	75.0
4	35.1	60.5	35.8	95.1	36.0	70.7	36.1	97.5	33.4	55.7
5	31.0	43.6	29.6	73.2	28.7	44.5	28.5	79.1	30.2	42.2

EXHIBIT 7.4: Data on PCO_2 (x) and Cerebral Blood Flow (y) for Five Regions of the Brain of Each of Five Chimpanzees
SOURCE: Tonse Raju, M.D., Department of Pediatrics, University of Illinois at Chicago.

if the variance of wim is indeed a function of sw and, if necessary, re-estimate β_0 and β_1.

4. Consider a model of the form $wim - sw = \alpha_0 + \alpha_1 t + \alpha_2 t^2$ where t is the axle order, i.e., $t = 1$ for axle 1, $t = 2$ for axle 23 and $t = 3$ for axle 45. Estimate the α's and obtain a confidence region for them.

Exercise 7.13: Exhibit 7.6 provides data on median family income for 34 Community Areas in the northern half of Chicago. Also given are the percentages of population who are black (PB), Spanish speaking (PS) and over 65 (PA). The contiguity matrix A is given in Exhibit 7.7. Both data sets were constructed by Prof. Siim Soot, Department of Geography, University of Illinois at Chicago.

1. Assume a first order spatial autoregressive model, estimate ρ and use it to estimate parameters in a model with income as the dependent variable and PB, PS and PA as independent variables. Are the coefficients of the independent variables significant?

2. Assume that the covariance matrix is of the form (7.26) with $q = 4$. Apply (7.9) to estimate parameters.

3. Under the same assumptions as part 1 above, apply (7.10) and (7.11) to estimate variances. Use these estimated variances to estimate parameters (making reasonable assumptions when the estimated variances turn out to be negative).

4. Try other values of q.

Test residuals (in each case) for the presence of spatial correlation and comment.

$sw^{(1)}$	$wim^{(1)}$	$sw^{(23)}$	$wim^{(23)}$	$sw^{(45)}$	$wim^{(45)}$
8660	5616	11920	15600	8520	10608
8820	6448	17040	19344	19480	18512
8780	6448	33040	34944	30040	29744
11180	9152	27840	33072	29040	31824
10680	10192	20200	30576	20300	30784
10820	7072	27200	29952	31900	29952
9500	7072	33200	34112	31820	29952
9660	6448	24000	19760	33200	16640
11320	9568	29520	33280	34100	34320
10460	6240	31640	30576	31620	27040
10720	5616	33560	42016	33720	35776
12740	9152	29720	30160	24380	23088
10920	9152	26560	29952	15280	14144
8720	6240	12880	14352	9920	11323
10380	6448	27260	29952	23020	20592
8980	5824	14160	13312	11340	11024
8780	7072	31280	37648	30000	37024
9160	6240	13900	13104	9660	9984
10220	7072	32820	36192	32740	35152
8820	6240	13940	13936	12300	12480
9460	6240	20640	26624	18680	19552
10120	8320	24700	18928	23200	20176
9500	6656	27240	35984	23460	25792
9820	7072	20740	20800	19280	17264
8760	7488	17660	21840	14500	16640
11360	8320	32380	37856	17480	17888

EXHIBIT 7.5: Data on Static Weights and Weight in Motion of Trucks
SOURCE: Saleh Mumayiz, Urban Transportation Center, University of Illinois at Chicago, who compiled the data from a data set provided by the Illinois Department of Transportation.

Obs #	Area Name	PB	PS	PA	Income
1	ROGERS PARK	9.41	11.92	14.96	18784
2	WEST RIDGE	0.72	3.71	22.68	25108
3	UPTOWN	15.06	23.26	13.88	14455
4	LINCOLN SQ.	0.54	11.31	17.98	20170
5	NORTH CENTER	1.17	19.01	13.61	19361
6	LAKE VIEW	6.93	18.80	14.77	20716
7	LINCOLN PARK	8.59	10.48	9.94	24508
8	NEAR N. SIDE	32.75	2.89	13.01	23395
9	EDISON PARK	0.00	0.96	18.64	27324
10	NORWOOD PARK	0.02	0.96	19.38	27595
11	JEFFERSON PK.	0.02	1.61	20.42	25082
12	FOREST GLEN	0.06	1.55	19.03	31651
13	NORTH PARK	0.94	5.54	17.95	25975
14	ALBANY PARK	0.61	19.69	11.14	19792
15	PORTAGE PARK	0.09	2.59	19.06	23402
16	IRVING PARK	0.13	8.62	15.59	21088
17	DUNNING	0.48	1.61	18.29	24445
18	MONTCLARE	0.00	1.66	20.47	24005
19	BELMONT	0.08	5.76	18.82	22245
20	HERMOSA	0.38	31.21	12.86	19118
21	AVONDALE	0.18	20.47	13.61	19144
22	LOGAN SQUARE	2.64	51.70	9.39	16224
23	HUMBOLT PARK	35.57	40.73	6.45	14461
24	WEST TOWN	8.99	56.72	8.56	12973
25	AUSTIN	72.45	5.90	6.10	16566
26	W GARFLD PK.	98.55	0.82	4.96	10922
27	E GARFLD PK.	99.00	0.83	7.81	9681
28	NEAR W. SIDE	76.45	9.96	8.14	7534
29	N. LAWNDALE	96.48	2.69	6.55	9902
30	LOWER W. SIDE	1.06	77.57	7.01	14486
31	LOOP	19.05	3.44	21.26	26789
32	NEAR S. SIDE	94.14	1.49	9.44	7326
33	ARMOUR SQUARE	25.35	4.83	14.77	15211
34	EDGEWATER	11.12	13.33	18.40	19859

EXHIBIT 7.6: Community Area Data for the North Part of the City of Chicago

|
|---|
| 0100000000000000000000000000000001 |
| 1001000000001000000000000000000001 |
| 0001110000000000000000000000000001 |
| 0110110000001101000000000000000001 |
| 0011011000000101000011000000000000 |
| 0011101000000000000000000000000000 |
| 0000110100000000000011010000000000 |
| 0000001000000000000000010001001000 |
| 0000000001000000000000000000000000 |
| 0000000010100000000000000000000000 |
| 0000000001010010000000000000000000 |
| 0000000000101111000000000000000000 |
| 0101000000010100000000000000000000 |
| 0001100000011011000000000000000000 |
| 0000000000110101101100000000000000 |
| 0001100000010110001110000000000000 |
| 0000000000000010011000000000000000 |
| 0000000000000000101000001000000000 |
| 0000000000000011110100101000000000 |
| 0000000000000011001011101000000000 |
| 0000101000000000100010100000000000 |
| 0000101000000000000110110000000000 |
| 0000000000000000001101011110000000 |
| 0000001100000000000001100011000000 |
| 0000000000000000011100100100100000 |
| 0000000000000000000000101010100000 |
| 0000000000000000000000110101100000 |
| 0000000100000000000000010010111100 |
| 0000000000000000000000001111010000 |
| 0000000000000000000000000001100110 |
| 0000000100000000000000000001000100 |
| 0000000000000000000000000001011010 |
| 0000000000000000000000000000010100 |
| 1111000000000000000000000000000000 |

EXHIBIT 7.7: The Contiguity Matrix A for the 34 Community Areas in Northern Part of Chicago

CHAPTER 8

Outliers and Influential Observations

8.1 Introduction

Sometimes, while most of the observations fit the model and meet G-M conditions at least approximately, some of the observations do not. This occurs when there is something wrong with the observations or if the model is faulty.

As far as the observations are concerned, there could have been a mistake in inputting or recording data. A few observations might reflect conditions or situations different from those under which other observations were obtained. For example, one or two of the observations from a chemical experiment may have been affected by chemical contamination or by equipment malfunction. If the data set on house prices were to include one or two observations where either the properties were particularly run-down or where for unusual circumstances they were sold at prices that did not reflect their 'true value', these points would most likely not belong to the model for houses.

Observations that do not fit the model might point also to deficiencies in the model. There could be an independent variable which should have been included in the model but was not. Such data points might also show us that the algebraic form of the model is incorrect; e.g., we might need to make some transformations to the data (Chapter 9) or consider forms which require the use of non-linear regression.

Because there are two main reasons for the existence of observations that do not belong to the model, there are two principal purposes in trying to identify them. One is obviously to protect the integrity of the model from the effects of points that do not belong to it. The other purpose is to identify shortcomings in the model. The latter point, which is particularly important, is often ignored by inexperienced analysts. In some sense the postulated model reflects what we already know or think we know; otherwise we would not have formulated the model the way we did. Points that do not conform to it are the surprises and can tell us things we did not know. This information can lead to substantial improvements in the model. It can also lead to discoveries which are valuable in themselves. As Daniel and Wood (1980, p. 29) put it, "Numerous patents have resulted from the recognition of outliers."

Observations that do not belong to the model often exhibit numerically large residuals and when they do they are called outliers. As described in Section 1.6, p. 11, outliers frequently have an inordinate influence on least squares estimates although not all influential points have large residuals. Although observations that do not fit the model do not necessarily have large residuals, nor are they necessarily influential, there is little hope of finding such non-conforming points if they do not stand out either as outliers or as influential points. Therefore we shall confine our attention in this chapter to outliers and influential points. Besides, a particularly influential point should be scrutinized simply because it has so much effect on the estimates.

A point has undue influence when it has a large residual or is located far away from other points in the space of the independent variables (Exhibit 1.10b, p. 12). A measure of this latter remoteness is the leverage h_{ii}, which we have already encountered in Section 5.3 (p. 106). Indeed, the residuals and the leverages constitute the blocks used in building virtually all measures to assess influence. Because of their importance, we shall examine leverages and residuals in the next two sections before proceeding to discuss how these measures might be used to identify influential points.

8.2 The Leverage

As stated in Section 5.3, the leverages h_{ii} are the diagonal elements of the hat matrix $H = X(X'X)^{-1}X'$. Consequently, an individual $h_{ii} = \boldsymbol{x}_i'(X'X)^{-1}\boldsymbol{x}_i$ where $\boldsymbol{x}_i'$ is a row of the design matrix X and, therefore, corresponds to a single observation. The key feature of a leverage h_{ii} is that it describes how far away the individual data point is from the centroid of all data points in the space of independent variables, i.e., how far removed $\boldsymbol{x}_i$ is from $\bar{\boldsymbol{x}} = n^{-1}\sum_{i=1}^n \boldsymbol{x}_i$. We shall show this in the subsection below.

In the case of a lever, greater influence can be generated at a point far removed from the fulcrum, than at a point closer to it. This is because for a given change at a point remote from the fulcrum the corresponding changes in close-in points are relatively small. The situation is similar for the leverage h_{ii} with the fulcrum at the centroid when $\beta_0 \neq 0$. When $\beta_0 = 0$ the fulcrum is obviously the origin and leverage measures the distance from the origin.

Since, as shown in Section 5.3, $\sum_{i=1}^n h_{ii} = k+1$, if we had the option of choosing independent variable values we would choose them so as to make each $h_{ii} = (k+1)/n$. However, such a choice is seldom ours to make. But if all h_{ii}'s are close to $(k+1)/n$ and if all the residuals turn out to be acceptably small, no point will have an undue influence. Belsley, Kuh and Welsch (1980, p. 17) offer $2(k+1)/n$ as a possible cut-off point for h_{ii}, but note that this criterion tends to draw attention to too many points.

A slight shortcoming of the leverage as a diagnostic tool is that it treats all independent variables the same regardless of how each one affects the dependent variable. For example, a variable that we might ultimately discard from the model because of its inefficacy (Chapter 11) would affect h_{ii} as much, or as little, as if it were the most important predictor. The DFFITS and DFBETAS described in Section 8.5 are more sensitive to the importance of an individual b_j.

8.2.1 *Leverage as Description of Remoteness

Let Z be the centered version of X, i.e., $Z' = (\boldsymbol{x}_1 - \bar{\boldsymbol{x}}, \ldots, \boldsymbol{x}_n - \bar{\boldsymbol{x}})$. Then the diagonal elements of $\tilde{H} = Z(Z'Z)^{-1}Z'$ are

$$\tilde{h}_{ii} = (\boldsymbol{x}_i - \bar{\boldsymbol{x}})'(Z'Z)^{-1}(\boldsymbol{x}_i - \bar{\boldsymbol{x}}). \tag{8.1}$$

Since $(n-1)^{-1}Z'Z$ is sample covariance matrix of the 'observations' $\boldsymbol{x}_i$, it is easy to see that (8.1) is a standardized form of squared distance between $\boldsymbol{x}_i$ and $\bar{\boldsymbol{x}}$. Apart from rendering (8.1) unit-free, this form of standardization also takes into account any relationship between independent variables.

Let $\bar{\boldsymbol{y}} = \bar{y}\mathbf{1}$ where $\bar{y} = n^{-1}\sum_{i=1}^{n} y_i$. Since the vector of predicted values is $\hat{\boldsymbol{y}} = H\boldsymbol{y}$,

$$\hat{\boldsymbol{y}} - \bar{\boldsymbol{y}} = H\boldsymbol{y} - n^{-1}\mathbf{1}\mathbf{1}'\boldsymbol{y} = (H - n^{-1}\mathbf{1}\mathbf{1}')\boldsymbol{y}.$$

On the other hand, since $\mathbf{1}'Z = \mathbf{o}$, it can be shown (using the centered model (2.40) on p.42) that $\hat{\boldsymbol{y}} - \bar{\boldsymbol{y}} = \tilde{H}\boldsymbol{y}$. It follows that

$$h_{ii} - n^{-1} = \tilde{h}_{ii}. \tag{8.2}$$

Therefore, h_{ii} also describes distance between $\boldsymbol{x}_i$ and $\bar{\boldsymbol{x}}$.

Equation 8.2 also shows that $n^{-1} \le h_{ii} \le 1$, which is why we took pains not to refer to h_{ii} as a distance.

8.3 The Residuals

For the purpose of detecting observations that do not belong to the model (and also influential points), more valuable than the residuals are the Studentized residuals, often called RSTUDENT and defined by

$$e_i^\star = \frac{e_i}{s_{(i)}\sqrt{1-h_{ii}}} \tag{8.3}$$

where e_i is, as before, a residual and $s_{(i)}$ is equivalent to s if least squares is run after deleting the ith case. Denote as $\boldsymbol{y}(i)$ and $X_{(i)}$ the results of removing the ith row from $\boldsymbol{y}$ and X and let $\boldsymbol{b}(i)$ be the least squares estimate of $\boldsymbol{\beta}$ based on $\boldsymbol{y}(i)$ and $X_{(i)}$, i.e., $\boldsymbol{b}(i) = (X_{(i)}'X_{(i)})^{-1}X_{(i)}'\boldsymbol{y}(i)$. Clearly,

$$(n-k-2)s_{(i)}^2 = \sum_{\substack{\ell=1\\ \ell\neq i}}^{n} [y_\ell - \boldsymbol{x}_\ell'\boldsymbol{b}(i)]^2$$

has essentially the same statistical properties as s^2, and, in particular, under G-M conditions $s^2_{(i)}$ is an unbiased and consistent estimate of σ^2. An alternative expression for it follows from

$$(n-k-2)s^2_{(i)} = (n-k-1)s^2 - e_i^2(1-h_{ii})^{-1} \tag{8.4}$$

which is proved in the appendix to this chapter. Obviously, (8.3) is most similar to the standardized residuals (5.4) and frequently will also be numerically similar. Since in either case, each residual is being divided by an estimate of its standard error, it is fairer to make comparisons between Studentized or standardized residuals than among the residuals themselves. However, analysts who have used residuals to identify outliers and watched the maximum of their numerical values decline as outliers are eliminated might find Studentized or standardized residuals a bit disconcerting. Since typically, s^2 or $s^2_{(i)}$ will decline as outliers are eliminated, the maximum of such residuals will not always decline. They are essentially relative measures.

The $e_i^\star$'s are fascinating quantities. It is shown in the appendix to this chapter (and slightly differently in Cook and Weisberg, 1982, p. 21) that if we append to our list of independent variables an additional one — an indicator variable z which is 1 for the ith case but is zero otherwise — then the t-value associated with this variable is exactly $e_i^\star$. This means that $e_i^\star$ has a t distribution when the errors are Gaussian and has a near t distribution under a wide range of circumstances (see Section 5.3; also see Exercise 8.2). With the presence of z in the model, the estimates of the coefficients of the other independent variables and the intercept are not affected by ith observation. Therefore, $e_i^\star$ is a standardized measure of the distance between the ith case and the model estimated on the remaining cases. Therefore, it can serve as a test statistic to decide if the ith point belongs to the model (see also Exercise 8.3).

8.4 Detecting Outliers and Points That Do Not Belong to the Model

Any one of the plots of the residuals — against $\hat{y}_i$'s or against any of the x_{ij}'s or even against the case numbers — will show us which residuals (if any) are large. And because these plots are easily drawn and also used for other purposes, this is possibly what is done most often.

However, for a search for points that might not fit the model, the Studentized residuals are much more useful. One could examine a listing or a plot of these against, say, case numbers. One can easily identify from these which $e_i^\star$ are significant at some given level, e.g., 5 per cent, and flag the corresponding observations for further scrutiny.

But this process can be pointlessly tedious. For example, if there are a hundred normally distributed observations all belonging to the model, then there would be naturally about five observations for which $e_i^\star$ would be significant at a 5 per cent level. Therefore, it would be useful to have methods for judging if outliers are present at all.

Several methods are possible. If the number of observations is small, the Bonferroni inequality can be used (see Section 3.8.4, p. 73). For larger numbers of observations the significance levels can be raised to reflect the number of observations, but this is not entirely satisfactory. A much better alternative is to use a normal plot (Section 5.2.1, p. 101) of the $e_i^\star$'s. If there are vertical jumps near either end of such a plot, or even if the plot turns sharply upwards or downwards near the ends, we might have points which should be flagged for further investigation.

As alternatives to the normal plot, some analysts use various other types of univariate displays like histograms, stem-and-leaf displays and box plots. It should be noted that causes other than outliers also can give disturbing shapes to normal plots (as well as other displays of Studentized residuals). One is heteroscedasticity; however, if the variance varies as a function of $E(y_i)$ or one of the independent variables, other plots can be used to detect its presence. But if the cause is a fluctuation in variance which is not a function of a known variable, or if the cause is an extremely long-tailed distribution of the y_i's, there is probably no simple way to distinguish their effects from those caused by the presence of outliers.

Some analysts treat each y_i as if it were a missing future observation and construct a confidence interval for it based on corresponding x_{ij}'s (as described in Chapter 3). If the observed y_i falls outside this interval, it could then be tagged for future study. Finally, the partial regression leverage plots described in Section 11.3.3, p. 243, have also been proposed as means for outlier detection.

8.5 Influential Observations

Despite the conclusions of Exercise 8.4 at the end of this chapter, not all influential points have large $e_i^\star$'s. But such points do need to be examined simply because they are influential and a bad observation that is influential will hurt the model more than a bad but less influential one. For such examinations, what better measures can we have than ones that tell us how much $\boldsymbol{b}$ or $\hat{\boldsymbol{y}}$ would change if a given point were to be deleted? Therefore, a crucial formula is

$$\boldsymbol{b} - \boldsymbol{b}(i) = \frac{(X'X)^{-1}\boldsymbol{x}_i e_i}{1 - h_{ii}}, \tag{8.5}$$

where, as in Section 8.2, $\boldsymbol{b}(i)$ is the least squares estimate of $\boldsymbol{\beta}$ obtained after deleting the ith case. A proof of (8.5) is given in the appendix to this chapter. The vector $\boldsymbol{b} - \boldsymbol{b}(i)$ is often called DFBETA$_i$.

Let

$$(X'X)^{-1}\boldsymbol{x}_i = (a_{0i}, \ldots, a_{ki})'.$$

Then, an individual component of DFBETA_i is

$$\mathrm{DFBETA}_{ij} = b_j - b_j(i) = \frac{a_{ji}e_i}{1-h_{ii}} \tag{8.6}$$

for $i = 1, \ldots, n$ and $j = 0, \ldots, k$. From (8.5) we also get

$$\mathrm{DFFIT}_i = \hat{y}_i - \hat{y}_i(i) = \boldsymbol{x}_i'\boldsymbol{b} - \boldsymbol{x}_i'\boldsymbol{b}(i) = h_{ii}e_i/(1-h_{ii}) \tag{8.7}$$

which tells us how much the predicted value $\hat{y}_i$, at the design point $\boldsymbol{x}_i$, would be affected if the ith case were deleted.

In order to eliminate the effect of units of measurement, standardized versions of these statistics are often used. Since by Theorem 2.2, p. 36, the covariance matrix of $\boldsymbol{b}$ is $\sigma^2(X'X)^{-1}$, the variance of an individual component b_j of $\boldsymbol{b}$ is the jth diagonal element $\sigma^2 q_{jj}$ of $\sigma^2(X'X)^{-1}$, where q_{ij} is the (i,j)th element of $(X'X)^{-1}$. It is appropriate to estimate σ^2 by $s_{(i)}^2$ since we are examining the ith observation and it may be suspect. Therefore, dividing the right side of (8.6) by the square root of $s_{(i)}^2 q_{jj}$ we get the standardized DFBETA_{ij}, which is

$$\mathrm{DFBETAS}_{ij} = \frac{a_{ji}e_i}{s_{(i)}(1-h_{ii})q_{jj}^{1/2}}. \tag{8.8}$$

Since the covariance matrix of $\hat{\boldsymbol{y}} = X\boldsymbol{b}$ is $\sigma^2 X(X'X)^{-1}X' = \sigma^2 H$, the variance of $\hat{y}_i$ may be estimated by $s_{(i)}^2 h_{ii}$. Hence, the standardized version of DFFIT_i is

$$\mathrm{DFFITS}_i = \frac{h_{ii}^{1/2}e_i}{s_{(i)}(1-h_{ii})}. \tag{8.9}$$

Obviously, these standardized forms are unit-free.

It is a relatively easy matter to express them as functions both of the leverage and the Studentized residuals. $\mathrm{DFBETAS}_{ij}$ can be written as $[q_{jj}(1-h_{ii})]^{-1/2}a_{ji}e_i^\star$ and DFFITS_i as

$$[h_{ii}/(1-h_{ii})]^{1/2}e_i^\star. \tag{8.10}$$

Therefore, if either the leverage increases or the Studentized residual increases, both measures of influence will increase.

A natural question that arises at this stage is how we use these measures to decide which of the data points is to be flagged for scrutiny. There are several possible approaches to answering this question and in a practical application one generally follows all of them; how one weights the evidence provided by each defines the analyst's style. If one is very familiar with the substantive application then one develops a feel for what is too large. Another approach is to run down the list of the measures of influence and identify those which are much larger than the others. Here, the types of plots that we described in connection with $e_i^\star$'s could be useful. It should be pointed out in this context that it is desirable to give some attention to DFBETAS's (or DFBETA's), tedious though it might be, since sometimes a point could be influential on a single b_j and not affect $\hat{y}_i$'s very much. A third approach is to use some criterion level. For DFFITS, Belsley, Kuh and Welsch (1980, p. 28) use $h_{ii} = (k+1)/n$ (see the paragraph just preceding Subsection 8.2.1) and $e_i^\star = 2$ (since under normality $e_i^\star$ has a t distribution and under a wide range of circumstances $e_i^\star$ has approximately a t distribution) in (8.10) to arrive at the criterion $2[(k+1)/(n-k-1)]^{1/2}$, or when n is much larger than k, the criterion $2[(k+1)/n]^{1/2}$. A larger value would tag the point for further examination. A criterion of $2n^{-1/2}$ for DFBETAS's has been suggested by Belsley, Kuh and Welsch (1980).

All four measures of influence examined in this section combine leverages and Studentized residuals to give expressions that are physically meaningful. Therefore, although the above-mentioned criteria are often useful, it is the physical meaning rather than criteria that is key here. In fact, we can write DFBETA$_{ij}$ as $\alpha_{ij}e_i$, where $\alpha_{ij} = a_{ji}/(1-h_{ii})$ is a function only of the independent variable values. Consequently, the standard error of DFBETA$_{ij}$ is $\alpha_{ij}\sigma(1-h_{ii})^{1/2}$. Therefore, if, instead of standardizing DFBETA by division by the standard error of b_j, we had standardized it by division by an estimate of its own standard deviation, we would have got exactly $e_i^\star$! The same is true of DFFIT.

8.5.1 Other Measures of Influence

The measures given in the last section are by no means the only ones available for detecting influential observations. Some measure the effect on the estimated covariance matrix of $\boldsymbol{b}$. One such measure is the ratio of the determinants of the estimated covariance matrix of $\boldsymbol{b}(i)$ and $\boldsymbol{b}$. This is called the covariance ratio and is given by

$$\text{Covariance Ratio} = \frac{\det[s_{(i)}^2(X_{(i)}'X_{(i)})^{-1}]}{\det[s^2(X'X)^{-1}]}. \tag{8.11}$$

A value of this ratio close to 1 would indicate lack of influence of the ith data point. It follows from (8.14) given in the appendix to this chapter and formulæ in Section A.8, p. 272, of Appendix A that

$$\begin{aligned}\det(X_{(i)}{}'X_{(i)}) &= \det(X'X - \boldsymbol{x}_i\boldsymbol{x}_i') = \det(X'X)\det[I - (X'X)^{-1}(\boldsymbol{x}_i\boldsymbol{x}_i')] \\ &= \det(X'X)\det[I - \boldsymbol{x}_i'(X'X)^{-1}\boldsymbol{x}_i] = (1-h_{ii})\det(X'X).\end{aligned}$$

Therefore, (8.11) can be written as $[s_{(i)}{}^2/s^2]^{k+1}(1-h_{ii})^{-1}$ and, by (8.4), it can be shown to equal the reciprocal of

$$(n-k-2+e_i^{\star 2})^{k+1}(1-h_{ii})(n-k-1)^{-k-1}$$

which is a function of Studentized residual and leverage.

Another set of measures, which we will also see in Chapter 11, Section 11.3.1, p. 239, are the PRESS residuals $e_{i,-1} = y_i - \hat{y}_i(i)$. Obviously, an individual $e_{i,-1}$ shows how well an observation is predicted by a model based on the other observations. From (8.7), it is easy to see that

$$e_{i,-1} - e_i = y_i - \hat{y}_i(i) - y_i + \hat{y}_i = \hat{y}_i - \hat{y}_i(i) = h_{ii}e_i/(1 - h_{ii}).$$

Hence

$$e_{i,-1} = h_{ii}e_i/(1 - h_{ii}) + e_i = e_i/(1 - h_{ii}). \tag{8.12}$$

A very frequently used measure of influence can be defined as the distance between the vectors $\boldsymbol{b}$ and $\boldsymbol{b}(i)$ 'standardized by' the estimated covariance matrix $s^2(X'X)^{-1}$ of $\boldsymbol{b}$. This distance, known as Cook's (1977) distance, is given by

$$(\boldsymbol{b} - \boldsymbol{b}(i))'X'X(\boldsymbol{b} - \boldsymbol{b}(i))/[(k+1)s^2]. \tag{8.13}$$

From (8.5) it follows that (8.13) equals

$$e_i^2 h_{ii}/[(k+1)s^2(1 - h_{ii})^2].$$

Hence, it is essentially the same as the square of the DFFITS_i. Another useful measure of distance has been given by Andrews and Pregibon (1978).

Actually, the number of measures available in the literature for identifying outliers and influential points verges on being mind-boggling. A partial list, with very readable explanations, is given in Chatterjee and Hadi (1986) and the discussion following it. Also see Cook and Weisberg (1980) and (1982), Cook (1986) and Lawrence (1988).

8.6 Examples

Example 8.1

Exhibit 8.1 shows some of the outlier and influential point diagnostics from a weighted regression (with weight n) of y on x using the data of Exhibit 6.10. From this and from the residual plot of Exhibit 8.2 we see that there is one very influential point (#29), which also has an unusually high RSTUDENT value. A possible reason for the great influence of the point is the large weight it receives, but this would not totally account for the RSTUDENT value. While weighting does affect RSTUDENT, it only compensates for the rather small variance one would expect from the mean of so many observations. Since a single value of slightly over 2 from among the 31 remaining observations is not so noteworthy, no other point seems to stand out.

Examination of a map showed that to make the bus trip corresponding to case #29, one has to transfer, i.e., change buses. This is a very plausible

Obs #	e_i	$e_i^\star$	$\Delta\hat{y}$	Obs #	e_i	$e_i^\star$	$\Delta\hat{y}$
1	3.28	0.32	0.04	17	3.41	0.79	0.22
2	9.43	0.95	0.16	18	-2.61	-0.53	-0.10
3	-4.22	-1.16	-0.29	19	-1.16	-0.20	-0.05
4	-4.74	-0.83	-0.18	20	0.02	0.00	0.00
5	4.64	0.66	0.10	21	-4.69	-1.30	-0.35
6	-10.14	-2.27	-0.74	22	-0.06	-0.02	-0.01
7	-3.82	-0.77	-0.15	23	-0.57	-0.12	-0.04
8	7.80	1.43	0.40	24	-1.91	-0.27	-0.04
9	-10.78	-1.09	-0.11	25	0.54	0.09	0.02
10	-0.59	-0.05	-0.01	26	3.09	0.44	0.07
11	-3.67	-1.29	-0.59	27	0.12	0.02	0.00
12	4.46	0.63	0.11	28	-1.06	-0.36	-0.21
13	5.07	0.50	0.09	29	8.06	5.27	3.05
14	8.27	1.19	0.19	30	-0.54	-0.08	-0.01
15	-7.38	-1.05	-0.14	31	3.09	0.31	0.03
16	-0.50	-0.09	-0.02	32	-4.46	-0.79	-0.18

EXHIBIT 8.1: Residuals (e), Studentized Residuals ($e^\star$) and DFFITS ($\Delta\hat{y}$) for Travel Time Example

reason for a large residual. If buses keep to schedule, half the time between buses (the headway) is the expectation of waiting time. But if they do not, the expected waiting time increases, since more people arrive during the longer gaps and consequently have to wait longer. In fact, it can be shown that the expected wait for randomly arriving buses is approximately the average headway. Moreover, waiting for buses is possibly so onerous that the time taken may be perceived as longer. Of course, one also waits for a bus at the beginning of a trip, but that affects all observations equally.

Unfortunately, case numbers 2 and 23 also require transfers. These points are not outliers. One possible reason is that the very influential point #29 may have pulled the regression line up so much that the residuals for points 2 and 23 were considerably smaller than they would otherwise have been (the x values are very close for all three points). At any rate, we found the argument of the last paragraph so compelling that we decided to explore the matter further.

On deleting case number 29 and rerunning the regression, the residual for observations 2 and 23 increased to about 13 and 3 respectively with RSTUDENT values of 1.9 and .89. While this was not a resounding confirmation of our conjecture, we nevertheless decided to go ahead and append to our model an indicator variable T which took the value of 1 for those cases which involved a bus transfer. Exhibit 8.3 provides parameter estimates and t-values while Exhibit 8.4 shows a residual plot. The R^2 value increased from about .66 to about .83 when the variable T was added. However, observation number 29 remained quite influential (DFFITS$_{29}$=2.8, $e_{29}^\star = 1.5$) although the residual $e_{29} = 1.2$ was quite small — not too surprising given the concentration of 17 trip makers (n — see Exhibit 6.10) at that point. It

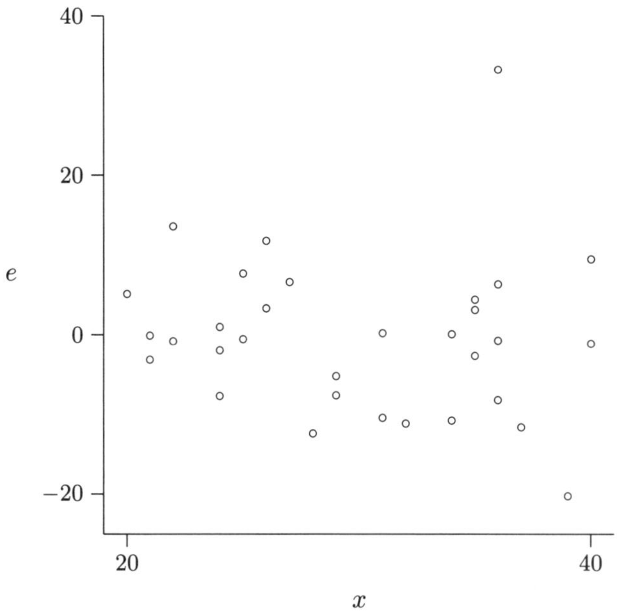

EXHIBIT 8.2: Residuals from a Weighted Regression of y on x Using Travel Time Data

Variable	b_j	s.e.(b_j)	t-value
Intercept	10.169	3.60	2.82
x	0.812	0.12	6.70
T	10.488	1.94	5.41

EXHIBIT 8.3: Parameter Estimates, Standard Error of Parameters and t-Values When the Variable T Was Included in Travel Time Model

would appear from these results that our course of action was reasonable. Indeed, the outlier led us to a conclusion which we might otherwise have overlooked.

If we examine Exhibit 8.4 using inter-quartile ranges (as in Example 6.1, p. 112), we might notice very slight heteroscedasticity. But we are getting a bit carried away at this stage! The model of Exhibit 8.3 is adequate for any purpose we can think of. ■

Example 8.2 (Continuation of Example 4.4, Page 92)

In Examples 5.2, p. 103, and 5.4, p. 106, we have already examined the residuals from the model that we constructed in Example 4.4, p. 92, of LIFE against a piecewise linear function of the log of income for 101 coun-

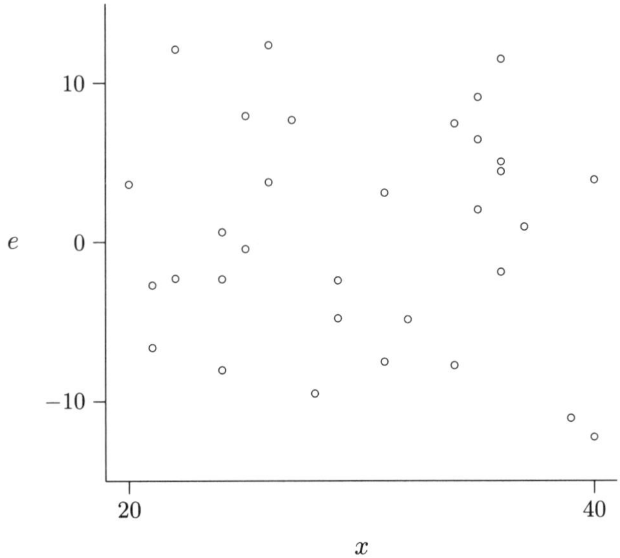

EXHIBIT 8.4: Residuals from a Weighted Regression of y on x and the Indicator Variable T Using Travel Time Data

tries. Exhibit 8.5 shows a normal plot of the Studentized residuals. (Readers preferring stem-and-leaf diagrams or other plots are invited to construct them.) Just like the plot in Exhibit 5.3, this plot also seems to show the possibility of outliers near both the top and the bottom. Exhibit 8.6 presents Studentized residuals $e_i^\star$'s, h_{ii}'s, DFFITS$_i$'s ($\Delta\hat{y}$) and DFBETAS$_{ij}$'s (Δb_j). Four observations — Iran (Observation 23), Libya (25), Saudi Arabia (27) and Ivory Coast (58) — have Studentized residuals less than -2 and two countries — Yugoslavia (49) and Sri Lanka (93) — have Studentized residuals above 2. Therefore, it is quite possible that some, though not necessarily all, of these observations do not fit the model (see Section 8.3). A point with a high Studentized residual does not necessarily have to be eliminated. Much depends on whether there are other reasons to believe that the point does not belong in the analysis.

While several of the h_{ii}'s are relatively large (e.g., greater than .06, using the criterion of Section 8.2), none is excessively so and for all six countries with numerically large Studentized residuals, the value of leverage is quite low. This is basically why only five countries have values of DFFITS exceeding in absolute value the cut-off mentioned in Section 8.5, which in this case works out to .345. All are among the six with large $e_i^\star$'s. The largest DFFITS corresponds to Saudi Arabia, which is indeed influential. Deleting it would change the predicted value at its GNP level by .67 standard error,

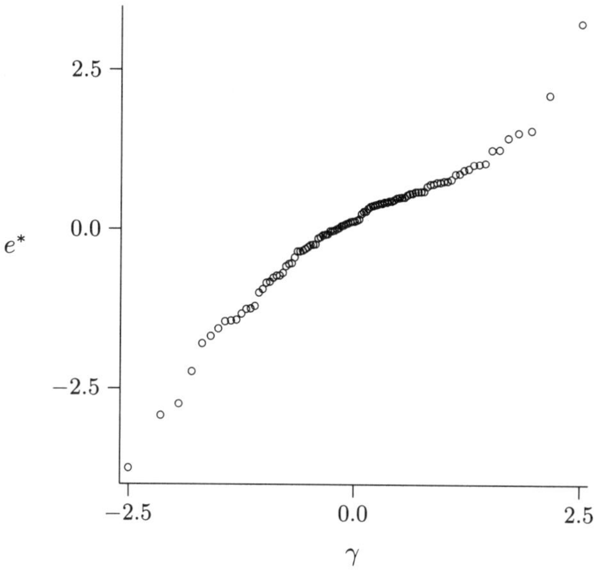

EXHIBIT 8.5: Normal Plot of Studentized Residuals from Life Expectancy Model

which can be shown to equal about 4.4 years; the change in the coefficients b_0, b_1 and b_2 would be about half a standard error each. No other point comes close to having so much influence, although the deletion of some would have a fair amount of effect on one or other of the estimate of β_j's. Most such points belong to the list of six given earlier. Some exceptions are Guinea, Laos, Uruguay and Portugal.

From the above discussion, it is obvious we should examine Saudi Arabia carefully. Its oil has made it a recently and suddenly rich country (note that the data are from the seventies). Its social services, education, etc., have not had time to catch up. This would also be true of Iran and Libya. Ivory Coast is another example of recent wealth although not because of oil (in the mid-70's). "Since the country attained independence from France in 1960, the Ivory Coast has experienced spectacular economic progress and relative political stability"(Encyclopædia Britannica, 1974, Macropædia, v. 11, p. 1181). In the early seventies this country of less than four million was the world's second largest producer of tropical hardwoods, the third largest producer of cocoa and a major producer of coffee.

All these countries have recently become rich. Therefore, we could introduce into the model per capita income from, say, twenty years before. Instead, we simply chose to eliminate these obviously unusual points since we decided that they should not be considered in an attempt to find a general relationship between average wealth and longevity.

Obs	$e_i^\star$	h_{ii}	$\Delta\hat{y}$	Δb_0	Δb_1	Δb_2	Obs	$e_i^\star$	h_{ii}	$\Delta\hat{y}$	Δb_0	Δb_1	Δb_2
1	.119	.044	.026	-.001	.001	.012	51	.402	.013	.046	.016	-.009	-.004
2	.048	.043	.010	-.001	.001	.005	52	-.689	.014	-.083	-.042	.030	-.004
3	.080	.043	.017	-.001	.001	.008	53	-1.456	.013	-.168	.011	-.037	.073
4	-.030	.077	-.008	-.001	.001	-.005	54	.736	.013	.083	.022	-.009	-.014
5	.120	.084	.036	.007	-.007	.025	55	1.022	.014	.123	-.025	.043	-.065
6	-.033	.042	-.007	.000	-.000	-.003	56	-1.686	.012	-.190	-.043	.013	.038
7	.325	.044	.070	-.002	.003	.033	57	-.250	.013	-.029	.002	-.007	.013
8	-.351	.084	-.106	-.019	.021	-.073	58	-2.923	.017	-.384	.143	-.197	.248
9	.561	.025	.090	-.050	.056	-.021	59	.017	.015	.002	-.000	.001	-.001
10	.423	.026	.069	-.029	.032	-.001	60	1.438	.015	.178	-.049	.075	-.104
11	.493	.041	.103	-.006	.007	.045	61	-.350	.013	-.040	-.013	.007	.004
12	.385	.060	.097	.008	-.009	.058	62	.001	.013	.000	-.000	.000	-.000
13	.058	.051	.013	.000	-.000	.007	63	-1.333	.021	-.197	.101	-.126	.143
14	.401	.060	.101	.009	-.010	.061	64	1.039	.015	.129	-.036	.055	-.076
15	.929	.046	.204	-.156	.174	-.171	65	.364	.012	.041	.006	-.000	-.011
16	.240	.098	.079	.017	-.019	.057	66	.093	.015	.011	-.003	.004	-.006
17	.420	.035	.080	-.012	.013	.027	67	1.256	.013	.142	.037	-.015	-.024
18	.558	.028	.095	-.031	.034	.010	68	-.146	.019	-.020	.009	-.012	.014
19	-.283	.097	-.093	-.020	.022	-.066	69	1.014	.018	.137	.092	-.076	.032
20	-.584	.019	-.081	.036	-.047	.056	70	-.244	.030	-.043	-.037	.033	-.020
21	-.089	.016	-.011	.003	-.005	.007	71	.671	.034	.127	.112	-.101	.063
22	.873	.022	.130	.099	-.084	.043	72	-.081	.037	-.016	-.014	.013	-.008
23	-2.238	.041	-.462	.360	-.400	.353	73	.136	.019	.019	.013	-.011	.005
24	-.827	.026	-.134	.080	-.096	.103	74	-1.253	.019	-.175	-.125	.104	-.049
25	-2.743	.036	-.530	.073	-.081	-.186	75	-.842	.036	-.163	-.145	.132	-.084
26	-1.443	.014	-.169	-.071	.046	.004	76	-.240	.031	-.043	-.037	.033	-.020
27	-3.751	.031	-.672	.496	-.551	.413	77	-.021	.032	-.004	-.003	.003	-.002
28	.353	.043	.075	-.059	.065	-.059	78	-1.799	.032	-.325	-.282	.253	-.155
29	.498	.046	.110	-.086	.096	-.088	79	-1.261	.024	-.198	-.158	.138	-.076
30	.948	.019	.131	-.058	.076	-.090	80	.155	.026	.025	.021	-.018	.011
31	.863	.027	.144	-.089	.106	-.113	81	.484	.014	.058	.028	-.020	.002
32	-1.430	.019	-.198	.088	-.114	.136	82	1.520	.036	.292	.260	-.235	.149
33	.585	.034	.110	-.076	.088	-.090	83	-.998	.020	-.141	-.102	.050	-.040
34	.587	.018	.079	-.032	.043	-.053	84	-.730	.018	-.099	-.067	.054	-.023
35	.439	.027	.073	-.048	.054	-.032	85	.439	.051	.102	.096	-.089	.060
36	-.338	.014	-.040	.006	-.012	.020	86	-.765	.014	-.090	-.041	.028	-.001
37	.458	.029	.078	-.024	.027	.010	87	.111	.027	.019	.016	-.014	.008
38	.780	.034	.147	-.102	.118	-.120	88	.533	.036	.103	.092	-.084	.053
39	.752	.014	.088	-.010	.024	-.042	89	1.557	.024	.242	.191	-.166	.090
40	.378	.032	.069	-.046	.054	-.055	90	.734	.042	.153	.140	-.129	.084
41	-.944	.023	-.144	.079	-.097	.108	91	-.534	.016	-.068	-.040	.031	-.010
42	-.097	.036	-.019	.013	-.015	.015	92	-.126	.029	-.021	-.018	.016	-.010
43	.274	.015	.033	-.008	.013	-.019	93	3.238	.015	.394	.204	-.149	.027
44	.501	.042	.104	-.081	.090	-.080	94	.703	.019	.097	.068	-.056	.025
45	.756	.042	.159	-.124	.138	-.123	95	-.313	.020	-.044	-.032	.027	-.013
46	.709	.035	.135	-.094	.108	-.109	96	-1.568	.015	-.191	-.101	.074	-.015
47	-.445	.019	-.062	.028	-.037	.043	97	.589	.017	.078	.052	-.042	.017
48	1.249	.038	.249	-.179	.204	-.205	98	-.730	.041	-.151	-.138	.126	-.082
49	2.112	.018	.285	-.116	.156	-.190	99	.277	.025	.045	.036	-.031	.018
50	-1.212	.014	-.144	.025	-.047	.074	100	.594	.032	.109	.095	-.085	.053
							101	-.547	.020	-.078	-.057	.048	-.023

EXHIBIT 8.6: Influence Diagnostics for Life Expectancy Model

On the positive residual side, both Sri Lanka and to a lesser extent Yugoslavia have a fair amount of influence. With over 300 hospitals in a country of 13 million (1 bed per 330), an excellent public health program, a literacy rate of over 70 per cent (80 per cent for males), and 'senility' as the major cause of death (all information taken from Encyclopædia Britannica, 1974, Micropædia, v. 9. pp. 506–507), Sri Lanka was rather unique among less developed countries of the early seventies. While Yugoslavia is not too influential, it is the only Socialist country on our list and for such countries the relationship between income and social services might be different from that for more free-enterprise countries. For this reason we decided to eliminate these two points as well.

Exhibit 8.7 shows a normal plot of the $e_i^\star$'s obtained after deletion of the points. It is fairly straight. There are still Studentized residuals with numerical values greater than 2 (there are 8 of them). One also finds large DFFITS's: the largest is −.44 for Ethiopia, followed, in order of absolute values, by −.41 for Cambodia (not shown in an exhibit). However, the value of s shrank from about 6.65 to about 4.95 when the outliers were deleted. Since except for h_{ii}, all statistics we are considering have $s_{(i)}$ in the denominator, the reduction of s implies, typically, a commensurate magnification. Therefore, a DFFITS$_i$ of .44 at this stage implies less effect on the b_j's than a .44 value before deletion of points. This is important to note, since after the deletion of points other points will appear with seemingly high influence. However, Cambodia did have rather a unique recent history and perhaps should be deleted.

But a more serious problem has also emerged. Exhibit 8.8, which shows a plot of residuals against predicteds after deletion of points, also shows the existence of more than moderate heteroscedasticity. In Exercise 8.11 the reader is requested to take necessary action to reduce the effect of heteroscedasticity and then to check to see if new outliers or influential points emerge. ■

Example 8.3 (Continuation of Example 6.6, Page 124)

Let us return to the dial-a-ride example introduced in Chapter 6 (see Example 6.6). To save space, we shall display only the diagnostics we refer to. Therefore, Exhibit 8.9 shows only the values of RSTUDENT, h_{ii} and DFFITS$_i$.

It can be seen that there are two very large $e_i^\star$'s (Cases 53 and 45) and two others (10 and 24) which are just under 2. Using the criterion mentioned in Section 8.2, which is $2(k+1)/n \approx .26$ (in this case), we find that h_{ii} exceeds this value in 8 cases, but in only four is it even higher than .3. The largest, .57, is for Case 1 (one of two services in Ann Arbor, Michigan). It is simply a large service with many vehicles serving a big population. The other three points with leverage greater than .3 are Case 24 (Benton Harbor, Michigan), Case 48 (in Buffalo) and Case 51 (in Detroit). The last

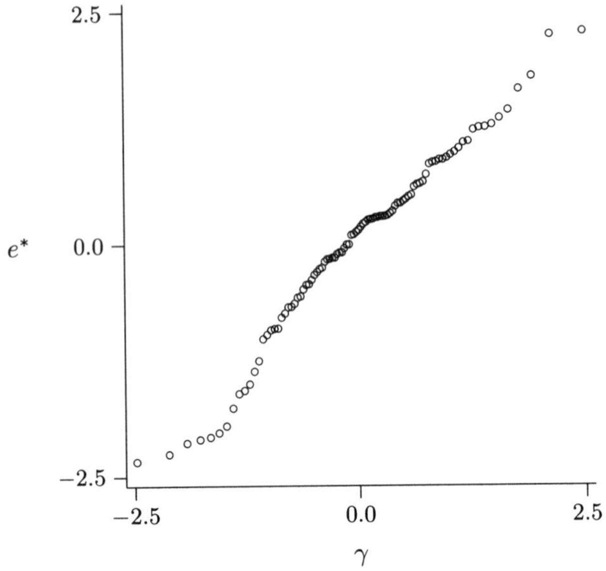

EXHIBIT 8.7: Normal Plot of Studentized Residuals from Life Expectancy Model Estimated After Deleting 6 Cases

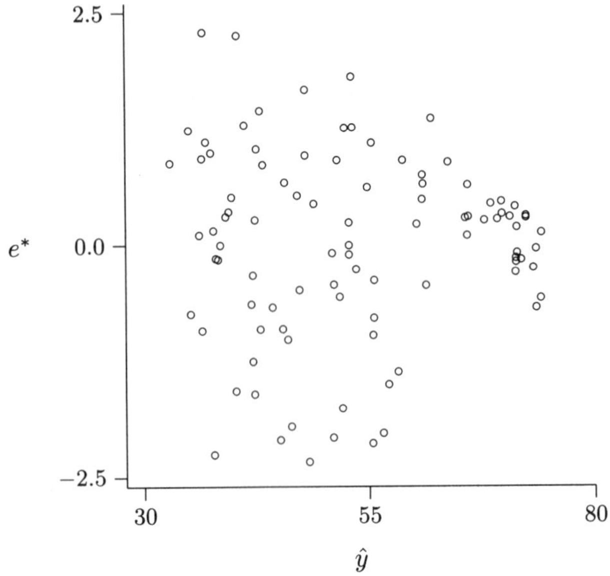

EXHIBIT 8.8: Plot of Studentized Residuals Against Predicteds for Life Expectancy Model with 6 Cases Deleted

Obs	$e^\star$	h_{ii}	$\Delta\hat{y}$	Obs	$e^\star$	h_{ii}	$\Delta\hat{y}$	Obs	$e^\star$	h_{ii}	$\Delta\hat{y}$
1	.263	.572	.305	19	.880	.064	.231	37	.102	.059	.025
2	1.026	.046	.226	20	-.282	.256	-.165	38	.474	.094	.153
3	1.029	.107	.357	21	.601	.047	.134	39	.539	.052	.127
4	-.101	.049	-.023	22	-1.026	.066	-.272	40	.407	.084	.123
5	.328	.158	.142	23	.217	.044	.047	41	-.855	.107	-.296
6	-.775	.116	-.281	24	-1.942	.366	-1.474	42	.211	.040	.043
7	.018	.208	.009	25	.607	.037	.118	43	-.178	.047	-.040
8	.213	.069	.058	26	-.381	.071	-.105	44	.101	.055	.024
9	-.929	.276	-.574	27	-.047	.024	-.007	45	2.945	.077	.849
10	-1.975	.248	-1.133	28	.420	.060	.106	46	-.553	.044	-.118
11	-.660	.101	-.221	29	-.868	.046	-.191	47	-.693	.200	-.347
12	.722	.080	.212	30	.392	.037	.076	48	-1.333	.362	-1.005
13	.244	.080	.072	31	.443	.037	.087	49	.571	.113	.204
14	-1.169	.070	-.321	32	-.880	.073	-.247	50	-1.299	.175	-.598
15	-.806	.075	-.230	33	.370	.101	.124	51	.065	.353	.048
16	.051	.040	.011	34	-1.018	.094	-.328	52	.262	.093	.084
17	.240	.170	.109	35	1.057	.294	.682	53	6.995	.291	4.483
18	-.489	.240	-.275	36	.454	.038	.090	54	.069	.291	.044

EXHIBIT 8.9: Influence Diagnostics for Dial-a-Ride Model

two charge no fares and since we set the fare at 1 cent for the purpose of taking logs, the value of this variable became a large negative number and contributed to the high leverage. Benton Harbor has an enormous service area. Although there are three much larger areas in the data, they are all counties, and since each had 3 or 4 vehicles, it might be conjectured that service was restricted mainly to their densest parts.

Using the criterion for DFFITS mentioned in Section 8.5, which is $2[(k+1)/n]^{1/2} = .72$, we see that five of the values shown are large, four of them being those we flagged above as having large $|e_i^\star|$'s. Of the points mentioned, Case 53 obviously stands out. It has by far the highest $e_i^\star$ and DFFITS$_i$. Its DFBETAS for IND (2.12) compared with the standard error of IND (.104) shows that its deletion would change the coefficient of IND (.80) by more than 20 per cent. Case 53 corresponds to Regina, Saskatchewan, which was well known in transportation planning circles as an extremely unusual service. Therefore, we had no difficulty in deleting the point. Case 45 was from Xenia, Ohio. and it too had a very high load factor — 225 trips per vehicle for a 12 hour day. But we decided to eliminate Regina only at this stage, since sometimes the deletion of one point can substantially change the influence statistics for other points and because Regina was such an unusual service. Ann Arbor (Case 1) does not have a large influence (DFFITS$_1$=.3), but because of its high leverage we should keep a close eye on it.

As feared, after Regina was eliminated, the DFFITS for Ann Arbor shot up. While the residual was low, RSTUDENT became 4.38 and DFFITS almost 6. We had to drop Ann Arbor because it just had too much influence and we were reluctant to let one point be so dominating. On the next round Xenia (RSTUDENT=6.1, DFFITS=1.7) was dropped. The normal plot of Studentized residuals after running a weighted least squares on the remaining points is shown in Exhibit 8.11 and appears to be fairly straight. RSTUDENT's and DFFITS's are given in Exhibit 8.10. While there are still fairly large DFFITS values, we did not seek to eliminate any other cases, largely because, after eliminating the three cases, the value of s dropped from 7.35 to 3.28 (see end of the discussion of Example 8.2).

Each of the three cases we eliminated represented a very unusual dial-a-ride system. If we assume that we are constructing this model in order to forecast ridership, we are implicitly assuming that these deleted points are sufficiently unusual that any system to which we would apply our model would not resemble them. Originally, these data were used to construct a model to predict ridership for yet to be started small dial-a-ride systems in the Chicago area. We were confident that such systems would not resemble the large and/or heavily used systems of Ann Arbor, Regina or Xenia.

Obs	$e^\star$	h_{ii}	$\Delta\hat{y}$	Obs	$e^\star$	h_{ii}	$\Delta\hat{y}$	Obs	$e^\star$	h_{ii}	$\Delta\hat{y}$
1	—	—	—	19	1.678	.070	.459	37	1.000	.066	.266
2	1.978	.068	.534	20	.558	.253	.324	38	.690	.091	.219
3	2.567	.113	.917	21	.680	.074	.192	39	.798	.064	.209
4	-.901	.059	-.225	22	-.996	.071	-.275	40	-.900	.169	-.407
5	1.100	.179	.514	23	-.512	.086	-.157	41	-.632	.103	-.215
6	-.048	.127	-.018	24	-.855	.438	-.755	42	-.455	.048	-.102
7	.288	.210	.148	25	.793	.052	.186	43	-.367	.050	-.084
8	.486	.068	.132	26	.053	.073	.015	44	.563	.060	.142
9	.861	.325	.598	27	-1.245	.054	-.298	45	—	—	—
10	-2.194	.263	-1.312	28	.221	.107	.077	46	-1.789	.047	-.398
11	-2.191	.110	-.769	29	-1.049	.050	-.241	47	1.599	.211	.828
12	1.487	.081	.444	30	.148	.063	.038	48	-1.686	.357	-1.258
13	-.068	.091	-.022	31	.234	.043	.049	49	.594	.125	.225
14	-.553	.083	-.166	32	-.586	.076	-.168	50	-.447	.169	-.202
15	-.339	.075	-.097	33	-1.384	.223	-.741	51	.366	.404	.301
16	-.340	.046	-.074	34	-.951	.093	-.305	52	.271	.114	.097
17	-.861	.229	-.469	35	1.214	.333	.858	53	—	—	—
18	.091	.268	.055	36	.284	.065	.075	54	.293	.301	.192

EXHIBIT 8.10: Influence Diagnostics for Dial-a-Ride Model After Deleting 3 Points

Other analysts might continue with the process of eliminating points or try new variables (e.g., an indicator variable for free fare services). We recommend to the reader that he or she retrace our steps and then try other alternatives. We also tried other alternatives, although to save space we did not report on them here. When we had used these data to construct a model to forecast ridership for new services in the Chicago area, we had deleted six cases.

We should mention that the computations for this exercise were made

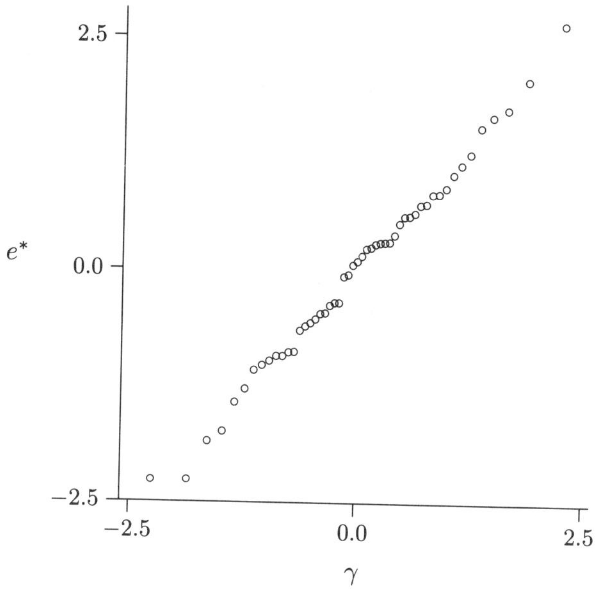

EXHIBIT 8.11: Normal Plot for Studentized Residuals from Dial-a-Ride Model Estimated After Deleting 3 Points

using the SAS NLIN procedure to compute the weights, which were as in Example 6.8, p. 125, followed by the REG procedure to get the influence diagnostics, the two procedures being put together in a single program. While the residuals printed out by SAS PROC REG are not weighted, the other statistics (RSTUDENT, DFFITS, DFBETAS) do take weighting into account. ■

The tedium of applications like these might be alleviated if we had a procedure that would remove alternative sets of cases in a single computer run and show us the effects of each alternative choice. Moreover, sometimes individual observations might not be too influential, but two or three together might turn out to have undue influence. Not much work has been done on this subject. One possibility is to use the indicator variable interpretation of Studentized residuals presented in Section 8.3. Then one would append an indicator variable corresponding to each of a limited number of observations and use a variable choice procedure (Chapter 11). Some techniques also exist for obtaining the combined influence of several data points. These procedures are described in detail in Belsley, Kuh and Welsch (1980), Cook and Weisberg (1980, 1982) and Hadi (1985; see also Chatterjee and Hadi, 1986, particularly the discussions following the paper).

Notice that in each of the three examples presented, we carefully examined the outlier or influential point. In one case, this led to the inclusion of an additional variable; in other cases we eliminated the points after deciding that they did not belong in the model we were seeking to construct. In 'real life' practical problems, it is not desirable to eliminate, without careful examination, every point with a large amount of influence or a large RSTUDENT value, although during diagnostic analysis one often needs to delete points temporarily and then bring them back in later. In textbook exercises, where the substantive background of data sets is not adequately known, it is also sometimes necessary to discard points that are suspected of not belonging to the model without fully understanding the underlying reasons.

Appendix to Chapter 8

8A Some Proofs

Proof of (8.5): By considering X and $X_{(i)}$ as partitioned matrices with each row forming a separate submatrix, we see that $X'X = \sum_{\ell=1}^{n} \boldsymbol{x}_\ell \boldsymbol{x}_\ell'$ and that

$$X_{(i)}'X_{(i)} = \sum_{\substack{\ell=1\\ \ell\neq i}}^{n} \boldsymbol{x}_\ell \boldsymbol{x}_\ell' = \sum_{\ell=1}^{n} \boldsymbol{x}_\ell \boldsymbol{x}_\ell' - \boldsymbol{x}_i \boldsymbol{x}_i' = X'X - \boldsymbol{x}_i \boldsymbol{x}_i'. \tag{8.14}$$

Therefore, from Theorem A.1, p. 275, in Appendix A, we get (see also Example A.6)

$$\begin{aligned}(X_{(i)}'X_{(i)})^{-1} &= (X'X - \boldsymbol{x}_i \boldsymbol{x}_i')^{-1}\\ &= (X'X)^{-1} + \frac{(X'X)^{-1}\boldsymbol{x}_i \boldsymbol{x}_i'(X'X)^{-1}}{1-h_{ii}}.\end{aligned}$$

Similarly,

$$X_{(i)}'\boldsymbol{y}_{(i)} = \sum_{j=1}^{n} \boldsymbol{x}_j y_j - \boldsymbol{x}_i y_i = X'\boldsymbol{y} - \boldsymbol{x}_i y_i. \tag{8.15}$$

From (8.14) and (8.15) we get

$$\begin{aligned}\boldsymbol{b}(i) &= \boldsymbol{b} + \frac{(X'X)^{-1}\boldsymbol{x}_i \boldsymbol{x}_i'\boldsymbol{b}}{1-h_{ii}} - (X'X)^{-1}\boldsymbol{x}_i y_i - \frac{(X'X)^{-1}\boldsymbol{x}_i \boldsymbol{x}_i'(X'X)^{-1}\boldsymbol{x}_i y_i}{1-h_{ii}}\\ &= \boldsymbol{b} + \frac{(X'X)^{-1}\boldsymbol{x}_i \boldsymbol{x}_i'\boldsymbol{b}}{1-h_{ii}} - (X'X)^{-1}\boldsymbol{x}_i y_i (1 + \frac{h_{ii}}{1-h_{ii}})\\ &= \boldsymbol{b} + \frac{(X'X)^{-1}\boldsymbol{x}_i (\boldsymbol{x}_i'\boldsymbol{b} - y_i)}{1-h_{ii}} = \boldsymbol{b} - \frac{(X'X)^{-1}\boldsymbol{x}_i e_i}{1-h_{ii}}\end{aligned}$$

and (8.5) follows. □

Proof of (8.4): Since H is an idempotent matrix we have $\sum_{\ell=1}^{n} h_{i\ell}^2 = h_{ii}$. From (2.13) it follows that $H\boldsymbol{e} = H[I-H]\boldsymbol{\epsilon} = \mathbf{o}$ and therefore, we have $\sum_{\ell=1}^{n} h_{i\ell}e_\ell = 0$. Applying these results along with (8.5) we get

$$\sum_{\substack{\ell=1\\ \ell\neq i}}^{n} [y_\ell - \boldsymbol{x}_\ell'\boldsymbol{b}(i)]^2 = \sum_{\substack{\ell=1\\ \ell\neq i}}^{n} [y_\ell - \boldsymbol{x}_\ell'\boldsymbol{b} + \boldsymbol{x}_\ell'\boldsymbol{b} - \boldsymbol{x}_\ell'\boldsymbol{b}(i)]^2$$

$$= \sum_{\substack{\ell=1\\ \ell\neq i}}^{n} [e_\ell + \boldsymbol{x}_\ell'(X'X)^{-1}\boldsymbol{x}_i e_i(1-h_{ii})^{-1}]^2 = \sum_{\substack{\ell=1\\ \ell\neq i}}^{n} [e_\ell + h_{i\ell}(1-h_{ii})^{-1}e_i]^2$$

$$= \sum_{\ell=1}^{n}[e_\ell + h_{i\ell}(1-h_{ii})^{-1}e_i]^2 - [e_i + h_{ii}(1-h_{ii})^{-1}e_i]^2$$

$$= \sum_{\ell=1}^{n} e_\ell^2 + (1-h_{ii})^{-2}e_i^2\sum_{\ell=1}^{n} h_{i\ell}^2 + 2e_i(1-h_{ii})^{-1}\sum_{\ell=1}^{n} e_\ell h_{i\ell} - (1-h_{ii})^{-2}e_i^2$$

$$= \sum_{\ell=1}^{n} e_\ell^2 - (1-h_{ii})^{-1}e_i^2,$$

and (8.4) follows. □

Proof of Indicator Variable Interpretation of RSTUDENT: For convenience of presentation and without loss of generality, we consider the case of the RSTUDENT for the first observation and show that it is the same as the t-value for an indicator variable z_i which takes the value 1 for the first case and is zero otherwise. This variable z_i is considered to be an additional independent variable, i.e., we append it to the original model

$$y_i = \beta_0 + \beta_1 x_{i1} + \cdots + \beta_k x_{ik} + \epsilon_i \tag{8.16}$$

and consider the model

$$y_i = \beta_0 + \beta_1 x_{i1} + \cdots + \beta_k x_{ik} + \delta z_i + \epsilon_i, \tag{8.17}$$

where $i = 1, \ldots, n$. This model, which has also been called the mean shift operator model, can be written as

$$\boldsymbol{y} = [\ \boldsymbol{1}^* \quad X\ ]\boldsymbol{\beta}^* + \boldsymbol{\epsilon},$$

where $\boldsymbol{1}^* = (1, 0, \ldots, 0)'$ and $\boldsymbol{\beta}^* = (\delta, \beta_0, \ldots, \beta_k)'$. Let

$$\left[\begin{pmatrix} \boldsymbol{1}^{*\prime} \\ X' \end{pmatrix}\begin{pmatrix} \boldsymbol{1}^* & X \end{pmatrix}\right]^{-1} = \begin{pmatrix} 1 & \boldsymbol{x}_1' \\ \boldsymbol{x}_1 & X'X \end{pmatrix}^{-1} = \begin{pmatrix} c_{11} & \boldsymbol{c}_{12}' \\ \boldsymbol{c}_{12} & C_{22} \end{pmatrix},$$

where $\boldsymbol{x}_1$ is the first row of X. Then, from the formulæ contained in Example A.8, on p. 276, we get $c_{11} = (1-h_{11})^{-1}$, $C_{22} = (X'X - \boldsymbol{x}_1\boldsymbol{x}_1')^{-1}$

and

$$\begin{aligned}\boldsymbol{c}_{12}' &= -\boldsymbol{x}_1'(X'X - \boldsymbol{x}_1\boldsymbol{x}_1')^{-1}\\ &= -\boldsymbol{x}_1'\left[(X'X)^{-1} + \frac{(X'X)^{-1}\boldsymbol{x}_1\boldsymbol{x}_1'(X'X)^{-1}}{1-h_{11}}\right] = -\frac{\boldsymbol{x}_1'(X'X)^{-1}}{1-h_{11}},\end{aligned}$$

where $h_{11} = \boldsymbol{x}_1'(X'X)^{-1}\boldsymbol{x}_1$. The least squares estimate of $\boldsymbol{\beta}^*$ is given by

$$\boldsymbol{b}^* = \begin{pmatrix} c_{11} & \boldsymbol{c}_{12}' \\ \boldsymbol{c}_{12} & C_{22} \end{pmatrix} \begin{pmatrix} (\boldsymbol{1}^*)' \\ X' \end{pmatrix} \boldsymbol{y} = \begin{pmatrix} c_{11} & \boldsymbol{c}_{12}' \\ \boldsymbol{c}_{12} & C_{22} \end{pmatrix} \begin{pmatrix} y_1 \\ X'\boldsymbol{y} \end{pmatrix}.$$

Therefore, the least squares estimate of δ is

$$\begin{aligned}\hat{\delta} &= c_{11}y_1 + \boldsymbol{c}_{12}'X'\boldsymbol{y}\\ &= (1-h_{11})^{-1}[y_1 - \boldsymbol{x}_1'(X'X)^{-1}X'\boldsymbol{y}] = (1-h_{11})^{-1}(y_1 - \hat{y}_1)\end{aligned}$$

where $\hat{y}_1 = \boldsymbol{x}_1'\boldsymbol{b}$ and $\boldsymbol{b} = (X'X)^{-1}X'\boldsymbol{y}$. Since $y_1 - \hat{y}_1$ is the usual residual e_1, its variance is $\sigma^2(1-h_{11})$. Consequently,

$$\hat{\delta}/\sqrt{\operatorname{var}(\hat{\delta})} = e_1/[\sigma(1-h_{11})^{1/2}].$$

Therefore, in order to show that $e_i^\star$ is the same as the t-value corresponding to the variable z_i in (8.17), all that remains to be done is to show that the usual unbiased estimate of σ^2 from the model (8.17) is the $s_{(1)}^2$ from the model (8.16).

Notice that

$$\begin{aligned}\boldsymbol{y}'\begin{pmatrix} \boldsymbol{1}^* & X \end{pmatrix}\boldsymbol{b}^* &= \begin{pmatrix} y_1 & \boldsymbol{y}'X \end{pmatrix} \begin{pmatrix} c_{11} & \boldsymbol{c}_{12}' \\ \boldsymbol{c}_{12} & C_{22} \end{pmatrix} \begin{pmatrix} y_1 \\ X'\boldsymbol{y} \end{pmatrix}\\ &= c_{11}y_1^2 + 2y_1\boldsymbol{y}'X\boldsymbol{c}_{12} + \boldsymbol{y}'XC_{22}X'\boldsymbol{y}\\ &= (1-h_{11})^{-1}y_1^2 + 2(1-h_{11})^{-1}y_1\boldsymbol{y}'X(X'X)^{-1}\boldsymbol{x}_1\\ &\qquad + \boldsymbol{y}'X(X'X - \boldsymbol{x}_1\boldsymbol{x}_1')^{-1}X'\boldsymbol{y}\\ &= (1-h_{11})^{-1}y_1^2 + 2(1-h_{11})^{-1}y_1\hat{y}_1\\ &\qquad + \boldsymbol{y}'X\left[(X'X)^{-1} + \frac{(X'X)^{-1}\boldsymbol{x}_1\boldsymbol{x}_1'(X'X)^{-1}}{1-h_{11}}\right]X'\boldsymbol{y}\\ &= (1-h_{11})^{-1}y_1^2 + 2(1-h_{11})^{-1}y_1\hat{y}_1 + \boldsymbol{y}'X\boldsymbol{b} + \hat{y}_1^2(1-h_{11})^{-1}\\ &= (1-h_{11})^{-1}(y_1 - \hat{y}_1)^2 + \boldsymbol{y}'X\boldsymbol{b}.\end{aligned}$$

Hence the residual sum of squares for (8.17) is

$$\begin{aligned}\boldsymbol{y}'\boldsymbol{y} - \boldsymbol{y}'\begin{pmatrix} \boldsymbol{1}^* & X \end{pmatrix}\boldsymbol{b}^* &= \boldsymbol{y}'\boldsymbol{y} - \boldsymbol{y}'X\boldsymbol{b} - (1-h_{11})^{-1}(y_1 - \hat{y}_1)^2\\ &= \sum_{i=1}^n e_i^2 - (1-h_{11})^{-1}e_1^2.\end{aligned}$$

It then follows from (8.4) that the usual unbiased estimate of σ^2 from the model (8.17) is the desired $s_{(1)}^2$. □

Problems

Exercise 8.1: Let

$$X'_{(i)} = (\boldsymbol{x}_1, \ldots, \boldsymbol{x}_{i-1}, \boldsymbol{x}_{i+1}, \ldots, \boldsymbol{x}_n)$$

and $P_{(i)} = X_{(i)}{}'X_{(i)}$. Show that

$$h_{ii} = 1 - [1 + \boldsymbol{x}_i' P_{(i)}{}^{-1} \boldsymbol{x}_i]^{-1}.$$

Thus, points with large values of $\boldsymbol{x}_i' P_{(i)}{}^{-1} \boldsymbol{x}_i$ have a large effect on the estimates.

Exercise 8.2: *Show directly (i.e., without using the indicator variable interpretation) that $e_i^\star$ has Student's t distribution with $n - k - 2$ degrees of freedom.
[**Hint:** Note that $\boldsymbol{b}(i)$ and $s_{(i)}{}^2$ and y_i are independent. Show that

$$\hat{y}_i = \boldsymbol{x}_i'[I + (X_{(i)}{}'X_{(i)})^{-1} \boldsymbol{x}_i \boldsymbol{x}_i']^{-1} \boldsymbol{b}(i) + h_{ii} y_i.$$

Hence show that e_i is independent of $s_{(i)}{}^2$.]

Exercise 8.3: *A method of determining if the first observation belongs to the model $y_i = \boldsymbol{x}_i'\boldsymbol{\beta} + \epsilon_i$ with $i = 2, \ldots, n$ is to test the hypothesis $\mathrm{E}[y_1] = \boldsymbol{x}_1\boldsymbol{\beta}$ against the alternative $\mathrm{E}[y_1] \neq \boldsymbol{x}_1\boldsymbol{\beta}$. Under the assumption that $\boldsymbol{\epsilon} \sim N(\mathbf{o}, \sigma^2 I)$, show that the likelihood ratio test would reject the hypothesis for large values of $(e_1^\star)^2$.

Exercise 8.4: Fit a line by least squares to the following points: (4,.9), (3,2.1), (2,2.9), (1,4.1) and (20,20). Obtain Studentized residuals and also plot the points and the estimated line. Does the point (20,20) appear as an outlier? Using a suitable indicator variable, numerically demonstrate the indicator variable interpretation of RSTUDENT's. Also demonstrate that DFFIT and the DFBETA's do indeed measure what has been claimed for them.

Exercise 8.5: Investigate the presence of outliers and influential points using h_{ii}, e_i^*, DFBETAS$_{ij}$ and DFFITS$_i$ for the model you fitted in Exercise 2.11, p. 50.

Exercise 8.6: Redo Exercises 1.12, p. 25, and 2.16, p. 53, after deleting the fourth case. Explain why, in both cases, the RSTUDENT value of the first observation changed so much on deleting the point.

Exercise 8.7: Run a least squares program on the data in Exhibit 1.11, p. 15, to obtain a regression of property crime rate on population. Examine various influence diagnostics and discuss what you would do. Regress violent crime rate on property crime rate and obtain influence diagnostics. State your conclusions. Do the same for a regression of violent crime rate against property crime rate and population.

Exercise 8.8: For the appropriately weighted version of the regression exercise in Example 4.5, p. 92, do there appear to be any outliers or unusually influential points?

Exercise 8.9: Consider the data in Exhibit 8.12 on male deaths per million in 1950 for lung cancer (y) and per capita cigarette consumption in 1930 (x).

1. Estimate a model expressing y as a linear function of x. Do any of the points look particularly influential? Delete the United States, rerun the model and check if the influences of Great Britain and Finland have been substantially altered. Now, put the U.S. back and delete Great Britain and examine the influence of points for the resultant model.

2. Try an appropriate broken line regression and examine the residuals. If you notice heteroscedasticity, run an appropriately weighted regression. Examine the data points for outliers or undue influence.

3. Do you think a plausible reason for using broken line regression is that the number of women who smoke might be much higher in countries with high per capita cigarette consumption?

4. Write a report discussing your various efforts and your final conclusion.

(A discussion of part 1 of this example is contained in Tufte, 1974, p. 78 *et seq.* The data was used in some of the earlier reports on *Smoking and Health* by the Advisory Committee to the U.S. Surgeon General. See reference to Doll, 1955.)

Country	y	x	Country	y	x
Ireland	58	220	Norway	90	250
Sweden	115	310	Canada	150	510
Denmark	165	380	Australia	170	455
United States	190	1280	Holland	245	460
Switzerland	250	530	Finland	350	1115
Great Britain	465	1145			

EXHIBIT 8.12: Data on Lung Cancer Deaths and Cigarette Smoking
SOURCE: Tufte (1974). Adapted by permission of Prentice Hall, Inc., Englewood Cliffs, New Jersey.

Exercise 8.10: Construct a model similar to that in Exercise 7.7, p. 146, but now do so ignoring serial correlation. Look for outliers and influential points. Write your conclusions.

Exercise 8.11: We ended our discussion of Example 8.2 without having taken it to a conclusion. Do so.

Exercise 8.12: Assuming you have completed Exercise 6.15, p. 130, check to see if some of the cases in the house price data set should have been deleted. Be careful; there are very few observations to work with and a very large number of independent variables.

Exercise 8.13: In the model used to carry out the test in Exercise 4.8, p. 97, do any of the RSTUDENT's or h_{ii}'s give cause for concern? Given that we have rather few observations, would you consider deleting any of them as unduly influential?

A	T	S	C	P	E	y
0	0	1.75	13.4	0.274	2	2.55
1	3	4.10	3.9	0.198	2	1.83
0	4	2.35	5.3	0.526	1	1.80
1	6	4.25	7.1	0.250	1	0.89
0	9	1.60	6.9	0.018	2	1.28
0	25	3.35	4.9	0.194	1	1.51
0	27	2.85	12.1	0.751	1	1.84
1	28	2.20	5.2	0.084	1	1.62
1	29	4.40	4.1	0.236	1	1.01
1	32	3.10	2.8	0.214	1	1.39
0	33	3.95	6.8	0.796	1	1.74
1	35	2.90	3.0	0.124	1	1.57
1	38	2.05	7.0	0.144	1	2.47
0	39	4.00	11.3	0.398	1	1.49
0	53	3.35	4.2	0.237	2	1.29
0	56	3.80	2.2	0.230	1	0.14
1	59	3.40	6.5	0.142	2	1.69
1	65	3.15	3.1	0.073	1	0.70
0	68	3.15	2.6	0.136	1	-0.19
1	82	4.01	8.3	0.123	1	0.08

EXHIBIT 8.13: Florida Cumulus Experiment Data
SOURCE: Woodley, *et al.* (1977). © 1977 by the AAAS. Reproduced with permission.

Exercise 8.14: The data in Exhibit 8.13 are from an experiment on the effects of cloud seeding by silver iodide crystals on precipitation. They were given in Woodley *et al.* (1978) and have also been analyzed by Cook and Weisberg (1980). Each of the cases represents a different day. The dependent variable y is the natural logarithm of precipitation in the target area in a 6-hour period (in 10^7 cubic meters). The independent variables include a dummy variable A on seeding (A $=1$ for seeded days and A $=0$

for days when no seeding was done), the number of days after the first day of the experiment (T), the percentage of cloud cover in the experimental area (C) and total rainfall (P) in the study area an hour before seeding (in 10^7 cubic meters). The independent variable S relates to heights of clouds and E is an indication of whether the radar echo was stationary (2) or moving (1). Using these six independent variables as well as the four additional ones obtained by taking products of the variable A with each of the variables S, C, P and E, obtain an OLS model for y. Identify influential points. Using only the information given in this chapter, give reasons why each of these points is influential.

Repeat the above exercise with the variable $A \times P$ replaced by $A \times \log[P]$.

Exercise 8.15: In the data set on hospital charges (Exhibit 4.13, p. 99), one case was left out, primarily to facilitate the use of the data in Chapter 5. This case represented a sixteen-year-old female patient who was treated by the doctor with ID number 730 for a severity level 2 condition. The charges totaled \$820. Would you include this case in your analysis of *log of charges*? Should any of the other observations be deleted?

Exercise 8.16: Using the same data as in Exercise 8.15, investigate the presence of outliers when 'charges' is the dependent variable. After deleting any outliers, examine plots of residuals against dependent and independent variables for both the 'logged' and the 'non-logged' model. Which one would you choose? If deletion of outliers (except for the point mentioned in the last exercise) was forbidden, what would your choice have been?

Exercise 8.17: For the models you originally constructed in Problem 2.20, p. 55, and refined in other exercises, look for outliers and influential points, and take whatever action you deem appropriate. (In the marriage rate example, can you explain the most prominent influential point?)

Exercise 8.18: Look for outliers and influential points in the model you constructed in Exercise 3.11, p. 77. Write your conclusions.

CHAPTER 9

Transformations

9.1 Introduction

Until now we have considered situations where the algebraic form of the model, i.e.,

$$\beta_0 + \beta_1 x_{i1} + \ldots + \beta_k x_{ik},$$

was approximately correct. Obviously, this will not always be so. The actual relationship may not be a linear function of the x_{ij}'s and sometimes not even of the β_j's. In some such cases we may still be able to do linear regression by *transforming* (i.e., using functions of) the independent and/or the dependent variables.

While we shall look at this subject in detail in the other sections of this chapter, let us consider two examples now. Suppose the true relationship between x_1, x_2 and y is $y = \beta_1 x_1^{1/2} + \beta_2 x_2^{1/2}$. Obviously, this is linear in β_1 and β_2 and should present no problem in itself if we know that we would have to take square roots of the x_j's. If we do not know this, then we will have to convince ourselves that there is a need to make some transformation and then try to find a good one by analyzing the data. As another example, consider a form that appears quite frequently in economics: $y = Ax_1^{\alpha} x_2^{\beta}$. By taking logarithms (the base is unimportant) of both sides, we get $\log(y) = \log(A) + \alpha \log(x_1) + \beta \log(x_2)$, and we *may* be able to use linear regression methods. Other examples as well as techniques for deciding on the need for transformations and for choosing them when needed are discussed in this chapter.

9.1.1 AN IMPORTANT WORD OF WARNING

If we transform the *dependent* variable y_i we would be changing var (y_i). We exploited this fact to our advantage in Chapter 6 where we transformed the y_i's to make previously unequal variances equal. Sometimes the same transformation of the y_i's will yield a good algebraic form and at the same time induce equality of variance. Unfortunately, often this will not be the case. The reader should be warned that if a transformation of a *dependent* variable is made, it may be necessary to take other steps to correct the heteroscedasticity induced. Transforming the dependent variable would also affect the normality of the ϵ_i's and, in addition, could give rise to bias, which would then have to be alleviated. (No such problems occur if only independent variables are transformed.)

Therefore, it has been recommended in the literature that nonlinear least squares (Appendix C) or direct maximization of the likelihood function is preferable as a means to estimate parameters for models with a transformed dependent variable and sometimes it is necessary to use these methods. Such models are sometimes called generalized linear models and have been treated in depth by McCullagh and Nelder (1983). However, as we shall see in the next section, under some specific circumstances, linear least squares may be used to estimate certain generalized linear models.

9.2 Some Common Transformations

Some types of transformations seem to appear more often than others. In this section we examine a few of them before embarking on a more general treatment.

9.2.1 Polynomial Regression

We have already encountered examples of polynomial regression in Chapter 1 (Section 1.2, p. 2), where we discovered that $y_i = \beta_0 + \beta_1 x_{i1} + \epsilon_i$ was not quite adequate to describe the data and we needed a model of the form

$$y_i = \beta_0 + \beta_1 x_{i1} + \beta_2 x_{i1}^2 + \epsilon_i.$$

Other, more complex polynomial models can also be used. For example,

$$y_i = \beta_0 + \beta_1 x_{i1} + \beta_2 x_{i1}^2 + \beta_3 x_{i2} + \beta_4 x_{i2}^2 + \beta_5 x_{i1} x_{i2} + \epsilon_i,$$

which is a second degree polynomial in two variables, and

$$y_i = \beta_0 + \beta_1 x_{i1} + \beta_2 x_{i1}^2 + \beta_3 x_{i1}^3 + \beta_4 x_{i1}^4 + \epsilon_i,$$

which is of fourth degree in one variable, may sometimes be employed. Polynomials, particularly those of second degree, are frequently used in fitting so-called response surfaces, which in turn are used to find the value of an independent variable or the combination of values of several independent variables which yield a maximum or a minimum value of the dependent variable (as in Exercise 2.18, p. 53).

However, if injudiciously done, polynomial regression can present two types of problems. First, the number of parameters increases rapidly with both the degree of the polynomial and the number of the (original) variables. For example, even a second degree polynomial with m variables can have as many as $1+2m+\frac{1}{2}m(m-1)$ parameters. Then for $m = 10$ we would have 66 parameters! Another problem, even with one independent variable, is that while high degree polynomials can be made to fit the data very well (with a high enough degree, an excellent fit can always be obtained even if no actual relationship exists!), such relationships usually mean little and are often worthless for predictive purposes.

9.2.2 Splines

There is no reason why broken line regression (Section 4.5, p. 89) cannot be generalized to broken curve regression. Then cases on either side of the break point, or knot, as it is sometimes called, would be described by curves given by, say, polynomials. An example of such a model which uses second degree polynomials and which is continuous across the knot is

$$y_i = \beta_0 + \beta_1 x_{i1} + \beta_2 x_{i1}^2 + \beta_3 (x_{i1} - x)\delta_i + \beta_4 (x_{i1} - x)^2 \delta_i + \epsilon_i,$$

where, as in Section 4.5, the value x of the independent variable x_1 represents the break point and

$$\delta_i = \begin{cases} 1 & \text{if } x_{i1} > x \\ 0 & \text{if } x_{i1} \leq x. \end{cases}$$

In fact, the model

$$y_i = \beta_0 + \beta_1 x_{i1} + \beta_2 x_{i1}^2 + \beta_3 (x_{i1} - x)^2 \delta_i + \epsilon_i \tag{9.1}$$

corresponds to a function of x_1 which is also differentiable. Such curves, consisting of polynomial curves coming together at knots, are sometimes called splines, although the word is more often reserved for the special case where, if the polynomials are of degree m, all derivatives of order $m-1$ or less exist across each knot. Model (9.1) is an example of the more restrictive definition.

When the break point(s) are known, the estimation of such broken curve models presents no difficulty. If they are not known, nonlinear least squares (Appendix C, see especially Example C.4, p. 313) could be used to obtain estimates of the break point and the other parameters. Splines are used in some computer 'graphics' packages to draw smooth curves through a set of points. For more on splines and broken curve regression, see Seber and Wild (1989, Ch. 9), Eubank (1984) and Smith (1979).

9.2.3 Multiplicative Models

By a multiplicative form of $x_1, \ldots, x_k$ we mean a function of the form

$$y = A \prod_{j=1}^{k} x_j^{\beta_j}, \tag{9.2}$$

where A and β_j's are parameters. Taking logs of both sides, we get

$$\log(y) = \beta_0 + \sum_{j=1}^{k} \beta_j \log(x_j), \tag{9.3}$$

where $\beta_0 = \log(A)$. With an appropriate error term added, (9.3) can frequently be estimated by linear least squares.

There are cases where multiplicative forms are suitable. Models based on such forms are in widespread use in econometrics where the well-known Cobb-Douglas model is just one example. The coefficients of multiplicative models have a very simple interpretation. Let Δy be a small change in y because of a small change Δx_j in x_j. From (9.3) it follows that $y^{-1}(\partial y/\partial x_j) = \beta_j x_j^{-1}$, i.e.,

$$\lim_{x_j \to 0} \frac{\Delta y/y}{\Delta x_j/x_j} = \beta_j.$$

Therefore, β_j is the limit of the ratio of the *percentage* change in y to the *percentage* change in x_j. In economics such a ratio is referred to as an *elasticity*. That β_j's have this interpretation might partially account for the popularity of multiplicative models.

While, as illustrated in Example 9.4, the need for a multiplicative model is difficult to diagnose using purely empirical means, the underlying situation can often give clues as to whether such a model is called for. We illustrate this in the example below.

Example 9.1 (Continuation of Example 6.6, Page 124)
Consider the dial-a-ride example. We expected the number of riders to be proportional to a function of the number of vehicles. We also expected it to be proportional to a function of the hours of service. Indeed, we felt that RDR was proportional to functions of each of the independent variables. Consequently, we wrote E[RDR] as a product of these functions.

We chose these functions to be power functions: VH^{β_1}, HR^{β_2} etc., partially based on convenience for least squares analysis, but also because they were intuitively reasonable. The result was a model of the form (9.2). ■

However, before we can estimate parameters in multiplicative models we need to decide how the error term should be introduced. Two possibilities are in common use. They imply somewhat different interpretations of the underlying substantive situation and are discussed in the next two subsections.

Multiplicative Errors

In the econometrics literature, the usual procedure is to use the model

$$y_i = A x_{i1}^{\beta_1} x_{i2}^{\beta_2} \ldots x_{ik}^{\beta_k} \epsilon_i. \tag{9.4}$$

Such models can arise in many ways. Suppose the real underlying model is $y_i = A \prod_{j=1}^{k} z_{ij}^{\beta_j}$, but for some reason, we do not know or cannot measure the z_{ij}'s. Instead, we rely on the surrogates $x_{ij} = z_{ij}\eta_{ij}$ where η_{ij} is an unobservable random variable. Then (9.4) results with $\epsilon_i = \prod_{j=1}^k \eta_{ij}^{-\beta_j}$. Another way the model (9.4) could occur is if the real underlying model contained a product $\prod_{j=k+1}^m x_{ij}^{\beta_j}$ of additional variables but these latter variables or their values were not known. Then, writing $x_{ij} = z_j \eta_{ij}$ for $j = k+1, \ldots, m$, we could absorb the z_j's into the A and the η_{ij}'s into the error term to get (9.4).

Taking natural logarithms of both sides of (9.4), we get a model akin to our usual regression model:

$$\log(y_i) = \log(A) + \beta_1 \log(x_{i1}) + \beta_2 \log(x_{i2}) + \cdots + \beta_k \log(x_{ik}) + \log(\epsilon_i). \quad (9.5)$$

However, this model has to be handled with some care. $\mathrm{E}(\log(\epsilon_i))$ will not usually be zero even if $\mathrm{E}(\epsilon_i)$ is assumed to be 1 — as is usually done. Econometricians frequently make the *additional* assumption that $\eta_i = \log(\epsilon_i)$ is normally distributed with mean, say, μ and variance σ^2, the same for all i. Then it can be shown that (Exercise B.1, p. 297) $1 = \mathrm{E}(\epsilon_i) = \mathrm{E}(e^{\eta_i}) = e^{\mu + \frac{1}{2}\sigma^2}$. Hence, $\mu + \frac{1}{2}\sigma^2 = 0$ and $\mathrm{E}[\log(\epsilon_i)] = \mu = -\sigma^2/2$. Therefore, we need to rewrite (9.5) as

$$\begin{aligned}\log(y_i) &= (\log(A) - \sigma^2/2) \\ &\quad + \beta_1 \log(x_{i1}) + \cdots + \beta_k \log(x_{ik}) + [\log(\epsilon_i) + \sigma^2/2].\end{aligned}$$

Now, $\mathrm{E}[\log(\epsilon_i) + \sigma^2/2] = 0$, and $\mathrm{var}\,[\log(\epsilon_i) + \sigma^2/2] = \mathrm{var}\,[\eta_i] = \sigma^2$. Thus if the y_i's are uncorrelated, we get a model for which the Gauss-Markov conditions are satisfied. Therefore, we can apply OLS and obtain estimates of $\beta_1, \ldots, \beta_k$ and also get the estimate b_0 of the intercept term $\beta_0 = (\log(A) - \sigma^2/2)$ and the estimate s^2 of σ^2. An estimate of A can be found from the last two estimates. One, proposed by Srivastava and Singh (1989), gives

$$A = e^{b_0 + \frac{1}{2}\gamma s^2}$$

where $\gamma = 1 - \boldsymbol{a}'(X'X)^{-1}\boldsymbol{a}$, X is the $n \times (k+1)$ design matrix

$$\begin{pmatrix} 1 & \log(x_{11}) & \ldots & \log(x_{1k}) \\ \multicolumn{4}{c}{\cdots\cdots\cdots\cdots\cdots\cdots\cdots} \\ \multicolumn{4}{c}{\cdots\cdots\cdots\cdots\cdots\cdots\cdots} \\ 1 & \log(x_{n1}) & \ldots & \log(x_{nk}) \end{pmatrix},$$

and $\boldsymbol{a}' = (1, 0, \ldots, 0)$.

Additive Errors

An alternative to (9.4) is

$$y_i = A x_{i1}^{\beta_1} x_{i2}^{\beta_2} \ldots x_{ik}^{\beta_k} + \epsilon_i \quad (9.6)$$

and its interpretation is fairly obvious. Assuming that $\mathrm{E}(\epsilon_i) = 0$, i.e., $\mathrm{E}(y_i) = Ax_{i1}^{\beta_1} \ldots x_{ik}^{\beta_k}$, and taking logarithms of both sides of (9.6), we have

$$\begin{aligned}\log[y_i] &= \log[\mathrm{E}(y_i) + \epsilon_i] = \log[\mathrm{E}(y_i)(1 + \epsilon_i/\mathrm{E}(y_i))] \\ &= \beta_0 + \beta_1 \log(x_{i1}) + \cdots + \beta_k \log(x_{ik}) + \log[1 + \epsilon_i/\mathrm{E}(y_i)].\end{aligned}$$

Since it is most unlikely that $\mathrm{E}[\log(1 + \epsilon_i/\mathrm{E}(y_i))]$ is even a constant (let alone zero), this is one of the situations where it has been recommended in the literature that means other than linear least squares be used to estimate the parameters of (9.6). However, in the frequently occurring case where y_i is counted and, therefore, has a Poisson distribution, linear least squares can be used if we are willing to tolerate a very slight bias. Then, by taking a Taylor's series expansion of

$$z_i = \log[y_i + \tfrac{1}{2}] - \log[\mathrm{E}(y_i)] = \log[1 + (\epsilon_i + \tfrac{1}{2})/\mathrm{E}(y_i)] \qquad (9.7)$$

it may be shown that for moderate to large values of $\mathrm{E}(y_i)$, $\mathrm{E}(z_i) \approx 0$. Exhibit 9.1 shows how good the approximation is. Thus, for say $\mathrm{E}[y_i] \geq 3$, adding a half to y_i before taking logs essentially eliminates bias. Corrections of this kind are sometimes called Anscombe's corrections (Anscombe, 1948, Rao, 1973, p.426). A similar procedure can be used to find bias-reducing corrections for several different transformations and a fair number of distributions of y_i for which the variance is a function of the mean.

$\mathrm{E}(y_i)$	$\mathrm{E}[z_i]$	$\mathrm{E}(y_i)$	$\mathrm{E}[z_i]$
2	.0259	5	-.0021
3	.0019	7	-.0012
4	-.0020	15	-.0002

EXHIBIT 9.1: Values of $\mathrm{E}[z_i]$ for Selected Values of $\mathrm{E}[y_i]$

Example 9.2 (Continuation of Example 9.1, Page 183)

It is obvious that for the dial-a-ride example we chose a model with additive errors. That is why we added a half to RDR before taking logs although, given the size of this variable (the smallest value of RDR was 56), the half makes very little difference. However, it is not obvious that multiplicative errors are entirely absent from the model.

Obviously variables are missing from the model, a very important one being the dedication and ability of key personnel. Another missing variable is the proportion of vehicles in operation per day. All we knew was the number of vehicles owned; we did not know how many were in the repair shop or sitting idle because there were no drivers. All these point towards the presence of a multiplicative error (in addition to an additive error).

This leads to a model of the form

$$\begin{aligned}\mathrm{RDR}_i &= C\,\mathrm{E}(\mathrm{RDR}_i)[1+{\epsilon_i}^{(1)}] + {\epsilon_i}^{(2)}\\ &= \mathrm{E}(\mathrm{RDR}_i)[1+{\epsilon_i}^{(1)} + ({\epsilon_i}^{(2)}/\,\mathrm{E}(\mathrm{RDR}_i))] = \mathrm{E}(\mathrm{RDR}_i)[1+\epsilon_i],\end{aligned}$$

where $\mathrm{E}({\epsilon_i}^{(1)}) = \mathrm{E}({\epsilon_i}^{(2)}) = \mathrm{E}(\epsilon_i) = 0$, ${\epsilon_i}^{(1)} + ({\epsilon_i}^{(2)}/\,\mathrm{E}(\mathrm{RDR}_i)) = \epsilon_i$ and C is a constant.

We surmised that variations in $\mathrm{RDR}_i/\,\mathrm{E}(\mathrm{RDR}_i)$ due to differences in key personnel, number of vehicles in use, etc., would decline with the size of the service, i.e., with $\mathrm{E}(\mathrm{RDR}_i)$. Therefore, we assumed that $\mathrm{var}\,({\epsilon_i}^{(1)})$ would be roughly proportional to $\mathrm{E}(\mathrm{RDR}_i)$ and therefore, $\mathrm{var}\,(\epsilon_i)$ would be roughly proportional to $\mathrm{E}(\mathrm{RDR}_i)$. This, of course, leads to the same weights as we would get for a purely additive errors model for a counted dependent variable. As mentioned above, the half that was added to RDR has little effect. Nor would there be any palpable difference in estimates if we replaced the half by other comparably sized numbers. Consequently, the course of action we followed would appear to be appropriate.

This example illustrates that choices to be made in applications of regression are frequently not clear-cut. However, we feel that we made a good choice. ■

9.2.4 The Logit Model for Proportions

We have already seen proportions of counts in Chapters 4 and 6. Such variables are of the form m_i/n_i, where m_i is the number of individuals out of n_i that possess a certain property. The variable m_i is a counted variable and n_i is usually treated as a fixed number. Since $0 \le m_i \le n_i$, it follows that $0 \le y_i \le 1$. In most applications where the dependent variable $y_i = m_i/n_i$ is a proportion of counts, it is desirable to set

$$\mathrm{E}(y_i) = f(Z_i) \text{ where } Z_i = \beta_0 + \beta_1 x_{i1} + \cdots + \beta_k x_{ik} \tag{9.8}$$

and f is a cumulative distribution function (since $0 \le f(z) \le 1$). An example of such a function is the logistic distribution function

$$f(z) = \exp(z)/[1+\exp(z)], \tag{9.9}$$

the shape of which is illustrated by the curve in Exhibit 9.2. When z is small, so is $f(z)$. The derivative of the curve is also small but it gradually increases as the cost difference increases. Between about $f(z) = 1/4$ and $f(z) = 3/4$ the curve is almost a straight line and then the derivative declines. The model (9.9) is often called a logit model.

Example 9.3

Suppose m_i is the number of travelers between two places who take the train and $n_i - m_i$ is the number of travelers who drive. Suppose we have

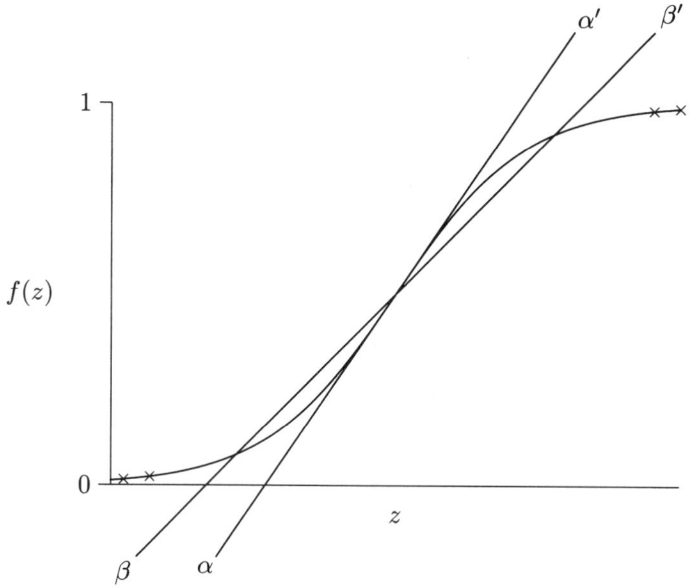

EXHIBIT 9.2: The Logistic Distribution Function

only one independent variable x_i which is the difference in travel times between the two modes, i.e., x_i = (train travel time − drive time). $\text{E}(y_i)$ (in (9.8)) is obviously the probability of choosing the train, given that either car or train is chosen. ∎

An interesting question we might pose here is: If we are primarily interested in the straight line part of $f(z)$, can we approximate the logistic distribution function by a straight line? The answer is usually no, as Exhibit 9.2 illustrates. Although we would want the line $\alpha - \alpha'$, we would get the line $\beta - \beta'$ because of the points marked with the $\times$'s, which are *not* idiosyncratic points. In fact, they are just about right on the logistic curve!

The case where the logistic distribution function is used in (9.8) is another one where the literature often recommends the use of nonlinear least squares (Appendix C) or direct maximization of the likelihood function (as can be done using SAS PROC LOGIST, described in SAS, 1986). However, it is frequently possible to apply linear least squares for this purpose. It may be verified that from (9.9) we get $f(z)/[1 - f(z)] = \exp(z)$, and, therefore, $\log[f(z)] - \log[1 - f(z)] = z$. Hence (see (9.8)),

$$\log[\,\text{E}(y_i)/(1 - \text{E}(y_i))] = Z_i.$$

If n_i's are fixed numbers and each $m_i = y_i n_i$ has a binomial distribution, then it may be shown, using a Taylor series approximation (essentially in the same way as on p. 185), that the expectation of the difference

$$\log[\mathrm{E}(y_i)/(1-\mathrm{E}(y_i))] - \log[(y_i+(2n)^{-1})/(1-y_i+(2n)^{-1})]$$

consists of terms of the order $[\mathrm{E}(m_i)]^{-2}$ and $[\mathrm{E}(n_i-m_i)]^{-2}$ and is small when $\mathrm{E}(m_i)$ and $\mathrm{E}(n_i-m_i)$ are large. Consequently, we may write our model as

$$\begin{aligned}&\log[(y_i+(2n)^{-1})/(1-y_i+(2n)^{-1})]\\ &\quad=\log(m_i+.5)-\log(n_i-m_i+.5)=Z_i+\epsilon_i\end{aligned} \tag{9.10}$$

where $\mathrm{E}[\epsilon_i]\approx 0$ for large values of $\mathrm{E}(m_i)$ and $\mathrm{E}(n_i-m_i)$. We shall call the function on the left side of (9.10) a *logit transformation* of y_i.

Obviously, (9.10) can be estimated by linear least squares. Since $n_i y_i$ has a binomial distribution, $\mathrm{var}\,[y_i+(2n_i)^{-1}]=\mathrm{var}\,[y_i]=n_i^{-1}\mathrm{E}(y_i)[1-\mathrm{E}(y_i)]$ and therefore the variance of the top line of (9.10) is, from (6.4)

$$\begin{aligned}&\mathrm{var}\,[\log(y_i+(2n_i)^{-1})-\log(1-y_i+(2n_i)^{-1})]\\ &\approx[\frac{1}{\mathrm{E}(y_i)}+\frac{1}{\mathrm{E}(1-y_i)}]^2\mathrm{var}\,(y_i)=\frac{\mathrm{E}(y_i)[1-\mathrm{E}(y_i)]}{n_i(\mathrm{E}(y_i)[1-\mathrm{E}(y_i)])^2}\\ &=[n_i\,\mathrm{E}(y_i)(1-\mathrm{E}(y_i))]^{-1}.\end{aligned} \tag{9.11}$$

Therefore, we need to weight (see Section 6.4, p. 118), and the weight should be the reciprocal of the last expression in (9.11). This procedure works quite well where most n_i's and m_i's are moderately large (notice that when $\mathrm{E}(m_i)$ or $\mathrm{E}(n_i-m_i)$ is small, the data point usually gets a relatively small weight). However, when this is not so, and particularly when all n_i's are 1 (so that y_i is either 0 or 1), other procedures (maximum likelihood or nonlinear least squares) must be used.

An alternative to the logistic distribution function, and practically indistinguishable from it, is the normal distribution function Φ (see Cox, 1970). Then, from (9.8), $\Phi^{-1}(\mathrm{E}(y_i))$ is linear in the parameters. Φ^{-1} is called a *probit transformation* and is often used in place of the logit.

9.3 Deciding on the Need for Transformations or Additional Terms

Perhaps the best starting point in any practical problem is to consider the underlying situation, as we did in Examples 9.1 and 9.2. Sometimes, of course, understanding the underlying situation is not sufficient. Then we must examine the data to see whether the chosen model

$$y_i=\beta_0+\beta_1 x_{i1}+\cdots+\beta_k x_{ik}+\epsilon_i, \tag{9.12}$$

is adequate or if transformations or additional terms involving the same x_{ij}'s would be helpful.

In this section we discuss several methods of doing this. They fall roughly into three categories. First we have graphical methods. Then there are methods where we append to the list of independent variables additional ones, which are nonlinear functions of those we had before, to see if the new variables have much value. If they do, we might wish to retain them, or seek some transformation of the original variables. Finally, we have near-neighbor techniques, where we compare variances estimated over small portions of the independent variable space with the variance estimated overall. Formal tests for deciding on the need for transformations or additional terms are sometimes loosely referred to as tests for nonlinearity.

9.3.1 Examining Residual Plots

By far the most frequently used method of checking for the need for transformations is to examine plots of residuals against each independent variable. We did this in Chapter 1. Exhibit 1.4, p. 5, showed such a plot. Clearly some action was needed. Exhibits 1.5 and 1.6 show that we have substantially improved the situation and that further efforts are unwarranted. We remind the reader that text-book examples are chosen carefully to illustrate a point. For many applications, plots are not nearly as obvious.

Unfortunately, this type of examination is marginal since we are considering one variable at a time. Therefore, we may not always be able to see the need for, say, product terms. A little reflection will convince the reader that the need for multiplicative models would be quite difficult to ascertain in this way. The example below provides such an illustration.

Example 9.4

Suppose a person knew quite a bit about regression but did not know the formula that the area (Y) of a rectangle is its length (X1) times width (X2). He suspected that length and width affect area; therefore he cut out a number of cardboard rectangles and measured their dimensions. He then weighed each and divided the weight by a known constant to obtain the area. However, because of a wide range of imperfections in his technique he realized that there is some error in this estimate and decided to use linear regression to estimate the (incorrect) model

$$\mathrm{Y}_i = \beta_0 + \beta_1 \mathrm{X1}_i + \beta_2 \mathrm{X2}_i + \epsilon_i.$$

Exhibit 9.3 shows his data. Actually, these numbers were generated using a pseudo-random number generator: X1 and X2 were obtained using a uniform distribution from 10 to 35 and Y_i was set to equal to $(\mathrm{X1}_i)(\mathrm{X2}_i)+\eta_i$ where $\eta_i \sim N(0, 1600)$.

His estimates for β_0, β_1 and $\beta 2$ were

$$b_0 = -423.8\ (47.4), \quad b_1 = 18.5\ (1.31), \quad b_2 = 22.6\ (1.38),$$

Y	X1	X2	Y	X1	X2	Y	X1	X2
457.7	17	27	582.6	30	19	539.6	19	29
568.2	22	27	590.4	34	19	516.8	19	25
452.8	16	24	131.6	16	13	791.2	27	29
783.7	34	23	415.0	33	11	516.9	15	31
629.4	30	21	606.3	28	21	493.8	30	18
216.9	13	17	628.7	20	31	366.3	32	12
1014.2	31	32	130.3	14	11	634.6	24	23
471.6	25	19	359.6	34	12	194.4	13	19
266.4	32	11	438.2	14	29	734.5	28	26
243.3	15	11	417.9	27	15	432.9	26	16
278.0	18	15	332.7	21	17	362.7	15	26
354.6	12	27	626.6	29	23	231.4	11	16
534.2	32	17	575.2	16	34	442.4	27	17
374.2	26	16	641.9	27	24	538.5	22	28
441.1	18	21	448.5	26	15	365.6	29	10
762.7	31	22	536.9	21	24	315.9	17	19
702.8	23	33	742.2	30	25			

EXHIBIT 9.3: 'Areas', Lengths and Widths of Rectangles

respectively, where the quantities within parentheses are the standard errors. The R^2 of .89 indicates a very good fit. Exhibits 9.4 and 9.5 are the residual plots one customarily examines. They do not appear to show any nonlinearity, and indeed our ill-informed analyst could easily emerge with a model from this exercise that area is a linear function of dimension. ■

9.3.2 Use of Additional Terms

One easy method of determining if nonlinearity exists is to insert nonlinear terms into the model and then test to see if they should have been included. Frequently they are polynomial terms. For example, suppose we are trying to determine if the model

$$y_i = \beta_0 + \beta_1 x_{i1} + \beta_2 x_{i2} + \epsilon_i$$

is adequate. Then we can consider the model

$$y_i = \beta_0 + \beta_1 x_{i1} + \beta_2 x_{i2} + \gamma_2 x_{i1}^2 + \gamma_2 x_{i1}^2 + \gamma_3 x_{i1} x_{i2} + \epsilon_1$$

and test the hypothesis that all γ_i's are zero against the alternative that at least one is not. Such a test would be an F-test as discussed in Chapter 3.

Frequently, if an analyst is unable to reject the hypothesis that a specific set of nonlinear terms is needed, he or she comes to the conclusion

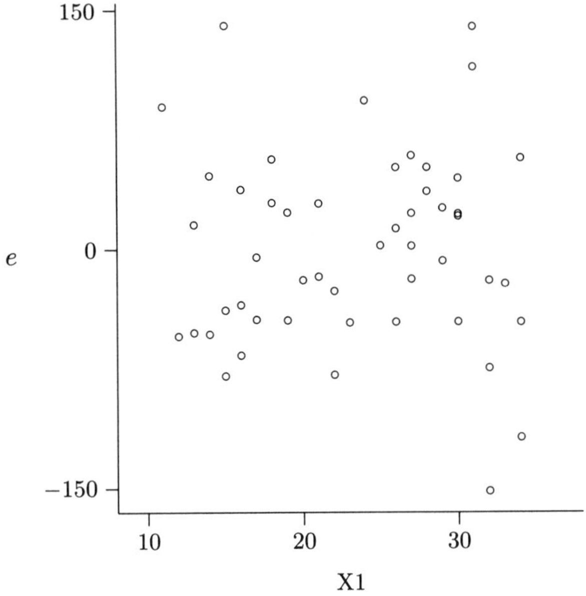

EXHIBIT 9.4: Plot of Residuals Against X1 for Data on 'Area' of Rectangles

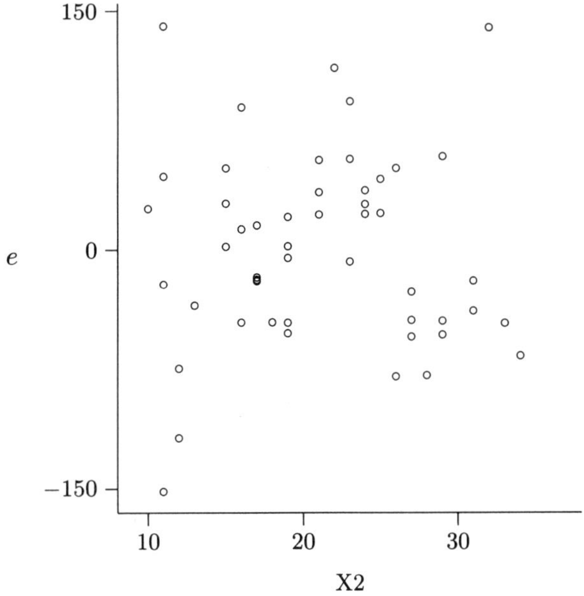

EXHIBIT 9.5: Plot of Residuals Against X2 for Data on 'Area' of Rectangles

that (9.3.2) is adequate and that no nonlinearity is indicated. A practical difficulty that often arises is that, if there are several variables and product terms are used, the number of terms in the augmented equation can get very large. In such cases, the user has to choose different selections of additional terms for different runs.

A related procedure is found in MINITAB (1988; see Burn and Ryan, 1983). Here, the model

$$y_i = \beta_0 + \sum_{\xi=1}^{k} \beta_\xi x_{i\xi} + \epsilon_i$$

is augmented to

$$y_i = \beta_0 + \sum_{\xi=1}^{k} \beta_\xi x_{i\xi} + \alpha_0 {z_i}^{(j)} + \sum_{\xi=1}^{k} \alpha_\xi {z_i}^{(j)} x_{i\xi} + \eta_i,$$

where ${z_i}^{(j)} = 0$ when $x_{ij} < \bar{x}_j$, ${z_i}^{(j)} = 1$ when $x_{ij} \geq \bar{x}_j$ and $\bar{x}_j = n^{-1}\sum_{i=1}^{n} x_{ij}$. In the augmented model, if the hypothesis $(\alpha_0, \alpha_j)' = \mathbf{o}$ is rejected, then a transformation of the variable x_j is indicated, and if $(\alpha_0, \ldots, \alpha_j - 1, \ldots, \alpha_j + 1, \ldots, \alpha_k)' = \mathbf{o}$ is rejected, perhaps interaction terms are needed. Each of the x_j's is considered in turn.

Tukey's Type of Nonadditivity

Tukey (1949) proposed a test for nonlinearity based on augmenting the model $\boldsymbol{y} = X\boldsymbol{\beta} + \boldsymbol{\epsilon}$ to

$$\boldsymbol{y} = X\boldsymbol{\beta} + \alpha \boldsymbol{f} + \boldsymbol{\epsilon},$$

where $\boldsymbol{f}$ is a known vector-valued function of $\boldsymbol{X\beta}$, $\mathrm{E}(\boldsymbol{\epsilon}) = 0$, and $\mathrm{cov}(\boldsymbol{\epsilon}) = \sigma^2 I$. Such tests are useful for determining if interactions (e.g., products of independent variables) should be included, without having to include a large number of terms and thereby giving up a large number of degrees of freedom. Let $\hat{\boldsymbol{f}}$ be the estimate of $\boldsymbol{f}$ obtained by substituting for $\boldsymbol{\beta}$ its estimate $\boldsymbol{b}$. Then a test for the hypothesis $\alpha = 0$ against $\alpha \neq 0$ is based on

$$\frac{(\boldsymbol{e}'\hat{\boldsymbol{f}})^2/\hat{\boldsymbol{f}}'\hat{\boldsymbol{f}}}{\boldsymbol{e}'\boldsymbol{e} - [(\boldsymbol{e}'\hat{\boldsymbol{f}})^2/\hat{\boldsymbol{f}}'\hat{\boldsymbol{f}}]},$$

which has an F distribution with 1 and $n-k-2$ d.f. Rejection would occur for large values of this statistic. Further details, extensions (to where $\boldsymbol{f}$ is matrix valued), and examples can be found in Tukey (1949), Milliken and Graybill (1970) and Andrews (1971a).

9.3.3 Use of Repeat Measurements

If for each of $i = 1, \ldots, n$ values $\boldsymbol{x}_i$ of the vector of independent variables, there are several observations $y_{i\ell}$ where $\ell = 1, \ldots, n_i$, then for each $\boldsymbol{x}_i$ we can compute a mean $\bar{y}_i$ and a variance s_i^2 for the corresponding $y_{i,1}, \ldots, y_{i,n_i}$ and $\mathrm{E}(s_i^2)$ is equal to σ^2 for all i. Therefore, by combining the s_i^2's we can get the following estimate of σ^2:

$$s_c^2 = \sum_{i=1}^{n}(n_i - 1)s_i^2 / \sum_{i=1}^{n}(n_i - 1).$$

This estimate would be immune from any nonlinearity. Now if we used the $\bar{y}_i$'s as the observations and ran a weighted regression with weights $w_i = n_i$, we would get another estimate s_w^2 of σ^2 with which we can compare s_c^2. If s_c^2 is much smaller than s_w^2, then we would reject the hypothesis of adequacy of our model (9.12). The ratio s_w^2/s_c^2 has an F distribution with $n-k-1$ and $\sum_{i=1}^{n}(n_i - 1)$ degrees of freedom.

Perhaps easier is to construct an indicator variable for each point $\boldsymbol{x}_i$ which is one for observations corresponding to it. These indicator variables can be appended to the list of independent variables and we could test if their coefficients are all zeros.

However, it is often difficult to find several observations for each combination of values of the predictor. It has been proposed that when several observations are not available for each $\boldsymbol{x}_i$, we can proceed, somewhat approximately, by dividing the range of each independent variable x_j into small intervals and considering the values of x_j to be the same in each interval. However, with many variables this immediately poses a problem. Suppose we divided the range of each of five independent variables into five intervals. Then we would get 5^5 cells, and if each cell were to contain several cases, a very large data set would be needed.

MINITAB (1988) has an option which performs such tests.

Theil's Approach

When there is only one independent variable, an alternative approach not requiring repeat observations is the following. Order the x_i's from smallest to largest (the sequence within ties does not matter) and let the index $i = 1, \ldots, n$ reflect the order, i.e.,

$$x_{11} \leq x_{21} \leq \cdots \leq x_{n1}.$$

Run a regression and compute residuals e_i and s^2 in the usual way. Obtain an alternative, approximately unbiased estimate of σ^2 from

$$s_s^2 = \sum_{i=2}^{n}(e_i - e_{i-1})^2/[2(n-1)]. \qquad (9.13)$$

Now we can compare (9.13) to s^2 or use s_s^2/s^2. As we saw in Chapter 7, the latter is also a test statistic for serial correlation and is readily available from packages. The reader can see, perhaps by referring back to Exhibit 1.9a on p. 12, that s_s^2 will usually be smaller than s^2 if transformations or additional terms are needed.

This approach cannot be directly extended to many dimensions, because there is no obvious way of ordering vectors. But it can be extended if we notice that what we are looking for in computing s_s^2 is not so much an ordered sequence of x_{i1}'s but proximity between pairs of x_{i1}'s. This idea has been exploited by Daniel and Wood (1980) and also in the approaches given in Section 9.3.5.

9.3.4 Daniel and Wood Near-Neighbor Approach

Daniel and Wood (1980, p. 133 *et seq.*) use the idea of estimating σ from pairs of $\boldsymbol{x}_i$'s which are close together. Instead of (9.13), they compute

$$.886 \sum\nolimits^{(\alpha)} |e_i - e_\ell| / \left(\alpha[(n-k-1)/n]^{1/2}\right) \tag{9.14}$$

where $\sum^{(\alpha)}$ is the summation over close-by $\boldsymbol{x}_i$'s and $\boldsymbol{x}_\ell$'s described below and α is the number of such pairs. For independently, normally distributed z_i's with common mean and variance σ^2 it can be shown that

$$\mathrm{E}(|z_i - z_\ell|/\sigma) = 1.128 = 1/.886.$$

Thus (9.14) is a reasonable estimator of σ, and can be compared with s. Obviously, e_i's are not independent, but with a large n assuming them to be independent is usually not too inappropriate. The $(n-k-1)/n$ approximately compensates for degrees of freedom.

To obtain close-by points, Daniel and Wood compute the distance $D_{i\ell}$ between $\boldsymbol{x}_i$ and $\boldsymbol{x}_\ell$ as

$$D_{i\ell} = \sum_{p=1}^{k} [b_p(x_{ip} - x_{\ell p})]^2 / s^2. \tag{9.15}$$

The role of the b_p is two-fold: to compensate for differences in units of measurement for different independent variables and to give less importance to less important independent variables which will presumably tend to have smaller b_p's. The s^2 is not critical; it is there to render the $D_{i\ell}$'s dimensionless.

The $D_{i\ell}$'s can be used to select a set of (i, ℓ) pairs that are close together. Because the total number of such pairs is quite large — $n(n-1)/2$ — Daniel and Wood use a two-step procedure. In the first step the $\hat{y}_i$'s are sorted or ranked as

$$\hat{y}_1 < \hat{y}_2 < \cdots < \hat{y}_n.$$

Then for each $\hat{y}_i$, the pairs $(i, i-1)$, $(i, i-2)$, $(i, i-3)$, $(i, i-4)$ are retained; this yields $4n - 10$ pairs. These $(4n - 10)$ pairs will include many of the pairs with small $D_{i\ell}$'s, since if a $D_{i\ell}$ is small, $|\hat{y}_i - \hat{y}_\ell|$ will be too. In the second step, the $D_{i\ell}$'s are computed and ranked for these $(4n - 10)$ pairs. For the smallest of these $D_{i\ell}$'s, (9.14) is computed with $\alpha = 1$, then (9.14) is computed for the $\alpha = 2$ smallest $D_{i\ell}$'s, then for the $\alpha = 3$ smallest $D_{i\ell}$'s and so on until $\alpha = 4n - 10$.

All $4n - 10$ values of (9.14) need to be examined. When α is small, the estimates of (9.14) may not be too reliable; when α is close to $4n - 10$, (9.14) may be based on points which are not too close. Daniel and Wood recommend choosing a set of consecutive values of (9.14) which are relatively constant and comparing them to s. A program which carries out this procedure is contained in the package that is a companion to Daniel and Wood (1980). Several examples of the procedure are given in that book.

9.3.5 Another Method Based on Near Neighbors

The Daniel and Wood approach helps us detect nonlinearity when it is quite serious and sometimes tells us when we do not need to worry about the adequacy of our model. However, it is difficult to decide what to do if s is somewhat larger, but not much larger, than (9.14). This is because the distribution of (9.14) is not known and is, moreover, difficult to obtain. Even if we replace (9.14) by a quantity the distribution of which is known, it is difficult to say much about a quantity which is being selected out of $4n - 10$ dependent values of it.

To remedy these problems, at least partially, we consider (see (9.13))

$$\sum_{i,\ell} w_{i\ell}(e_i - e_\ell)^2 / ws^2, \tag{9.16}$$

where $w = \sum_{i,\ell} w_{i\ell}$, $w_{i\ell} = g(d_{i\ell})$, $w_{i\ell} = w_{\ell i}$ and $w_{ii} = 0$ for all i and ℓ. The g is some function (e.g., $g(d_{i\ell}) = d_{i\ell}^{-2}$) which is positive and monotonically *fairly rapidly* decreasing and $d_{i\ell}$ is a measure of distance between $\boldsymbol{x}_i$ and $\boldsymbol{x}_\ell$. A simple example of such a measure is $||\boldsymbol{x}_i - \boldsymbol{x}_\ell|| = [(\boldsymbol{x}_i - \boldsymbol{x}_\ell)'(\boldsymbol{x}_i - \boldsymbol{x}_\ell)]^{1/2}$. Alternatively, and preferably, we could invoke the centered model of Section 2.10, p. 42, and since $Z'Z$ is the sample covariance matrix of the independent variable values, use

$$(\boldsymbol{z}_i - \boldsymbol{z}_\ell)'(Z'Z)^{-1}(\boldsymbol{z}_i - \boldsymbol{z}_\ell) = \tilde{h}_{ii} + \tilde{h}_{\ell\ell} - 2\tilde{h}_{i\ell} \tag{9.17}$$

where $\boldsymbol{z}_i$ is a row of Z and $\tilde{H} = (\tilde{h}_{i\ell}) = Z(Z'Z)^{-1}Z'$. Another possibility is

$$h_{ii} + h_{\ell\ell} - 2h_{i\ell}. \tag{9.18}$$

Note that, as may easily be verified, this last expression is $\sigma^{-2}\text{var}\,[\hat{y}_i - \hat{y}_\ell]$.

Formally, (9.16) is the same as Geary's statistic for spatial correlation (Section 7.7.1, p. 143). As in that case, an alternative to (9.16) is

$$\Omega = \sum_{i,\ell} (w_{i\ell}/w) e_i e_\ell / s^2. \tag{9.19}$$

As before, probably the most convenient way to use (9.16) and (9.19) when n is large is to invoke the fact that both (9.16) and (9.19) are asymptotically normal with means and variances as given in Section 7.7. For the readers' convenience these formulæ are reproduced below. Note that both (7.19) and (7.20) can be written in the form $c\boldsymbol{e}'V\boldsymbol{e}/s^2$ where V is a suitable $n \times n$ matrix. Under normality, the mean of $c\boldsymbol{e}'V\boldsymbol{e}/s^2$ is $c\,\text{tr}B$ and its variance is $2c^2(n-k+1)^{-1}[(n-k-1)\,\text{tr}B^2 - (\text{tr}B)^2]$, where $B = M'VM$ and $M = I - X(X'X)^{-1}X'$. When the errors are not normal, under some mild conditions the mean and variance are, asymptotically, $c\,\text{tr}B$ and $2c^2\,\text{tr}B^2$.

For reasons that are fairly obvious, the method works better when a large proportion of the independent variables require transformations, or additional terms involving a large proportion of the variables are called for. This is fortunate, since the need to transform fewer variables is typically easy to detect by other means.

A fundamental difference between (9.15) and the other measures is that the latter do not include the b_j's. Of course, there is nothing to prevent us from including b_j's in the measures that do not contain them or from removing them from the one that does. However, a choice needs to be made. The reason for including b_j's was to reduce the effects of relatively unimportant variables. It is not clear that this is always accomplished, since b_j may not necessarily be too good a measure of importance. The alternative position is that when we use these measures, we should merely focus on nonlinearity and disregard variable importance. After finding and more or less correcting for nonlinearity, we are in a better position to judge the importance of variables.

Example 9.5 (Continuation of Example 9.4, Page 189)
For the data described in Example 9.4 with $g(d_{i\ell})$ set as $||\boldsymbol{x}_i - \boldsymbol{x}_\ell||^2$, the approximate value of $(\Omega - \text{E}(\Omega))/(\text{var}\,\Omega)^{1/2}$ was computed using the formulæ for the means and variances given above for the normal case. This value was 3.73, showing that the linear model used in that example should be rejected. Therefore, the test helps us take a correct decision even though the plots were quite unhelpful. ■

Example 9.6 (Continuation of Example 9.2, Page 185)
Consider the dial-a-ride data of Exhibit 6.8, p. 125, without making any transformations. Because the independent variables are in vastly different units of measurement, we used (9.17) as distance and set its reciprocal as the $w_{i\ell}$'s.

The possible presence of heteroscedasticity required that we consider weighted residuals. Therefore, we set the vector of residuals as $W^{1/2}(\boldsymbol{y} - X\boldsymbol{b}_{WLS})$ where $\boldsymbol{b}_{WLS} = (X'WX)^{-1}X'W\boldsymbol{y}$, $\boldsymbol{y} = (y_1, \ldots, y_n)'$ is the vector of values of the dependent variable RDR (number of riders), X is the design matrix including all the remaining variables and W is the diagonal matrix of weights, which are estimates of $1/\mathrm{E}[y_1]$ since the y_i's are counted variables. These estimates were simply taken to be $\hat{y}^{-1}$ from an ordinary least squares exercise. When any of the weights were negative it was replaced by a zero. Notice that weighting also requires replacing X by $W^{1/2}X$ in the expression for B in the formulæ for the mean and variance of (9.19).

Computation of $(\Omega - \mathrm{E}(\Omega))/(\mathrm{var}\,\Omega)^{1/2}$, again using the formulæ for the mean and variance of Ω under an assumption of normality, yielded 3.7, which strongly recommends transformations. A similar exercise after taking logs yielded .78. ■

An alternative which is easier to apply than the one mentioned above and detects certain kinds of nonlinearity is described in Burn and Ryan (1983) and is available in MINITAB (1988). In this method the sum of squares of residuals — call it SSE_C — is computed from a regression based only on the m design points for which $h_{ii} \leq 1.1(k+1)/n$. From the discussion of Section 8.2.1, p. 156, it is easily seen that these points would be close to the centroid of all the design points. If, now, SSE is the error sum of squares when all points are used, it is easily seen that

$$(n-m)^{-1}\mathrm{SSE}_{LOF}/[(m-k-1)^{-1}\mathrm{SSE}_C], \tag{9.20}$$

where

$$\mathrm{SSE}_{LOF} = \mathrm{SSE} - \mathrm{SSE}_C,$$

would be high in the presence of substantial nonlinearity. It may also be seen that SSE_{LOF} is independent of SSE_C and has a chi-square distribution with $(n-m)$ degrees of freedom. Therefore, (9.20) has an F distribution under the hypothesis of nonlinearity.

9.4 Choosing Transformations

In the last section we examined methods for deciding whether a transformation was necessary. A few of the techniques can also give us a hint as to what transformations are called for. However, in general, these methods simply identify need and leave the analyst to find the suitable transformations. In this section we shall discuss some methods to help us make this choice. Subsections 9.4.1 and 9.4.2 present essentially graphical methods, while the remaining subsections of this section are devoted to analytical procedures.

9.4.1 Graphical Method: One Independent Variable

When there is only one independent variable, we can examine a plot of the dependent variable against it. Sometimes this plot will immediately suggest a course of action — e.g., the use of the logit function examined earlier in this chapter, or of broken line regression, as in Chapter 4. If no obvious course is apparent, the following method given in Mosteller and Tukey (1977) has been found useful by the authors.

Divide the range of the independent variable into three portions, making a good compromise between getting equal numbers of data points in each portion and making the three portions roughly equal. For each of the three sets of data points thus created, find a point (which may or may not be one of the data points) which is a good representative of the set. For each set, a good choice is the point whose coordinates are the medians of the x and y values for the points in the set. Find the slope of the line joining the first two points (going from left to right) and the slope for the line joining the last two. If the two slopes are equal, then the data points should be describing a straight line. If not, the middle of three points will be below (the convex case) or above (the concave case) the line joining the other two.

$\vdots$		$\vdots$
$-1/y^2$		$\vdots$
$-1/y$	$\Uparrow$	x^5
$-1/y^{1/2}$	IF CONVEX	x^4
$\log(y)$	UP THE LADDER	x^3
$y^{1/2}$		x^2
y	◁ YOU ARE HERE ▷	x
y^2		$x^{1/2}$
y^3	IF CONCAVE	$\log(x)$
y^4	DOWN THE LADDER	$-1/x^{1/2}$
y^5	$\Downarrow$	$-1/x$
$\vdots$		$-1/x^2$
$\vdots$		$-1/x^3$
		$\vdots$

EXHIBIT 9.6: Ladder of Transformations

We may now transform either the dependent variable y or the independent variable x using the ladder of power transformations given in Exhibit 9.6. If the three points are in a convex configuration we move up the ladder; if they are in a concave configuration we move down. In either case we apply the transformations to the chosen coordinates (x or y) of the three points. If the two slopes mentioned above become roughly equal we stop; if the configuration changes from convex to concave or *vice versa* we have gone too far.

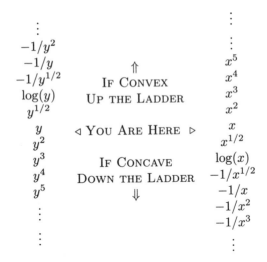

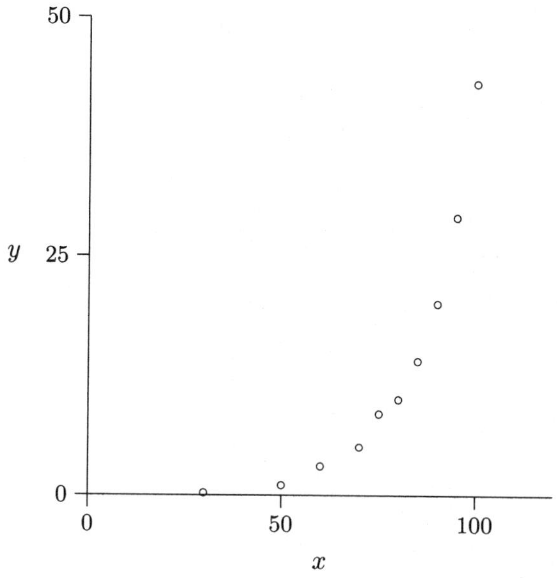

EXHIBIT 9.7: Plot of Stevens' Data

Example 9.7

Consider the data of Exhibit 1.15, p. 23. Suppose we do not know that Stevens took logarithms or we wish to check if his action was appropriate. Exhibit 9.7 shows a plot of y against x. Consider the three points with coordinates

$$(50, 1),\ (77.5, 9.25) \quad \text{and} \quad (95, 29).$$

The slopes between the first two and last two points are

$$8.25/27.5 = .3 \quad \text{and} \quad 19.75/17.5 = 1.13,$$

which are far from equal. Suppose we decide to transform the dependent variable. Going up the ladder to $y^{1/2}$, we take square roots of the y coordinates and get the slopes

$$(3.04 - 1)/27.5 = .074 \quad \text{and} \quad 2.35/17.5 = .134.$$

They are getting closer. Now, moving on to $\log(y)$, we get the slopes

$$(.966 - 0)/27.5 = .035 \quad \text{and} \quad .496/17.5 = .028.$$

These are quite close, and, indeed, we may have gone too far. Nevertheless, let us try $-1/y^{1/2}$; the slopes are .025 and .0082; the situation is getting worse. We can deem the slopes for $\log(y)$ close enough and go with it (after all, the points chosen were somewhat arbitrary), or we could look for some small positive power for y (e.g., $y^{(.1)}$, for which, incidentally, the slopes are nearly equal). ■

Sometimes, particularly when some values of the dependent variable y are negative, an analyst might wish to add a quantity c to y before raising it to a power. A graphical method for choosing c and the power has been given in Dolby (1963).

9.4.2 Graphical Method: Many Independent Variables

The primary problem with graphical methods is that usually we have only two dimensions to work with. Therefore, we need to find a way to examine two dimensional cross-sections of an essentially many dimensional situation. Examining plots of the dependent variables against each independent variable is sometimes not very useful, since the clutter introduced by the effects of other independent variables makes the identification of transformations difficult. For example, the points $(y, x_1, x_2) =$ (2,1,3), (1,2,1), (3,2,3), (1,3,0), (3,3,2), (2,4,0) all lie on the plane $y = -3 + x_1 + x_2$, but a plot of y against just x_1 will show the six points arranged in a regular hexagon. Wood (1973) has given an alternative method which is usually much more useful.

Component Plus Residual Plots

Consider the estimated model

$$y_i = b_0 + b_i x_{i1} + \cdots + b_k x_{ik} + e_i,$$

where $i = 1, \ldots, n$, the b_j's are estimates of β_j's and the e_i's are the residuals. Then

$$y_i - b_0 - \sum_{j=1:j\neq m}^{k} b_j x_{ij} = b_m x_{im} + e_i.$$

Thus $b_m x_{im} + e_i$ is essentially y_i with the linear effects of the other variables removed. Plotting $b_m x_{im} + e_i$ against x_{im} eliminates some of the clutter mentioned above. Wood (1973) called these plots *component plus residual plots* since $b_m x_{im}$ may be seen as a component of $\hat{y}_i$ (see also Daniel and Wood, 1980). The plot has also been called a *partial residuals plot*. The method of the last section can be applied to each of these plots, but now we would be transforming only the independent variables.

Obviously, the value of such plots depends on the quality of the estimate of b_j's. If the independent variables are related (see Chapter 10), b_j's could be so far away from the appropriate β_j that component plus residual plots could become misleading. Moreover, if, say, one of several highly related independent variable requires a transformation, the appearance of bending can be shared by several of the plots. Then also such plots are of limited value. A modification of component plus residual plots which includes nonlinear terms in the component has been proposed by Mallows (1985).

R	I	S	R	I	S	R	I	S
99	6.5	1	670	26.0	2	400	18.0	3
125	11.0	1	820	44.0	2	640	22.0	3
200	17.0	1	325	8.0	2	468	6.0	4
550	37.0	1	366	15.0	2	668	22.0	4
100	9.0	1	325	9.0	2	850	24.0	4
250	6.0	2	411	7.0	3	825	27.0	4
400	14.0	2	580	14.0	3	950	27.0	5
475	18.0	2	580	17.0	3	1000	31.0	6
250	12.0	2	580	16.5	3	780	12.0	6

EXHIBIT 9.8: Data on Monthly Rent (R dollars), Annual Income (I $\times$ 1000 dollars) and Household Size (S).

Example 9.8

Generally, the rent (R) a household pays for an apartment is related to its income (I). It is conceivable that rents would also be related to household size (S), since larger households would require more space. On the other hand, larger households would have less money available for rent. The data in Exhibit 9.8 was extracted from a much larger set by one of the authors. The original data set was collected about 20 years ago from several sources, principally successful lease applications. The purpose of the original study was to determine rent levels appropriate for low income housing.

Exhibits 9.9 and 9.10 show component plus residual plots for the two independent variables. $\mathcal{C}(\mathrm{I})$ and $\mathcal{C}(\mathrm{S})$ denote respectively the component *plus* residuals for I and S. Obviously, the plot corresponding to income requires no action. In the case of the plot for S, some action seemed necessary. The nature of the plot made a search for three (see Section 9.4.1) typical points difficult. Therefore, we chose the four points (1,60), (2,280), (3,410), (5,580). Taking the square roots of S and then computing slopes between pairs of transformed points, we get approximately 531, 409 and 337. We have not gone far enough. With logs we get 317, 320 and 332, showing that a log transformation may be about right for S. ■

Example 9.9

Consider the data set of Exhibit 9.11 where the dependent variable PCS is the percentage savings that occurred when the operation of some transit routes was given over to private companies. The independent variables are

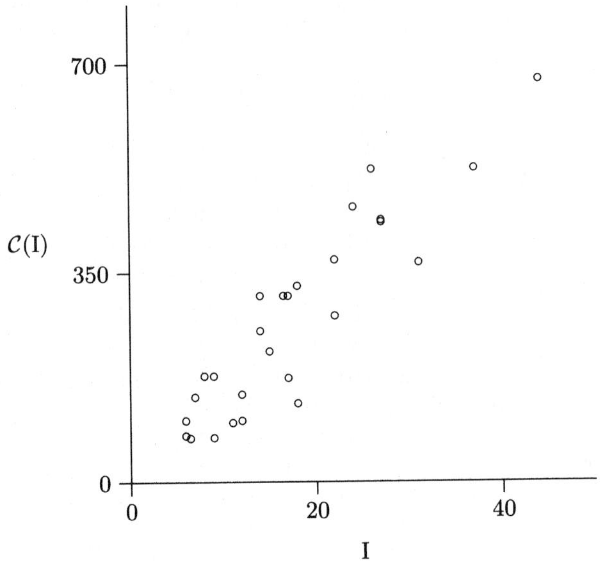

EXHIBIT 9.9: Component Plus Residual Plot for Income in Rent Model

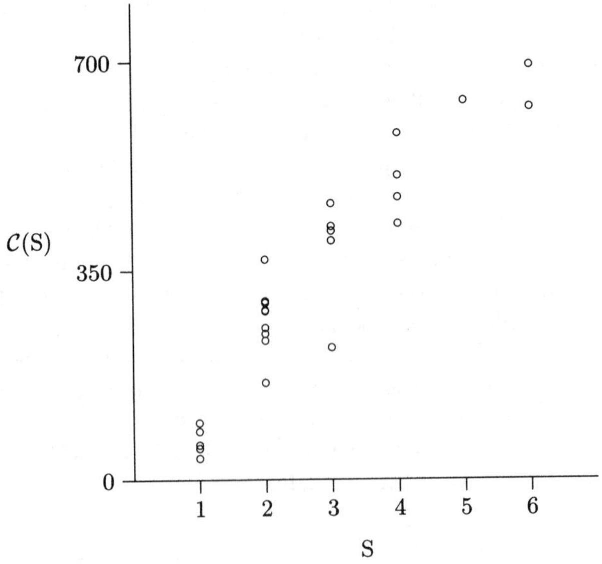

EXHIBIT 9.10: Component Plus Residual Plot for Household Size in Rent Model

V1	V2	V3	V4	V5	V6	V7	V8	V9	PCS
55.0	0.10	17.95	379.45	15.0	1327.02	4.16	0.00	10.68	19.8
67.0	1.00	11.94	767.52	20.0	16120.02	0.12	1.00	7.21	29.0
20.0	1.00	17.77	92.40	15.0	3060.24	15.58	1.00	11.56	50.0
66.0	0.89	20.18	810.98	0.0	1861.85	3.33	1.00	8.02	48.0
35.0	1.00	7.57	220.00	0.0	806.69	4.38	1.00	6.11	30.0
69.0	0.45	19.57	468.22	16.0	1014.00	3.52	0.00	7.28	30.0
35.0	0.43	11.00	184.30	0.0	35.78	0.76	1.00	6.14	0.1
48.0	1.00	13.94	206.44	0.0	24.15	1.20	1.00	7.37	27.0
45.0	0.83	14.21	255.00	0.0	38.99	0.47	0.88	7.34	3.1
45.0	0.80	13.32	145.00	0.0	27.38	0.47	0.67	7.19	0.1
26.0	1.00	12.63	173.00	0.0	25.08	0.47	1.00	6.09	30.2
20.0	1.00	19.03	405.10	0.0	1606.77	9.57	0.00	9.28	45.9
45.0	0.00	29.93	74.41	15.0	1327.02	4.16	0.00	8.34	49.0
45.0	0.00	33.29	38.80	25.0	1327.02	4.16	1.00	6.38	50.0
60.0	0.75	10.13	297.20	12.0	7869.54	16.68	1.00	7.80	43.0
45.0	0.71	16.83	360.27	25.0	5179.78	12.36	1.00	8.84	22.9
9.0	1.00	13.90	27.93	0.0	514.62	2.00	0.00	6.94	10.0

V1 Average capacity of buses in service
V2 Ratio of buses in use during non-peak period to those in use in peak period [Peak Periods are those when most people travel to or from work]
V3 Average speed
V4 Vehicle-miles contracted
V5 Distance from center of metropolitan area
V6 Population of metropolitan area
V7 Percentage of work trips in the metropolitan area that are made by transit
V8 Buses owned by sponsor ÷ buses owned by contractor
V9 Per capita income for metropolitan area
PCS Per cent savings

EXHIBIT 9.11: Data on Transit Privatization
SOURCE: Prof. E.K. Morlok, Department of Systems Engineering, University of Pennsylvania.

those that the analyst (Morlok) initially thought were important determinants of it. While several of the independent variables seemed to require transformations, to save space, we have shown only the component plus residual plot for V2 in Exhibit 9.12. The axes labels are similar to those in Example 9.8. It is possible to see a very convex pattern here, but quite clearly the transformations of Exhibit 9.6 without alteration are not appropriate. We tried the two independent variables V2 and $(\mathrm{V2})^2$. We also replaced V6 by $\mathrm{V6}^{-1/2}$ for reasons the reader is invited to explore in Exercise 9.11. The resultant component plus residual plot for V2, presented in Exhibit 9.13, shows that we did not do too poorly. ■

9.4.3 Analytic Methods: Transforming the Response

A good transformation should make residuals smaller. However, when we transform the dependent variable so that y_i becomes $f(y_i)$, we also change scales. Therefore, we cannot simply compare the s^2's after making various transformations. We need to make some adjustments to them. Such adjustments are used in the methods described below.

The Box and Cox Method

For situations where all $y_i > 0$, Box and Cox (1964) considered the following family of transformations:

$$y_i^{(\lambda)} = \begin{cases} (y_i^\lambda - 1)/\lambda & \text{when } \lambda \neq 0 \\ \log(y_i) & \text{when } \lambda = 0, \end{cases} \tag{9.21}$$

where $i = 1, \ldots, n$. [Notice that, by the use of L'Hospital's Rule, it can be shown that

$$\lim_{\lambda \to 0} (y_i^\lambda - 1)/\lambda = \log(y_i).]$$

Define $\boldsymbol{y}^{(\lambda)} = (y_1^{(\lambda)}, \ldots, y_n^{(\lambda)})'$.

To find the appropriate transformation from the family (9.21), all we need do is maximize $-\frac{1}{2}n\log[s^2(\lambda)]$ given by

$$-\tfrac{1}{2}n\log[s^2(\lambda)] = (\lambda - 1)\sum_{i=1}^{n} \log(y_i) - \tfrac{1}{2}n\log[\hat{\sigma}^2(\lambda)], \tag{9.22}$$

where $\hat{\sigma}^2(\lambda) = n^{-1}\boldsymbol{y}^{(\lambda)\prime}[I - H]\boldsymbol{y}^{(\lambda)}$; i.e., $\hat{\sigma}^2(\lambda)$ is the sum of squares of the residuals divided by n when $y_i{}^{(\lambda)}$'s are used as observations (see (2.13)). $s^2(\lambda)$ is essentially the s^2 for the transformed model adjusted for change of scale for the dependent variable $y^{(\lambda)}$. The maximization is best carried out by simply computing (9.22) for several values of λ. We shall sometimes

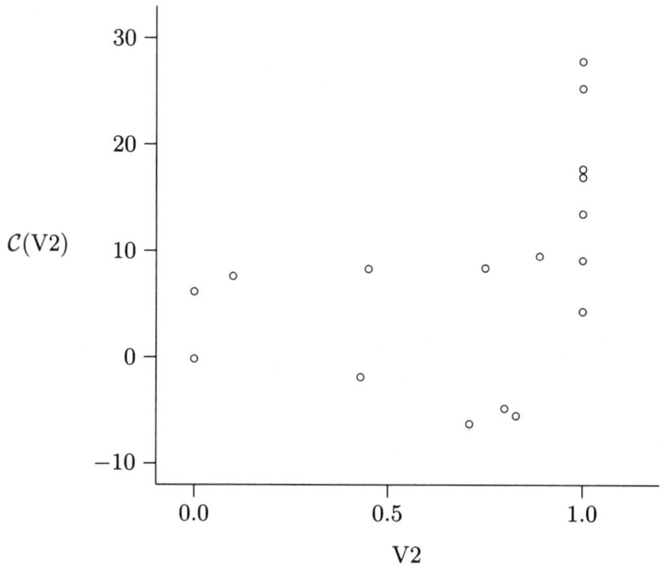

EXHIBIT 9.12: Component Plus Residual Plot for V2

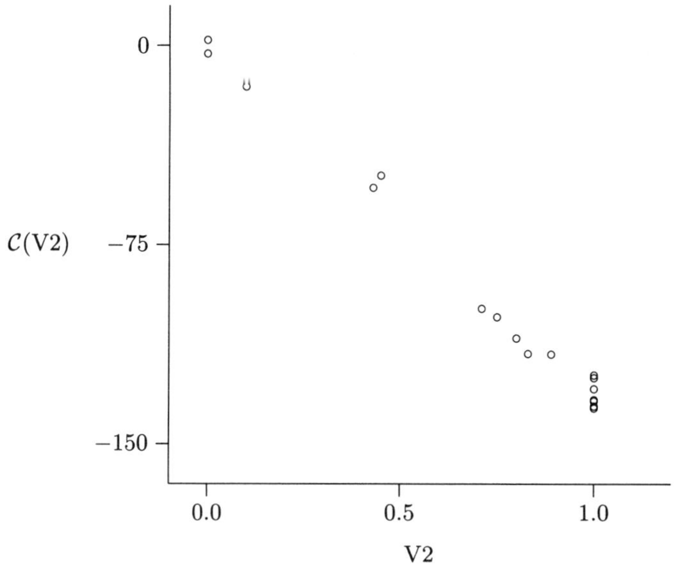

EXHIBIT 9.13: Component Plus Residual Plot for V2 after including $V2^2$

call $-\frac{1}{2}n\log[s^2(\lambda)]$ the Box-Cox objective function. A justification for the procedure is given in the subsection below.

Let λ_{max} be the value of λ which maximizes (9.22). Then, under fairly general conditions, for any other λ,

$$n\log[s^2(\lambda)] - n\log[s^2(\lambda_{max})] \tag{9.23}$$

has approximately the chi-square distribution with 1 degree of freedom. This follows from the fact, which will be obvious from the discussion in the next subsection, that (9.23) is *twice* the logarithm of a likelihood ratio, which is known to have such a distribution.

The Box-Cox transformations were originally introduced to reduce non-normality in data. However, as an examination of (9.22) will show, it should be well suited to our purpose of reducing nonlinearity. It should be pointed out that the approach will attempt to find transformations which will try to reduce the residuals associated with outliers and also reduce heteroscedasticity, particularly if there was no acute nonlinearity to begin with. Andrews (1971b) has noted the fairly severe effect of outliers and influential points on choice of the Box-Cox transformations and since then a number of methods have been proposed to combat such situations; e.g., see Cook and Wang (1983), Carroll (1982), Carroll and Ruppert (1985), Atkinson (1985, 1986). A discussion of this subject is, unfortunately, beyond the scope of this book.

Example 9.10 (Continuation of Example 9.7, Page 199)
Let us return to Stevens' data. After using a set of λ's separated by .1 (Exhibit 9.14) and surmising that the maximum of $-\log[s^2(\lambda)]$ occurred for λ in the interval (0, .1), we used a more finely separated set of values of λ to obtain Exhibit 9.15. The largest value of $-(n/2)\log[s^2(\lambda)]$ occurs around .07. This gives us the transformation $[y^{.07} - 1]/.07$. Actually, we would just use $y^{.07}$, which is about what we found in Example 9.7.

λ	0	.1	.2	.3	.4	.5	.6	.7	.8
$-\frac{1}{2}n\log[s^2(\lambda)]$	1.99	5.23	-.698	-8.40	-10.89	-12.99	-14.85	-16.58	-18.21

EXHIBIT 9.14: Values of Box-Cox Objective Function Over a Coarse Grid: Stevens' Data

λ	.01	.02	.03	.04	.05	.06	.07	.08	.09
$-\frac{1}{2}n\log[s^2(\lambda)]$	2.74	3.48	4.17	4.78	5.28	5.63	5.80	5.78	5.58

EXHIBIT 9.15: Values of Box-Cox Objective Function Over a Fine Grid: Stevens' Data

Notice that the difference $n\log[s^2(0)] - n\log[s^2(\lambda_{max})] = 2 \times 3.81 = 7.62$ is significant at even a 1 per cent level (the 10, 5, 1 and .001 per cent critical values are, respectively, 2.71, 3.84, 6.63 and 10.83). Nevertheless, because .07 is so close to zero, the obvious intuitive appeal of logs in this case overrides this consideration, at least in the judgment of the authors.■

Example 9.11 (Continuation of Example 5.5, Page 107)
Application of the Box-Cox method to seek an appropriate transformation of the dependent variable 'charges' in the hospital charge data given in Exhibit 4.13, p. 99, yielded Exhibit 9.16. The maximum occurs for $\lambda = .2$ and both the log transformation ($\lambda = 0$) and the square root are significant using (9.23). The reader who has completed Exercises 8.15 and 8.16 will recognize that the main shortcoming of the model using simply 'charges' as the dependent variable is the presence of the outlier. Therefore, to some extent, the Box-Cox procedure chose the transformation it did in order to take care of it.

The original data set from which our set was extracted had about 2000 cases covering several diagnostic categories (ailments). In that set, the point we have called an outlier did not appear to be so. The space between the isolated outlier and other points was 'filled in' with observations. For this larger data set, Feinglass found that a transformation between the log and square root was most appropriate and chose the log transformation because the coefficients then have a fairly simple explanation in terms of percentages (see Exercise 9.1). He also did not want to give the appearance of undue fastidiousness resulting from a transformation like, say, $y^{.17}$! ■

λ	-.5	-.1	0	.1	.2	.3	.4	.5	1.0
$-\frac{1}{2}n\log[s^2(\lambda)]$	-420.6	-404.5	-403.6	-400.82	-400.19	-400.37	-401.3	-402.9	-418.5

EXHIBIT 9.16: Values of Box-Cox Objective Function: Hospital Charge Data

*Derivation of Box-Cox Objective Function

Since $\partial y_i{}^{(\lambda)}/\partial y_\ell = 0$ when $i \neq \ell$, the Jacobian of the transformation from $y_i^{(\lambda)}$ to y_i is given by (see Section B.6, p. 293)

$$\prod_{i=1}^{n}(\partial y_i^{(\lambda)}/\partial y_i) = (\prod_{i=1}^{n} y_i)^{\lambda-1}. \tag{9.24}$$

Assuming that $\boldsymbol{y}^{(\lambda)}$ has a normal distribution with mean $X\boldsymbol{\beta}$ and covariance matrix $\sigma^2 I$, the density function of $\boldsymbol{y}$ is proportional to

$$(\prod_{i=1}^{n} y_i^{\lambda-1})(\sigma^2)^{-\frac{1}{2}n}\exp[-\frac{1}{2\sigma^2}[(\boldsymbol{y}^{(\lambda)} - X\boldsymbol{\beta})'(\boldsymbol{y}^{(\lambda)} - X\boldsymbol{\beta}]].$$

Therefore, the log-likelihood function is, ignoring constants,

$$\begin{aligned}\log[\mathcal{L}(\boldsymbol{\beta},\sigma^2,\lambda)] &= (\lambda-1)\sum_{i=1}^{n}\log[y_i] \\ &\quad -\tfrac{1}{2}n\log(\sigma^2)-\tfrac{1}{2}\sigma^{-2}[(\boldsymbol{y}^{(\lambda)}-X\boldsymbol{\beta})'(\boldsymbol{y}^{(\lambda)}-X\boldsymbol{\beta})].\end{aligned} \tag{9.25}$$

Maximizing (9.25) by differentiating it with respect to $\boldsymbol{\beta}$ and σ^2 and equating to zero, we get, for a given λ, the maximum likelihood estimates of $\boldsymbol{\beta}$ and σ^2. These are $\hat{\boldsymbol{\beta}}(\lambda) = (X'X)^{-1}X'\boldsymbol{y}^{(\lambda)}$ and $\hat{\sigma}^2(\lambda) = n^{-1}(\boldsymbol{y}^{(\lambda)})'[I-H]\boldsymbol{y}^{(\lambda)}$. Substituting these estimates into equation (9.25), we get

$$\log[\mathcal{L}(\hat{\boldsymbol{\beta}},\hat{\sigma}^2,\lambda)] = -\tfrac{1}{2}n\log[s^2(\lambda)]$$

as given in (9.22)

*Other Families of Transformations for Dependent Variables

Sometimes it is useful to 'start' (Mosteller and Tukey, 1977) our transformations by adding a constant to y_i before applying the transformation. We saw an example of this when we added Anscombe's correction. If the variable has a natural non-zero lower bound, particularly if it is negative, such starts often work well. If the value a of the start is unknown, the following family of transformations can be useful:

$$y_i^{(\lambda)} = \begin{cases}[(y_i+a)^\lambda - 1]/\lambda, & \text{when } \lambda \neq 0,\\ \log(y_i+a), & \text{when } \lambda = 0.\end{cases} \tag{9.26}$$

Then, as before, it can be shown that the maximum likelihood estimate of (λ, a) maximizes

$$-\tfrac{1}{2}n\log[s_1^2(\lambda)] = (\lambda-1)\sum_{i=1}^{n}\log(y_i+a) - \tfrac{1}{2}n\log[\sigma^2(\lambda)], \tag{9.27}$$

where $\hat{\sigma}^2(\lambda) = n^{-1}\boldsymbol{y}^{(\lambda)\prime}[I-H]\boldsymbol{y}^{(\lambda)}$, with $\boldsymbol{y}^{(\lambda)} = (y_1^{(\lambda)},\ldots,y_n^{(\lambda)})'$ defined by (9.26). Therefore, except for the addition of a and the fact that the search for the maximum must now be conducted over two variables, the procedure is essentially the same as the one given earlier. The corresponding chi-square statistic has two degrees of freedom.

Since both approaches just described try to minimize sums of squares, they might not always work well if the distribution of residuals from the untransformed model has a fairly symmetric appearance. In such cases the following family of 'modulus' transformations, due to John and Draper (1980), might be appropriate.

$$y_i^{(\lambda)} = \begin{cases}(\text{sign})[\{|y_i-b|+1\}^\lambda - 1]/\lambda, & \text{when } \lambda \neq 0,\\ (\text{sign})\log[\{|y_i-b|\}+1], & \text{when } \lambda = 0,\end{cases} \tag{9.28}$$

where sign=sign of $(y_i - b)$ and b is some *preselected* value such as the arithmetic or geometric mean. Estimation of λ proceeds analogously to that for the previous two families: maximize

$$-\tfrac{1}{2}n\log[s_1^2(\lambda)] = (\lambda-1)\sum_{i=1}^{n}\log(|y_i-b|+1) - \tfrac{1}{2}n\log[\sigma^2(\lambda)], \qquad (9.29)$$

where $\hat{\sigma}^2(\lambda)$ is now defined in terms of (9.28).

9.4.4 Analytic Methods: Transforming the Predictors

Analogous to the Box-Cox family of transformations for the dependent variable is the family of transformations given by Box and Tidwell (1962) for independent variables:

$$z_{ij} = \begin{cases} [x_{ij}^{\alpha_j} - 1]/\alpha_j, & \text{when } \alpha_j \neq 0 \\ \log(x_{ij}), & \text{when } \alpha_j = 0, \end{cases} \qquad (9.30)$$

for $i = 1, \ldots, n$ and $j = 1, \ldots, k$. We need to estimate the α_i's. Since the dependent variable is not being transformed, we need not worry about changes of scale, and simply minimize

$$\sum_{i=1}^{n}(y_i - \beta_0 - \beta_1 z_{i1} - \cdots - \beta_k z_{ik})^2.$$

This minimization is probably most easily carried out using nonlinear least squares (Appendix C). An alternative method requiring repeated applications of linear least squares is given in Box and Tidwell (1962).

Example 9.12 (Continuation of Example 9.8, Page 201)
Since the apartment rent model has two independent variables I and S, we applied nonlinear least squares to

$$R = \beta_0 + \beta_1\frac{\mathrm{I}^{\alpha_1}-1}{\alpha_1} + \beta_2\frac{\mathrm{S}^{\alpha_2}-1}{\alpha_2}.$$

The estimates are shown in Exhibit 9.17. It would appear that I requires no action. The course of action for S is less clear. While its power α_2 has an estimate of .378, the standard error is so large that it includes all reasonable options! Based on our previous work (Example 9.8), we stayed with a log transformation. ∎

9.4.5 Simultaneous Power Transformations for Predictors and Response

We could also simultaneously obtain appropriate power transformations for the dependent variable as well as for the independent variables. For sim-

Parameter	Estimate	Std. Error
β_0	-18.441	44.986
β_1	16.101	5.623
β_2	246.049	121.788
α_1	0.988	0.110
α_2	0.378	1.127

EXHIBIT 9.17: Parameter Estimates and Corresponding Standard Errors for Nonlinear Least Squares Applied to Rent Model

plicity in presentation we consider only the case where all the observations (dependent and independent) are positive; other cases can be handled in a manner similar to those given in the subsections near the end of Section 9.4.3. The transformations are given by

$$y_i^{(\lambda)} = \begin{cases} (y_i^\lambda - 1)/\lambda & \text{when } \lambda \neq 0 \\ \log(y_i) & \text{when } \lambda = 0 \end{cases}$$

$$x_{ij}^{(\gamma_j)} = \begin{cases} (x_{ij}^{\gamma_j} - 1)/\gamma_j & \text{when } \gamma_j \neq 0 \\ \log(x_{ij}) & \text{when } \gamma_i = 0 \end{cases}$$

for $i = 1, \ldots, n$, and $j = 1, \ldots, k$. Let $X^{(\Upsilon)}$ denote the matrix $(x_{ij}^{(\gamma_j)})$ and let $H^{(\Upsilon)} = X^{(\Upsilon)}(X^{(\Upsilon)\prime}X^{(\Upsilon)})^{-1}X^{(\Upsilon)\prime}$. Then it may be shown that the maximum likelihood estimate of (λ, γ) is obtained by maximizing (9.22) but now $\hat{\sigma}^2 = n^{-1}\boldsymbol{y}^{(\lambda)\prime}[I - H^{(\Upsilon)}]\boldsymbol{y}^{(\lambda)}$. This is a problem in nonlinear optimization which can also be handled by some nonlinear least squares procedures where user-supplied 'loss functions' can be used in place of the sum of squares 'loss' required by least squares (e.g., SAS PROC NLIN described in SAS, 1985b and SYSTAT[1] NONLIN described in Wilkinson, 1987). A modified Newton algorithm is given in Spitzer (1982) for the case in which λ and the γ_j's are all the same.

[1]SYSTAT is a trademark of SYSTAT, Inc., Evanston, IL.

Appendix to Chapter 9

9A Some Programs

The lines of program (albeit æsthetically rather crude) given below compute $(\Omega - E(\Omega))/(\text{var}\,\Omega)^{1/2}$ of Section 9.3.5, using formulæ for means and variances derived under the normality assumption. They are written in SAS PROC MATRIX, which is described in SAS, 1982b. It is assumed that a data set has already been defined in which the first variable is the dependent variable and the remaining ones are the independent variables. The data set name needs to be inserted over the underlines and the lines given below inserted after definition of the data.

```
PROC MATRIX; FETCH X DATA=_______;
N=NROW(X);KCOL=NCOL(X); Y=X(,1); Z=J(N,1,1); X(,1)=Z;
ADJ=I(M)-X*(INV(X'*X))*X'; ADJ2=I(N)-((Z*Z')#/N);
E=ADJ*Y; D=J(M,M,0);
DO KK=1 TO KCOL;
   SS=ADJ2*X(,KK); SD=SS'*SS#/N;
   SD1=(SD)##(.5);X(,KK)=X(,KK)#/SD1;
END;
DO K=1 TO M; DO L=1 TO M;
   DO KK=1 TO KCOL;
      D(K,L)=D(K,L)+(X(K,KK)-X(L,KK))**2;
   END;
   IF K=L THEN D(K,L)=0; ELSE D(K,L)=1#/D(K,L);
END;END;
B=ADJ*D*ADJ; NUM=E'*D*E; DEN=E'*E; DEN=DEN#/(M-KCOL);
ST=NUM#/DEN; MEAN=TRACE(B);
VAR=2*((M-3)*TRACE(B**2)-(TRACE(B)**2) )#/(M-1);
ST=(ST-MEAN)#/SQRT(VAR); PRINT ST;
```

While there are several programs written to carry out the Box-Cox computations of Section 9.4.3 (e.g., see Stead, 1987, and other appropriate entries in SAS, 1988), unless there is severe multicollinearity present, adequate and simple programs can be written using MINITAB or SAS PROC MATRIX or PROC IML. The lines below illustrate such a program in SAS PROC MATRIX. One needs to adjust the 'DO' statement for the grid to be searched over. Otherwise, the method of use is the same as for the program lines given above.

```
PROC MATRIX; FETCH X DATA=_______;
N=NROW(X); Y=X(,1); Z=J(N,1,1); X(,1)=Z;
H=X*INV(X'*X)*X'; M=I(N)-H; M=M#/(N);
YLOG=LOG(Y);G=SUM(YLOG);
DO L=.0 TO .1 BY .01;
```

```
   IF L=0 THEN YL=YLOG; ELSE YL=(Y##L-Z)#/L;
   S=YL'*M*YL; S=LOG(S); S=S#N#/2;GG=(L-1)#G;
   S=-S+GG; PRINT L S;
END;
```

Problems

Exercise 9.1: Suppose you have completed the study recommended in Exercise 4.9, p. 97 (with the dependent variable logged), and you need to explain the meaning of the estimates of β_j's to your client. What would you say? Suppose, in addition to the independent variables already included, you had an additional one: severity times age. How would you explain the meaning of the coefficient of this variable to your client?

Discuss the merits and shortcomings of logging the variable 'charges', pretending that you have not yet seen the data.

Exercise 9.2: Derive Anscombe's correction for $\log(y_i)$ when y_i is Poisson (p. 185). If the dependent variable y_i were a proportion of counts (no logs), what would be the appropriate Anscombe's correction?

Exercise 9.3: Suppose someone comes to you for help with a regression problem where the dependent variable is a proportion of counts. The independent variable values are such that the expectations of the dependent variable values for each of them must lie between .25 and .75. He will use the model only to make predictions which are expected to be in the same range. He does not wish to get involved with logits. Can he get by without transforming the dependent variable?

Exercise 9.4: Derive the results of Section 9.4.5 by writing down the likelihood equation and maximizing with respect to β and σ^2. Give details.

Exercise 9.5: Using the weights you did in Exercise 8.11, p. 178, and deleting the outliers we deleted in Example 8.2, p. 163, fit a model which is a second degree polynomial in LINC to the data of Exhibit 4.7, p. 90. Examine residuals to see if you prefer this model over the broken line model.

Exercise 9.6: Fit a model of the form (9.1) to the data of Problem 9.5, choosing as knot the point LINC= 7.

Exercise 9.7: A generalization of the Cobb-Douglas model given in Part 1 of Exercise 2.19, p. 54, is the translog model

$$\log[V_t] = \beta_0 + \beta_1 k_t + \beta_2 l_t + \tfrac{1}{2}\beta_3 k_t^2 + \tfrac{1}{2}\beta_4 l_t^2 + \beta_5 k_t l_t + \epsilon_t,$$

where lower case Latin letters are logs of corresponding capital letters, but otherwise the notation is the same as in Exercise 2.19. Estimate the parameters of this translog model for each of the sectors 20, 36 and 37 under the assumption that Gauss-Markov conditions are met. Test the hypothesis that the Cobb-Douglas model is adequate against the alternative that the translog model is needed. For the Cobb-Douglas model of Part 1 of Exercise 2.19, give an estimate of α.

t_a	t_r	m_a	m_r	t_a	t_r	m_a	m_r	t_a	t_r	m_a	m_r
360	459	71	61	360	422	48	36	650	265	11	29
860	410	29	87	840	408	21	51	850	229	9	43
450	487	41	41	918	432	38	227	319	395	35	24
800	505	46	97	280	307	29	41	300	350	11	16
70	200	41	20	507	375	40	119	900	370	13	34
915	339	29	237	235	540	75	40	1000	388	14	79
795	324	60	135	217	250	35	35	900	388	14	69
862	547	147	313	450	600	23	12	335	500	26	27
100	120	13	11	1178	513	27	184	450	440	19	30
186	450	33	17	300	300	12	12	377	464	35	43
850	450	62	205	185	350	24	12	1050	425	13	54
992	388	42	265	650	362	9	16	1000	437	14	148
100	120	13	11	172	173	18	26	350	600	13	16
150	200	25	24	1100	465	12	72	950	450	13	36
141	250	43	32	1200	471	29	148	1053	403	30	80
809	435	27	44	220	400	5	10	900	460	17	29
800	420	16	47	100	200	19	14	1000	409	10	81

EXHIBIT 9.18: Data Travel Times and Usage for Automobiles and Public Transportation

SOURCE: Selected by Robert Drozd from Census UTP Package.

Exercise 9.8: Exhibit 9.18 provides data on travel time by car (t_a), travel time by public transportation (t_r), the number of people who used any kind of public transportation (m_r) and number of those who used a car or van either as driver or passenger (m_a) to travel between pairs of points. Travel times, which are in tenths of minutes, are averages of those reported by the travelers. The data set was *selected* (not randomly) from the Bureau of Census Urban Transportation Planning Package for the Chicago area and then modified by one of the authors. The primary modification was the addition to auto travel times of an amount of time which reflected the cost of parking. For downtown zones this amounted to about sixty minutes. Obtain a logit model for the proportion of auto users corresponding to each of the three sets of independent variables:

1. The difference in travel times
2. The two travel times
3. Square roots of the two travel times.

Which model do you think is best? Why?

Based on what you have found, formulate and estimate a model that you think might be better than the ones you have already fitted.

Using each of the models you constructed, obtain an estimate of the proportion of public transportation users among 50 commuters whose travel times are $t_a = 500$ and $t_r = 500$. Treating these proportions as future observations, give 95 per cent confidence intervals for them.

Exercise 9.9: Apply as many of the procedures of Section 9.3.5, p. 195, as you can to both the 'logged' and 'unlogged' models of the dial-a-ride data (Exhibit 6.8, p. 125). Selectively drop some of the independent variables and rerun the procedure. State any conclusions you reach.

ACC	WHP	SP	G	ACC	WHP	SP	G	ACC	WHP	SP	G
8.0	20.5	7.5	0	8.0	40.0	7.5	0	1.0	84.5	22.5	0
5.0	20.5	22.5	0	5.0	40.0	22.5	0	1.0	84.5	30.0	0
5.0	20.5	30.0	0	2.0	40.0	30.0	0	0.6	84.5	40.0	0
4.0	20.5	40.0	0	1.8	40.0	40.0	0	0.2	84.5	50.0	0
3.0	20.5	50.0	0	1.5	40.0	50.0	0	1.6	84.5	7.5	2
2.0	20.5	60.0	0	0.7	40.0	60.0	0	0.6	84.5	22.5	2
7.8	20.5	7.5	2	7.8	40.0	7.5	2	0.6	84.5	35.0	2
4.6	20.5	22.5	2	4.6	40.0	22.5	2	0.2	84.5	45.0	2
4.2	20.5	35.0	2	1.6	40.0	35.0	2	0.7	84.5	7.5	6
3.4	20.5	45.0	2	1.4	40.0	45.0	2	2.0	257.0	7.5	0
2.4	20.5	55.0	2	1.0	40.0	55.0	2	1.0	257.0	22.5	0
6.7	20.5	7.5	6	6.7	40.0	7.5	6	0.8	257.0	30.0	0
3.7	20.5	22.5	6	3.7	40.0	22.5	6	0.4	257.0	40.0	0
3.4	20.5	35.0	6	0.7	40.0	35.0	6	1.6	257.0	7.5	2
2.5	20.5	45.0	6	0.5	40.0	45.0	6	0.6	257.0	22.5	2
1.5	20.5	55.0	6	0.2	40.0	55.0	6	0.3	257.0	35.0	2
				2.0	84.5	7.5	0	0.7	257.0	7.5	6

EXHIBIT 9.19: Acceleration Data
SOURCE: Raj Tejwaney, Department of Civil Engineering, University of Illinois at Chicago.

Exercise 9.10: Exhibit 9.19 gives observations on the acceleration (ACC) of different vehicles along with their weight-to-horsepower ratio (WHP), the speed at which they were traveling (SP), and the grade (G; G $= 0$ implies the road was horizontal).

1. Run a regression using ACC as your dependent variable without making any transformations, and obtain component plus residual plots.

2. Obtain a good fitting model by making whatever changes you think are necessary. Obtain appropriate plots to verify that you have succeeded.

3. The component plus residual plot involving G in 1 (above) appears to show heteroscedasticity. If you have been successful in 2, the appearance of any serious heteroscedasticity should vanish without your having to weight or transform the dependent variable. Explain why you think this happens.

Exercise 9.11: Explore transformations of other independent variables and complete building a model for PCS that we started in Example 9.9.

Exercise 9.12: Exhibit 9.20 gives cleanup costs (C) corresponding to sales at hot dog stands (R_{HD}) and at beer stands (R_B) for a medium-sized sports facility. (The authors have lost all records other than the data; consequently the units of measurement are not known.) Use component plus residual plots and also the Box-Tidwell method to find suitable transformations of the independent variables. Can you reconcile the difference in results you get from the two approaches? [You may need to use weighted least squares.]

C	R_{HD}	R_B	C	R_{HD}	R_B	C	R_{HD}	R_B
10.4	579	302	8.2	191	409	11.0	240	513
10.6	661	442	12.2	680	540	14.3	701	814
8.5	273	308	9.6	376	389	14.7	675	917
8.9	840	400	7.7	120	276	8.9	408	300
6.7	207	97	7.5	113	248	9.2	411	517
						8.7	272	312

EXHIBIT 9.20: Stadium Cleanup Data

Exercise 9.13: Apply the Box-Cox approach to Model 2 of Exercise 2.15, p. 53, to find an appropriate transformation for the dependent variable.

Exercise 9.14: Find appropriate transformations of independent variables in the models considered in Exercise 2.20, p. 55 (now properly weighted and with perhaps some particularly influential points deleted. In at least one case we saw no need for transformations.) Rerun the transformed models and check if fit has improved.

Exercise 9.15: The only time the market value or, equivalently, the depreciation of a piece of real property is observable is when it is sold. However, for various reasons, including the assessment of property taxes, it is often necessary to estimate its value at other times on the basis of selling prices of other similar properties.

The data set in Exhibit 9.21 provides the age and corresponding depreciation for 11 similar properties. After fitting an appropriate curve, estimate the depreciation at age zero; i.e., the depreciation right after the facility was first built. Notice that since such large factories are designed for specific, highly specialized production processes, if some other manufacturer were to buy it they would have to make major changes in order to adapt it to their use. Consequently, the depreciation at year zero need not be zero.

Clearly, the estimate of depreciation you make will depend on the transformation you use. Therefore, it is important to choose it carefully.

Obtain 95 per cent confidence intervals for the prediction you make, as well as for the corresponding future observation. Discuss which of these two confidence intervals is appropriate if your interest is in estimating 'a fair market value.'

Age	Depr.	Age	Depr.	Age	Depr.
7	60.7	22	68.7	35	90.9
17	83.5	22	81.3	48	96.9
19	69.2	28	94.5	21	87.7
20	70.5	28	80.8		

EXHIBIT 9.21: Depreciation in Market Value of Large Factories
SOURCE: Diamond-Star Motors, Normal, IL. We are grateful to Gary Shultz, General Counsel, for making these data available.

CHAPTER 10

Multicollinearity

10.1 Introduction

Until this point, the only difficulties with least squares estimation that we have considered have been associated with violations of Gauss-Markov conditions. These conditions only assure us that least squares estimates will be 'best' for a given set of independent variables; i.e., for a given X matrix. Unfortunately, the quality of estimates, as measured by their variances, can be seriously and adversely affected if the independent variables are closely related to each other. This situation, which (with a slight abuse of language) is called *multicollinearity*, is the subject of this chapter and is also the underlying factor that motivates the methods treated in Chapters 11 and 12.

10.2 Multicollinearity and Its Effects

If the columns of X are linearly dependent, then $X'X$ is singular and the estimates of $\boldsymbol{\beta}$, which depend on a generalized inverse of $X'X$, cannot be unique. In practice, such situations are rare, and even when they do occur, practical remedies are not difficult to find. A much more troublesome situation arises when the columns of X are nearly linearly dependent. Just as linear dependency leads to the singularity of $X'X$, near linear dependency leads to near singularity. Since singularity may be defined in terms of the existence of a unit vector $\boldsymbol{c}$ (i.e., with $\boldsymbol{c}'\boldsymbol{c} = 1$) such that $X\boldsymbol{c} = 0$ or equivalently $\boldsymbol{c}'X'X\boldsymbol{c} = 0$, we may characterize near singularity in terms of the existence of a unit vector $\boldsymbol{c}$ such that $||X\boldsymbol{c}||^2 = \boldsymbol{c}'X'X\boldsymbol{c} = \delta$ is small. This is equivalent to saying that, for some $\boldsymbol{c} = (c_0, \ldots, c_k)'$ of unit length, the length of $\sum_{j=0}^{k} c_j \boldsymbol{x}_{[j]}$ is small where $X = (\boldsymbol{x}_{[0]}, \ldots, \boldsymbol{x}_{[k]})$. (The presence of square brackets in the subscript of $\boldsymbol{x}$ emphasizes that we are discussing a column of X and distinguishes it from $\boldsymbol{x}_i$ which we have used to denote a row of X.)

When near singularity exists, the variance of estimates can be adversely affected, as we now show. From the Cauchy-Schwartz inequality

$$\begin{aligned}1 = [\boldsymbol{c}'\boldsymbol{c}]^2 &= [\boldsymbol{c}'(X'X)^{1/2}(X'X)^{-1/2}\boldsymbol{c}]^2 \\ &\le \boldsymbol{c}'(X'X)\boldsymbol{c}\boldsymbol{c}'(X'X)^{-1}\boldsymbol{c} = \delta\boldsymbol{c}'(X'X)^{-1}\boldsymbol{c}.\end{aligned}$$

Consequently, $\text{var}\,(\boldsymbol{c}'\boldsymbol{b}) = \sigma^2\boldsymbol{c}'(X'X)^{-1}\boldsymbol{c} \ge \sigma^2/\delta$, which will be large when

δ is small in comparison to σ^2. This will usually result in some of the b_j's have large variances. We could also get counter-intuitive results, especially in signs of b_j's. Moreover, near singularity can magnify effects of inaccuracies in the elements of X. Therefore, it is most desirable to detect the presence of near singularity and to identify its causes when it is there.

Multicollinearity is the special case of near singularity where there is a (linear) near relationship between two or more $\boldsymbol{x}_{[j]}$'s; i.e., it requires that the length of $\sum_{j=0}^{k} c_j \boldsymbol{x}_{[j]}$ be small with at least two $\boldsymbol{x}_{[j]}$'s and corresponding c_j's not too small.

Since $\sum_{j=0}^{k} c_j \boldsymbol{x}_{[j]}$ is affected by the units in which the variables are measured, when assessing smallness it is desirable to scale X; i.e., instead of $\boldsymbol{y} = X\boldsymbol{\beta} + \boldsymbol{\epsilon}$ consider the equivalent model

$$\boldsymbol{y} = X_{(s)}\boldsymbol{\beta}_{(s)} + \boldsymbol{\epsilon}, \tag{10.1}$$

where $X_{(s)} = XD_{(s)}{}^{-1}$, $\boldsymbol{\beta}_{(s)} = D_{(s)}\boldsymbol{\beta}$ and

$$D_{(s)} = \operatorname{diag}(||\boldsymbol{x}_{[0]}||, \ldots, ||\boldsymbol{x}_{[k]}||). \tag{10.2}$$

Since $(X_{(s)}{}'X_{(s)})^{-1} = D_{(s)}(X'X)^{-1}D_{(s)}$, it is easy to see that, if $\boldsymbol{b}_{(s)}$ is the least squares estimator of $\boldsymbol{\beta}_{(s)}$, then

$$\boldsymbol{b}_{(s)} = D_{(s)}\boldsymbol{b} \text{ and } \operatorname{cov}(\boldsymbol{b}_{(s)}) = D_{(s)} \operatorname{cov}(\boldsymbol{b}) D_{(s)}.$$

A consequence of this scaling is that it removes from consideration near singularities caused by a single $\boldsymbol{x}_{[j]}$ being of small length, i.e., near singularities which are not multicollinearities.

For $\boldsymbol{d} = D_{(s)}\boldsymbol{c}$,

$$\boldsymbol{c}'X'X\boldsymbol{c} = \boldsymbol{c}'D_{(s)}D_{(s)}{}^{-1}X'XD_{(s)}{}^{-1}D_{(s)}\boldsymbol{c} = \boldsymbol{d}'X_{(s)}{}'X_{(s)}\boldsymbol{d} \geq d_{min}||\boldsymbol{d}||^2$$

(see Example A.9, p. 277), where d_{min} is the smallest characteristic root of $X_{(s)}{}'X_{(s)}$. Therefore, if multicollinearity is present ($\boldsymbol{c}'X'X\boldsymbol{c}$ is small with $||\boldsymbol{d}||$ not too small), d_{min} will be small. Conversely, if we have a small eigenvalue of $X_{(s)}{}'X_{(s)}$, and if γ_0 is the corresponding eigenvector, then $\gamma_0'X_{(s)}{}'X_{(s)}\gamma_0$ is small and it may easily be shown that multicollinearity then would be present. Moreover, since the eigenvectors are mutually orthogonal, each small eigenvalue represents a different near relationship.

Belsley, Kuh and Welsch (1980) give a very thorough discussion of multicollinearity from a slightly different starting point and cover in greater depth most of the material of this chapter (see also Stewart, 1987, and the comments following that article). While in much of our discussion we use just the scaled model (10.1), several authors recommend a model that is both centered and scaled (Section 2.11, p. 44). In fact, there is quite a controversy on the subject — see Belsley (1984, 1986) and commentaries following Stewart (1987).

The example below illustrates some of the difficulties caused by multicollinearity.

Example 10.1
Consider the highly multicollinear values of the independent variables x_1 and x_2 given in Exhibit 10.1. These data were made up by the authors. The dependent variables $y^{(1)}$, $y^{(2)}$ and $y^{(3)}$ may be considered as different samples. They were obtained by adding a $N(0, .01)$ pseudo-random number to

$$x_1 + \frac{1}{2}x_2 \tag{10.3}$$

and it is easily seen that corresponding values of the dependent variables are much alike.

x_1	x_2	$y^{(1)}$	$y^{(2)}$	$y^{(3)}$
2.705	2.695	4.10	4.10	4.06
2.995	3.005	4.34	4.73	4.39
3.255	3.245	4.95	4.81	5.02
3.595	3.605	5.36	5.30	5.23
3.805	3.795	5.64	5.75	5.57
4.145	4.155	6.18	6.26	6.50
4.405	4.395	6.69	6.61	6.65
4.745	4.755	7.24	7.13	7.26
4.905	4.895	7.46	7.30	7.48
4.845	4.855	7.23	7.32	7.39

EXHIBIT 10.1: Multicollinear Data

To keep matters confined to the highly multicollinear x_1 and x_2, let us consider the model $y_i = \beta_1 x_{i1} + \beta_2 x_{i2} + \epsilon_i$. The least squares estimate $\boldsymbol{b}$ of $\boldsymbol{\beta}$ for the dependent variable $y^{(1)}$ is

$$b_1 = 5.21\ (2.00), \quad b_2 = -3.70\ (-1.42) \quad [s = .082],$$

where the quantities within parentheses are the t-values. For the dependent variables $y^{(2)}$ and $y^{(3)}$ the corresponding estimates are

$$b_1 = -1.40\ (-.47), \quad b_2 = 2.90\ (.98) \quad [s = .094]$$

and

$$b_1 = .54\ (.12), \quad b_2 = .97\ (.21) \quad [s = .147].$$

These fairly wild fluctuations in the value of $\boldsymbol{b}$ are due to the fact that

$$(X'X)^{-1} = \begin{pmatrix} 1001.134 & -1001.05 \\ -1001.05 & 1000.972 \end{pmatrix}.$$

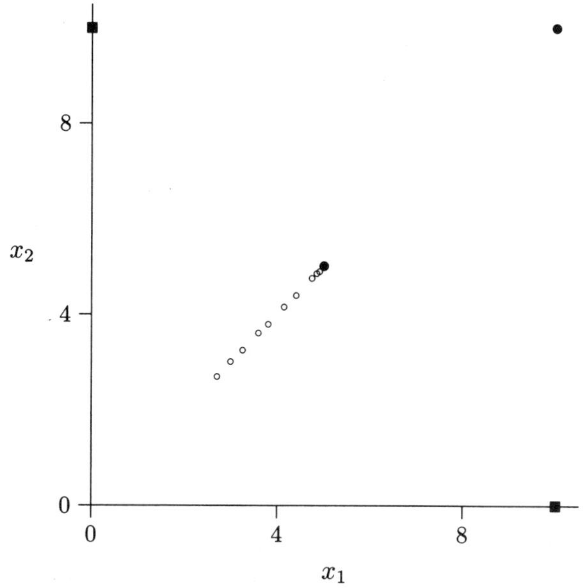

EXHIBIT 10.2: Plot of Independent Variable Values for Multicollinear Data. [A small circle denotes points corresponding to 'observations', a bullet • corresponds to forecast points along the direction of the data points and a solid box ■ to forecast points in a direction orthogonal to the direction of the data points.]

Since $\boldsymbol{b} = (X'X)^{-1}X'\boldsymbol{y}$, the slightest change in $X'\boldsymbol{y}$ is magnified by the matrix $(X'X)^{-1}$, leading to these fluctuations in the value of $\boldsymbol{b}$. This is mirrored in the size of the covariance matrix $\sigma^2(X'X)^{-1}$ of $\boldsymbol{b}$, which remains large even though σ and its estimate s are quite small.

The eigenvalues of $(X'X)$ are 341.6 and .0007 and the corresponding eigenvectors are

$$(.707\ .707)' \text{ and } (.707\ \ -.707)'. \tag{10.4}$$

Therefore, the length of the linear combination $\boldsymbol{z}_2 = .707\boldsymbol{x}_{[1]} - .707\boldsymbol{x}_{[2]}$ is very small compared to that of $\boldsymbol{z}_1 = .707\boldsymbol{x}_{[1]} + .707\boldsymbol{x}_{[2]}$, where $\boldsymbol{x}_{[j]}$ is the vector of values of the independent variable x_j.

Now let us see how this multicollinearity affects prediction. For this purpose we shall simply focus on estimates obtained using $y^{(1)}$ as dependent variable. For $(x_1, x_2) = (5, 5)$ or $(x_1, x_2) = (10, 10)$, which lie along the direction of the data points (i.e., along $\boldsymbol{z}_1$ — see Exhibit 10.2), the corresponding $\hat{y}$'s are 7.52 and 15.04. Inserting these values of (x_1, x_2) into (10.3) we get 7.5 and 15, so that our predictions are fairly good even though the estimates of individual β_j's are clearly poor. The corresponding standard errors of $\hat{y}$ are .0325 and .0650, which are also quite small. Therefore, estimates along the $\boldsymbol{z}_1$ direction are quite good. This is also true of predictions made for points close to the $\boldsymbol{z}_1$ direction and, in particular, would be true for all predictions corresponding to the design points used to estimate the model, i.e., the $\hat{y}_i$'s.

For $(x_1, x_2) = (0, 10)$ and $(x_1, x_2) = (10, 0)$, which lie in a direction orthogonal to the direction of the points (i.e., in the $\boldsymbol{z}_2$ direction), $\hat{y} = -37$ and 52, which is obviously nonsense. Moreover, s.e.$(\hat{y}) = 26$ in both cases. Therefore, a prediction in this direction or close to it is perhaps best not made at all.

In situations like this, many analysts might delete one of the independent variables. The basic idea is that if x_1 and x_2 are related, then $c_1x_1 + c_2x_2 \approx 0$ and we can solve this approximate equation and express one of the variables (say x_2) in terms of the other. If we now substitute for x_2 in the original model we get a model without x_2.

When we remove x_2, the estimate of the sole coefficient (continuing to use $y^{(1)}$ as the dependent variable) turns out to be 1.5 with a standard error of .007. The 1.5 is easy to explain using (10.3): In the absence of x_2, x_1 is playing the role of both x_1 and x_2. But now that x_2 is missing from the equation, the forecast for the point $(x_1, x_2) = (0, 10)$ would be 0, which is obviously incorrect. Moreover, the standard error is a 'reassuringly small' 0. From the discussion of Section 11.2.1 we shall see that the deletion of x_1 has caused the model to become quite biased. If we deleted x_1 from the model, we would get a terrible estimate for the point $(x_1, x_2) = (10, 0)$. An alternative to deleting one or the other variable is to use a composite of the two. One such possibility is to use $x_1 + x_2$. The estimated coefficient of this composite independent variable is .75, which also does not yield good forecasts for the two points along the eigenvector $\boldsymbol{z}_2$ corresponding to the small eigenvalue. The fact is that in this problem we simply do not have enough information in the $\boldsymbol{z}_2$ direction to make a good forecast.

As we shall see in the next two chapters, sometimes it may be reasonable to trade off some of the high variance for bias. However, note that for this purpose bias is usually computed for the predicteds $\hat{y}_i$, which under acute multicollinearity, might not reflect the high levels of bias we would encounter in estimating future observations for independent variable values which are not related in much the same way as the design points. ■

10.3 Detecting Multicollinearity

10.3.1 Tolerances and Variance Inflation Factors

One obvious method of assessing the degree to which each independent variable is related to all other independent variables is to examine R_j^2, which is the value of R^2 between the variable x_j and all other independent variables (i.e., R_j^2 is the R^2 we would get if we regressed x_j against all other

x_ℓ's). The *tolerance* TOL_j is defined as

$$\text{TOL}_j = 1 - R_j^2 \qquad (10.5)$$

and, obviously, TOL_j is close to one if x_j is not closely related to other predictors. The *variance inflation factor* VIF_j is given by

$$\text{VIF}_j = \text{TOL}_j^{-1}. \qquad (10.6)$$

Clearly, a value of VIF_j close to one indicates no relationship, while larger values indicate presence of multicollinearity.

Notice that neither R_j^2 nor the method below will detect the situation where an independent variable has nearly constant values — i.e., it is related to the constant term. However, this case will be identified by the method of Section 10.3.3.

Perhaps the most frequently used device for detecting multicollinearity is the correlation matrix $\mathcal{R}$ of the x_j's mentioned in Section 2.11. The difficulty with $\mathcal{R}$ is that while it shows relationships between individual pairs of variables, it is difficult to see from it the strength of relationships between each x_j and all other predictors. However, it may be shown (Exercise 10.2) that the diagonal elements of $\mathcal{R}^{-1}$ are exactly the variance inflation factors. Since if no multicollinearity is present these diagonal elements are 1, VIF_j's show to what extent the variance of an individual b_j has been inflated by the presence of multicollinearity. Stewart (1987) has defined *collinearity index* as the square root of the variance inflation factors and, in that context, has presented a number of interesting properties of VIF_j's.

SAS provides, on request, all the multicollinearity detection devices mentioned above and other packages provide at least a few from which others can be readily obtained.

10.3.2 Eigenvalues and Condition Numbers

We have already seen in Section 10.2 the value of eigenvalues and the desirability of using $X'_{(s)}X_{(s)}$ rather than $X'X$. Moreover, since the sum of eigenvalues is equal to the trace, and each diagonal element of $X'_{(s)}X_{(s)}$ is 1,

$$\sum_{j=0}^{k} \lambda_j = \text{tr}(X'_{(s)}X_{(s)}) = k + 1 \qquad (10.7)$$

where λ_j's are the eigenvalues of $X'_{(s)}X_{(s)}$. This provides the immediate ability to make the comparisons we alluded to in Section 10.2. We need only consider what fraction of $k+1$ any given eigenvalue is or even what fraction of $k+1$ is accounted for by any collection of eigenvalues.

Another method of judging the size of one eigenvalue in relation to the others is through the use of the *condition number* η_j, which also has some other pleasant properties described in Belsley, Kuh and Welsch (1980) and Stewart (1987). It is defined as

$$\eta_j = \sqrt{\lambda_{max}/\lambda_j} \tag{10.8}$$

where $\lambda_{max} = \max_{0 \le j \le k} \lambda_j$. Belsley, Kuh and Welsch (1980) suggest that an eigenvalue with $\eta_j > 30$ be flagged for further examination.

10.3.3 Variance Components

If we wish to go further than the mere detection of multicollinearity to determine which linear combinations of columns of X are causing it, we can use the eigenvectors (Section 10.2). A more frequently used approach involves the variance of the coefficients of x_j's. Since it is usually applied to $X_{(s)}$, in this section we will be discussing variances of the components ${b_j}^{(s)}$ of $\boldsymbol{b}_{(s)} = D_{(s)}\boldsymbol{b}$ (see (10.1) and (10.2)).

Since $X'_{(s)}X_{(s)}$ is symmetric, it can be written as $\Gamma D_\lambda \Gamma'$ where $D_\lambda = \text{diag}\,(\lambda_0, \ldots, \lambda_k)$ and

$$\Gamma = \begin{pmatrix} \gamma_{00} & \gamma_{01} & \cdots & \gamma_{0k} \\ \gamma_{10} & \gamma_{11} & \cdots & \gamma_{1k} \\ \cdots & \cdots & \cdots & \cdots \\ \cdots & \cdots & \cdots & \cdots \\ \gamma_{k0} & \gamma_{k1} & \cdots & \gamma_{kk} \end{pmatrix}$$

is an orthogonal matrix. Hence, $\text{cov}(\boldsymbol{b}_{(s)}) = \sigma^2(X'_{(s)}X_{(s)})^{-1} = \sigma^2\Gamma D_\lambda^{-1}\Gamma'$ and, therefore,

$$\text{var}\,[{b_j}^{(s)}] = \sigma^2 \sum_{\ell=0}^{k} \lambda_\ell^{-1}\gamma_{j\ell}^2. \tag{10.9}$$

The $\lambda_\ell^{-1}\gamma_{j\ell}^2$'s are called *components of variance* of ${b_j}^{(s)}$, and

$$\phi_{\ell j} = \lambda_\ell^{-1}\gamma_{j\ell}^2 / \sum_{\ell=0}^{k} \lambda_\ell^{-1}\gamma_{j\ell}^2, \tag{10.10}$$

is called the *proportion of variance* of the jth coefficient ${b_j}^{(s)}$ corresponding to the ℓth eigenvector.

Notice that, except for the standardization to make $\sum_{\ell=0}^{k} \phi_{\ell j} = 1$, $\phi_{\ell j}$ is simply the reciprocal of the eigenvalue times the square of a component of the eigenvector. Therefore, it provides information very similar to that given by the matrix of eigenvectors, Γ. However, now we know something about how the variances of b_j's are being affected, which, of course, is valuable. Much of the damage done by multicollinearity is to variances and the proportions of variance tell us which relationship is doing how much damage to which variance.

The proportions of variance are usually presented in a table with the rows corresponding to the ℓ's and the columns to j's, (or, in some programs, with the rows corresponding to the j's and the columns to the ℓ's). Examination of such a table typically consists of identifying rows corresponding to small eigenvalues and then, in such rows, identifying $\phi_{\ell j}$'s which are large. Notice that since we are considering scaled variables, near singularity cannot occur unless there is multicollinearity. Therefore, every small eigenvalue is caused by the relationship between at least two independent variables. Also notice that since each column obviously adds to one, if the same independent variable features prominently in several different linear combinations of small length then $\phi_{\ell j}$ cannot be too large for some combinations. One must compensate for this fact in one's analysis of the tables.

Some analysts prefer to use the centered and scaled version $Z_{(s)}$ of X described in Section 2.11 instead of merely the scaled version $X_{(s)}$. The eigenvalues of $Z'_{(s)}Z_{(s)}$ can be obtained and components of variance computed in the same way as for $(X'_{(s)}X_{(s)})$. They are also available from several packages.

x_1	x_4	x_2	x_3	x_5	y	x_1	x_4	x_2	x_3	x_5	y
10.0	8.9	13.9	14.0	18.8	17.9	9.7	9.9	13.5	14.0	20.0	17.2
7.4	6.0	10.4	11.1	13.1	16.2	10.0	5.9	14.0	14.0	12.7	16.6
8.9	5.8	12.5	13.4	12.4	16.8	7.5	8.3	10.4	11.2	17.8	17.1
6.6	8.4	9.1	9.8	17.2	13.3	7.7	7.1	10.9	11.6	15.1	16.7
5.7	8.3	8.1	8.7	16.9	15.8	5.0	5.3	6.9	7.5	11.6	16.2
4.7	9.1	6.7	7.1	18.9	12.6	8.5	4.6	11.8	12.7	10.2	16.7
9.1	9.5	12.8	13.7	20.0	17.6	4.9	7.4	6.8	7.3	15.4	15.4
8.7	6.3	12.0	13.0	13.3	16.8	7.3	9.4	10.1	11.0	20.0	15.9
8.7	6.5	12.1	13.0	14.3	16.6	3.8	8.4	5.3	5.7	17.4	16.3
6.0	6.8	8.4	9.0	14.7	15.1	5.3	4.9	7.3	7.9	10.2	14.9
5.8	5.9	8.0	8.6	12.7	16.1	7.5	9.0	10.6	11.3	18.7	17.5
7.0	5.3	9.9	10.5	11.0	17.0	6.2	8.7	8.7	9.2	18.4	16.4
7.9	5.8	10.9	11.8	12.7	17.3	8.0	7.6	11.3	12.0	16.1	16.1
7.9	8.3	10.9	11.7	17.5	16.3	9.9	5.5	13.7	14.0	12.3	17.9
8.9	5.2	12.4	13.4	10.9	15.7	6.7	6.6	9.4	10.1	13.6	16.0

EXHIBIT 10.3: Supervisor Rating Data

10.4 Examples

Example 10.2

The data given in Exhibit 10.3 are part of a larger data set which was gathered for purposes other than that illustrated in this example. The six variables are each composites obtained from responses to several questions on a questionnaire. The dependent variable y is a composite of responses to questions on attitudes towards the respondent's supervisor and on job satisfaction. The highest possible score of 20 would be obtained if a respondent showed a strong liking for the supervisor and also was most satisfied with his or her job. The predictor x_1, which measures the level of social contact each respondent felt he or she had with the supervisor, was based on questions such as 'Do you see your supervisor outside of your work place?'. The predictor x_2 measures the perceived level of the supervisor's personal interest in the employee's personal life, based on questions like 'Would you discuss a personal problem with your supervisor?' while x_3 measures the level of support the employee feels from the supervisor, based on questions like 'Is your supervisor supportive of your work?' Variables x_4 and x_5 measure the drive of the supervisor. The former is based directly on the employee's perception of this drive and the latter on such questions as 'Does your supervisor encourage you to learn new skills?'

Since the maximum value of y was 20 and most scores were close to it, we used as dependent variable $-\log[1 - y/20]$, which is part of the logit transformation (Section 9.2.4, p. 186) and seems to work quite well. There was also no apparent heteroscedasticity and no weighting was done. However, it should be noted that none of the collinearity diagnostics depend on the dependent variable.

Variable	b_j	s.e.	TOL
Intercept	0.830	0.318	
x_1	-0.012	0.647	0.002
x_2	0.199	0.483	0.001
x_3	-0.117	0.178	0.010
x_4	-0.367	0.294	0.008
x_5	0.186	0.147	0.009

EXHIBIT 10.4: Parameter Estimates, t-Values and Tolerances for Supervisor Rating Data

The tolerances shown in Exhibit 10.4 and the eigenvalues and condition numbers presented in Exhibit 10.5 all tell us that we have a multicollinear disaster on our hands. The last row of the table in Exhibit 10.5 shows one of the offending linear combinations — that involving x_1 and x_2. The second last row shows another such linear combination involving x_4 and x_5. The third row from the bottom is more interesting. The only large variance component in it is associated with x_3. But the .0425 in the var (b_1) column accounts for most of the variance of var (b_1) not accounted for by the smallest eigenvector and a similar situation exists in the var (b_2) column. Therefore, there is linear combination involving x_1, x_2 and x_3

Eigen-values	Condition Number	Proportions of var (b_0)	var (b_1)	var (b_2)	var (b_3)	var (b_4)	var (b_5)
5.867	1	.0005	.0000	.0000	.0000	.0000	.0000
.114	7.16	.0021	.0001	.0001	.0005	.0010	.0008
.0182	17.96	.7905	.0001	.0001	.0002	.0015	.0010
.00031	137.4	.1482	.0425	.0294	.9851	.0000	.0001
.00019	175.3	.0530	.0021	.0023	.0002	.9341	.9442
.000037	400.5	.0057	.9552	.9680	.0139	.0633	.0540

EXHIBIT 10.5: Eigenvalues, Condition Numbers and Variance Proportions for Supervisor Rating Data

which also contributes to a small eigenvalue. The first three rows do not have eigenvalues small enough to concern us too much. However, row 3 is interesting because the variance of the intercept has a large component in it. It seems that the intercept is somewhat related to several of the variables. We shall shortly see another and slightly more obvious example of this.

At this stage many analysts would confirm their findings from the examination of variance components by running regressions (e.g., of x_1 against x_2, x_3 against x_1 and x_2, etc.). However, we did not do this because the relationships were fairly obvious from an examination of the data.

The only purpose of the methods of this chapter is to identify multicollinearity and its causes. However, in this example, because of the somewhat arbitrary way in which the independent variables were created, there is a fairly simple way to reduce multicollinearity. Since x_1, x_2 and x_3 seem highly related, we decided to combine them. (Actually, we did this in two steps: first we combined x_1 and x_2 and then found that this combination was highly related to x_3. But to save space we have not shown this intermediate step.) Since the variables are highly related and also because their scales are somewhat alike, we simply added the three to get a new variable z_1. We might have handled this differently but this simple expedient seems adequate. Similarly, we added x_4 and x_5 to get a variable z_2. We can interpret z_1 as a measure of favorable 'people orientation' and z_2 as a measure of high 'task' or 'mission' orientation.

Variable	b_j	t	TOL
Intercept	0.816	2.904	
z_1	0.029	4.478	0.9986
z_2	0.003	0.348	0.9986

EXHIBIT 10.6: Parameter Estimates, t-Values and Tolerances for Supervisor Rating Data After Combining Variables

Eigen-values	Condition Number	Proportions of var (b_0)	var (b_1)	var (b_2)
2.939188	1	0.0025	0.0053	0.0047
0.046307	7.97	0.0008	0.5400	0.4226
0.014505	14.2	0.9967	0.4546	0.5727

EXHIBIT 10.7: Eigenvalues, Condition Numbers and Variance Proportions for Supervisor Rating Data After Combining Variables

The value of s hardly changed (.231 vs. .228) when we went from using the x_j's to using the z_j's as independent variables. The tolerances in Exhibit 10.6 show that multicollinearity is almost gone. None of the condition numbers are small enough to concern us much.

However, the difference between what the tolerances show (virtually no multicollinearity) and what the condition numbers show (mild multicollinearity) is perhaps worthy of attention. The R_j^2's (since they are essentially R^2's) are defined with $\bar{y}$ subtracted out. In that sense they are centered measures and relationships involving the intercept term play no role in them or in the tolerances. However, since the variance components and the eigenvalues are computed with a model that has not been centered, relationships involving the intercept do play a role. The somewhat small eigenvalue in the last row of Exhibit 10.7 is due to such a relationship. An examination of the values of z_1, z_2 would have revealed it. Both independent variables have values which cluster around moderately large positive numbers. The 'variable' associated with the intercept term takes only one value — 1. Therefore, it is most predictable that there would be a linear combination of small length that involves them.

The t-values given in Exhibit 10.6 show that we might wish to delete z_2. It would appear that employees are quite indifferent to the task orientation of their supervisors. ■

Example 10.3 (Continuation of Example 4.3, Page 88)
Let us return to the house price example. As a result of working out the related exercises, the reader has probably already come to the conclusion that outliers are present and that some weighting is necessary. We deleted the 20th case and also weighted the regression. For the latter procedure we ran the regression and set weights at $w_i = 1$ if $\hat{y}_i < 50$ and $w_i = .25$ if $\hat{y}_i \geq 50$, and then ran the regression again with these weights. (This was perhaps an overly conservative approach. Several other possibilities exist. Weighting with $\text{E}[y_i]$ is a possibility. A bolder move would have been to

use

$$w_i = \begin{cases} 1 & \text{if } \hat{y}_i < 50, \\ .25 & \text{if } 50 \le \hat{y}_i < 60, \\ .06 & \text{if } 60 \le \hat{y}_i, \end{cases} \qquad (10.11)$$

and also remove one or two more outliers.)

Variable	b	t-value	TOL
Intercept	19.253	4.87	
FLR	0.019	7.25	0.34
RMS	2.535	2.33	0.13
BDR	-5.848	-4.37	0.14
BTH	2.350	1.58	0.62
GAR	1.687	1.58	0.48
LOT	0.282	3.36	0.72
FP	5.952	2.27	0.69
ST	9.978	5.29	0.81
L1	0.745	0.38	0.37
L2	3.874	2.30	0.53

EXHIBIT 10.8: Parameter Estimates, t-Values and Tolerances for House Price Model After Weighting and Eliminating an Outlier

Eigen-value	Cond. Num.	Proportions of Variance of Coefficients of Int.	FLR	RMS	BDR	BTH	GAR	LOT	FP	ST	L1	L2
7.27	1	.0004	.0008	.0001	.0003	.0012	.0029	.0007	.0020	.0032	.0024	.0018
1.29	2.4	.0005	.0001	.0001	.0001	.0002	.0346	.0004	.0899	.0375	.0352	.1013
.836	3.0	.0000	.0001	.0000	.0000	.0002	.0058	.0002	.4437	.2406	.0166	.0236
.750	3.1	.0003	.0005	.0001	.0002	.0003	.0035	.0007	.0950	.5387	.0126	.1067
.430	4.1	.0003	.0012	.0000	.0014	.0022	.4214	.0001	.0935	.0069	.1534	.0001
.239	5.5	.0033	.0009	.0001	.0005	.0028	.1563	.0290	.0845	.1058	.2925	.4682
.067	1.4	.0218	.2110	.0053	.0376	.2564	.0440	.0377	.0004	.0169	.3267	.0674
.061	1.9	.0427	.0007	.0000	.0182	.6295	.0297	.1444	.0293	.0269	.1180	.0345
.034	14.5	.0015	.6889	.0163	.1587	.0753	.0922	.0154	.0001	.0008	.0052	.0045
.024	17.4	.4007	.0060	.0143	.0304	.0010	.0777	.7321	.1614	.0107	.0256	.1711
.005	39.8	.5287	.0898	.9636	.7525	.0308	.1319	.0392	.0003	.0122	.0118	.0208

EXHIBIT 10.9: Eigenvalues, Condition Numbers and Variance Proportions for House Price Model After Removing Outlier and Weighting

Exhibit 10.8 shows the parameter estimates, t-values and tolerances. Obviously, there is a fair amount of multicollinearity, particularly involving RMS and BDR. Exhibit 10.9 shows eigenvalues, condition numbers and variance proportions. Only one condition number is very large. Corresponding variance proportions show a pattern we have seen in the last example. The intercept, RMS and BDR dominate a linear combination of small length. The reason is easy to see from the raw data (Exhibit 2.2, page 32). RMS–BDR is very frequently 3. The situation would become a little better if RMS were replaced by

$$\text{COR} = \text{RMS} - \text{BDR} - \text{Mean of } (\text{RMS} - \text{BDR}).$$

The reader is invited to do this and find that tolerances and standard errors are quite improved. The largest condition number drops to about 17.2.

The other rows of Exhibit 10.9 do not show any particularly small linear combinations. However, the reader is invited to verify that there is a slight and fairly harmless relationship between the intercept and LOT, between FLR and BDR and between BTH and LOT. ■

Problems

Exercise 10.1: Let

$$\boldsymbol{x}_{[1]} = (10.1,\ 9.9,\ 5.1,\ 4.9)' \text{ and } \boldsymbol{x}_{[2]} = (9.9,\ 10.1,\ 4.9,\ 5.1)',$$

and let $\boldsymbol{w}_1 = \boldsymbol{x}_{[1]} + \boldsymbol{x}_{[2]}$ and $\boldsymbol{w}_2 = \boldsymbol{x}_{[1]} - \boldsymbol{x}_{[2]}$. Consider a design matrix $W = (\boldsymbol{w}_1\ \boldsymbol{w}_2)$. Assuming $\text{var}(\epsilon_i) = 1$ in the model $\boldsymbol{y} = W\boldsymbol{\beta} + \boldsymbol{\epsilon}$, obtain the standard errors of the estimates of β_1 and β_2.

Suppose we now substitute for $\boldsymbol{w}_1$ and $\boldsymbol{w}_2$ and consider the model $\boldsymbol{y} = X\boldsymbol{\alpha} + \boldsymbol{\epsilon}$ where $X = (\boldsymbol{x}_{[1]}\ \boldsymbol{x}_{[2]})$. Obtain the standard errors of the estimates of α_1 and α_2 from those of β_1 and β_2.

Exercise 10.2: Show that $r^{jj} = \text{VIF}_j$ where $r^{ij}$is the (ij)th element of $\mathcal{R}^{-1}$.
[**Hint:** Write $\mathcal{R} = (Z_{(s)})'Z_{(s)}$ as

$$\begin{pmatrix} c_{11} & \boldsymbol{c}_{12}' \\ \boldsymbol{c}_{12} & C_{22} \end{pmatrix}.$$

Now show that the top left hand element of $\mathcal{R}^{-1}$ is $(1 - \boldsymbol{c}_{12}'C_{22}^{-1}\boldsymbol{c}_{12})^{-1}$ and then use the material from Section B.7, p. 295.]

Exercise 10.3: Using the table in Exhibit 10.10, can you identify the principal causes of multicollinearity?

Eigen-values	Condition Number	Proportions of var (b_0)	var (b_1)	var (b_2)	var (b_3)	var (b_4)	var (b_5)
5.7698	1	.0009	.0033	.0000	.0001	.0001	.0000
0.1313	6.629	.0007	.7599	.0008	.0004	.0003	.0001
0.0663	9.332	.2526	.0484	.0038	.0027	.0057	.0001
0.0268	14.69	.3761	.0003	.0012	.0019	.0695	.0000
.00573	31.73	.0402	.1705	.0634	.2585	.0004	.0001
.00017	186.0	.3295	.0176	.9307	.7364	.9241	.9997

EXHIBIT 10.10: Eigenvalues, Condition Numbers and Variance Proportions for Exercise

Exercise 10.4: In Example 10.3 we suggested using a variable COR in place of RMS. Show that this in fact does improve matters. Also identify the linear combinations associated with the smaller eigenvalues in this changed model.

Exercise 10.5: For the regression model using all independent variables in Exercise 3.14, p. 79, investigate the multicollinearity structure. Be sure to examine the data for outliers and heteroscedasticity before you begin.

Exercise 10.6: The data set presented in Exhibit 10.11, given in Longley (1967), has been used by him and subsequently by others to check the numerical accuracy of regression programs. Consider a model where total employment (Total) is the dependent variable, and the remaining variables are the independent variables. These latter variables are a price index (Def), gross national product (GNP), unemployment rate (Unemp), employment in the armed forces (AF), noninstitutional population (Population) and the year. Identify the major causes of multicollinearity in these data.

The estimate of the coefficient of the variable 'Def' should be 15.062. Does the computer package you are using give this value? You might also wish to test other computer packages with this data set.

Def	GNP	Unemp	AF	Population	Year	Total
83.0	234289	2356	1590	107608	1947	60323
88.5	259426	2325	1456	108632	1948	61122
88.2	258054	3682	1616	109773	1949	60171
89.5	284599	3351	1650	110929	1950	61187
96.2	328975	2099	3099	112075	1951	63221
98.1	346999	1932	3594	113270	1952	63639
99.0	365385	1870	3547	115094	1953	64989
100.0	363112	3578	3350	116219	1954	63761
101.2	397469	2904	3048	117388	1955	66019
104.6	419180	2822	2857	118734	1956	67857
108.4	442769	2936	2798	120445	1957	68169
110.8	444546	4681	2637	121950	1958	66513
112.6	482704	3813	2552	123366	1959	68655
114.2	502601	3931	2514	125368	1960	69564
115.7	518173	4806	2572	127852	1961	69331
116.9	554894	4007	2827	130081	1962	70551

EXHIBIT 10.11: Longley's Data

SOURCE: Reproduced from the *Journal of the American Statistical Association*, **62**, with the permission of the ASA.

Exercise 10.7: Investigate the multicollinearities in the unconstrained translog model of Exercise 9.7

Exercise 10.8: Investigate multicollinearity in each of the models in Exercise 2.20, p. 55.

CHAPTER 11

Variable Selection

11.1 Introduction

Frequently we start out with a fairly long list of independent variables that we suspect have some effect on the dependent variable, but for various reasons we would like to cull the list. One important reason is the resultant parsimony: It is easier to work with simpler models. Another is that reducing the number of variables often reduces multicollinearity. Still another reason is that it lowers the ratio of the number of variables to the number of observations, which is beneficial in many ways.

Obviously, when we delete variables from the model we would like to select those which by themselves or because of the presence of other independent variables have little effect on the dependent variable. In the absence of multicollinearity, as, say, with data from a well-designed experiment, variable search is very simple. One only needs to examine the b_j's and their standard errors and take a decision. Multicollinearity makes such decisions more difficult, and is the cause of any complexity in the methods given in this chapter. Notice that in the last chapter, when we examined the possibility of relationships among the independent variables, we ignored the effects any of them might have on the dependent variable.

Ideally, given the other variables in the model, those selected for removal have *no* effect on the dependent variable. This ideal situation is not likely to occur very often, and when it does not we could bias the regression (as we shall see in Section 11.2). Moreover, as we shall also show in Section 11.2, on the average, s^2 will tend to increase with a reduction in the variable list. Thus the practice of variable search is often a matter of making the best compromise between keeping s^2 and bias low and achieving parsimony and reducing multicollinearity.

During variable selection, one frequently finds, clustered around the chosen model, other models which are nearly 'as good' and not 'statistically distinguishable'. As with many other decisions in the practice of regression, the decisions involved in variable selection are seldom obvious. More often there is no unique choice and the one that is made reflects the analyst's best judgment at the time.

There is yet another problem with variable search procedures. Suppose we apply such a procedure to twenty independent variables constructed entirely of random numbers. Some of these variables, by sheer chance, may appear to be related to the dependent variable. Since we are picking the 'best' subset from our list, they appear in our short list and we end up with a nonsense relationship. An illustration can be found in Wilkinson (1979), and his findings are startling.

Despite these problems, variable search procedures can be invaluable if carried out judiciously. However, some further caveats are in order. We should be careful not to drop an important variable. For example, if the purpose of the study is to determine the relationship between the price of something and its sales, it would be silly to drop price just because our search procedure recommends it unless we are sure that price has virtually no effect on sales. In fact, the lack of effect might be the key finding here or may only be due to a poor design matrix where the corresponding column had a very short length or was highly related to some other column. Researchers whose primary interest is forecasting should also make sure that they are not (without very strong reasons) dropping an easy-to-predict independent variable in favor of a more troublesome one. Finally, if theoretical considerations or intuitive understanding of the underlying structure of the relationship suggest otherwise, the results of the mechanical procedures given below in Section 11.3 should play second fiddle. Ultimately, it is the researcher who should choose the variables — not the 'computer'!

11.2 Some Effects of Dropping Variables

Assume that

$$y_i = \beta_0 + \beta_1 x_{i1} + \cdots + \beta_k x_{ik} + \epsilon_i \tag{11.1}$$

is the correct model and consider

$$y_i = \beta_0 + \beta_1 x_{i1} + \cdots + \beta_{p-1} x_{i(p-1)} + \epsilon_i^{(p)} \tag{11.2}$$

which includes only the first $p-1 < k$ independent variables from (11.1). In this section we discuss the effects of considering the incorrect abbreviated model (11.1). Since we have 'starred' several of the subsections of this section, we describe below some of the principal results obtained in the subsections.

As mentioned earlier, deleting some independent variables usually biases the estimates of the parameters left in the model. However, no bias occurs if the values of the deleted variables are orthogonal to those of the remaining variables, and then the estimates of $\beta_0, \ldots, \beta_{p-1}$ are exactly the same whether $x_{ip}, \ldots, x_{ik}$ are included or not. Deletion of variables usually increases the value of the expectation of s^2 and decreases (in the sense that the difference is non-negative definite) the covariance matrix of the estimates of $\beta_0, \ldots, \beta_{p-1}$. Note that we are referring to the covariance matrix, which is defined in terms of σ^2, not an estimate of it, which would frequently behave differently. Because estimates of $\beta_0, \ldots, \beta_{p-1}$ are biased, it is not surprising that predicted values usually become biased. One measure of this bias is called Mallows' C_p. While a definition of C_p is postponed to

Section 11.2.4, the key property for applications is that if (11.2) does not lead to much bias in the predicteds, then

$$\mathrm{E}(C_p) \approx p.$$

Therefore, if one is considering several candidate models, one can look at the corresponding s_p^2, R_p^2 and C_p values (the first two are the familiar s^2, R^2 measures applied to possibly truncated p variable model; all are provided by most packages) and judge which models are relatively bias-free. These become the ones from which a selection can be made.

Since coefficients become biased, the deletion of variables can cause residuals of the abbreviated model to have non-zero expectations (Section 11.2.4). This leads to plots of residuals against predicteds which sometimes show a pattern as if the residuals were related to the predicted values. When one sees such a plot, it is reasonable to suspect that a variable has been left out that should have been included. However, not much value should be placed on an apparent absence of pattern since even in quite biased models, such patterned plots do not occur with any regularity.

11.2.1 Effects on Estimates of β_j

Assume that the correct model is

$$\boldsymbol{y} = X\boldsymbol{\beta} + \boldsymbol{\epsilon}, \tag{11.3}$$

where $\mathrm{E}(\boldsymbol{\epsilon}) = 0$ and $\mathrm{cov}(\boldsymbol{\epsilon}) = \sigma^2 I$. Let $X = (X_1, X_2)$, and $\boldsymbol{\beta}' = (\boldsymbol{\beta}_1' \ \boldsymbol{\beta}_2')$ where X_1 is $n \times p$ dimensional, $\boldsymbol{\beta}_1$ is a p-vector and the other dimensions are chosen appropriately. Then

$$\boldsymbol{y} = X_1\boldsymbol{\beta}_1 + X_2\boldsymbol{\beta}_2 + \boldsymbol{\epsilon}. \tag{11.4}$$

While this is the correct model, suppose we leave out $X_2\boldsymbol{\beta}_2$ and obtain the estimate $\boldsymbol{b}_1$ of $\boldsymbol{\beta}_1$ by least squares. Then $\boldsymbol{b}_1 = (X_1'X_1)^{-1}X_1'\boldsymbol{y}$, which not only is usually different from the first p components of the estimate $\boldsymbol{b}$ of $\boldsymbol{\beta}$ obtained by applying least squares to the full model (11.3), but also is usually biased, since

$$\begin{aligned}\mathrm{E}(\boldsymbol{b}_1) &= (X_1'X_1)^{-1}X_1'\,\mathrm{E}(\boldsymbol{y}) \\ &= (X_1'X_1)^{-1}X_1'(X_1\boldsymbol{\beta}_1 + X_2\boldsymbol{\beta}_2) = \boldsymbol{\beta}_1 + (X_1'X_1)^{-1}X_1'X_2\boldsymbol{\beta}_2.\end{aligned}$$

Thus, our estimate of $\boldsymbol{\beta}_1$ obtained by least squares after deleting $X_2\boldsymbol{\beta}_2$ is biased by the amount $(X_1'X_1)^{-1}X_1'X_2\boldsymbol{\beta}_2$.

Notice that the bias depends both on $\boldsymbol{\beta}_2$ and X_2. For example, if X_2 is orthogonal to X_1, that is, if $X_1'X_2 = 0$, then there is no bias. In fact if $X_1'X_2 = 0$,

$$(X'X)^{-1} = \begin{pmatrix} X_1'X_1 & 0 \\ 0 & X_2'X_2 \end{pmatrix}^{-1} = \begin{pmatrix} (X_1'X_1)^{-1} & 0 \\ 0 & (X_2'X_2)^{-1} \end{pmatrix}$$

(see Example A.8, p. 276) and since $X'\boldsymbol{y} = (X_1'\boldsymbol{y}\ X_2'\boldsymbol{y})$, it follows that $\boldsymbol{b}_1$ is actually the same as the first p components of $\boldsymbol{b}$.

11.2.2 *Effect on Estimation of Error Variance

The estimate of σ^2 based on the full model is given by

$$s^2 = \text{RSS}_{k+1}/(n-k-1) \equiv (n-k-1)^{-1}\boldsymbol{y}'[I-H]\boldsymbol{y}.$$

When we delete $X_2\boldsymbol{\beta}_2$, an estimate of σ^2 based on p independent variables will be given by

$$s_p^2 = \text{RSS}_p/(n-p) = (n-p)^{-1}(\boldsymbol{y}-\hat{\boldsymbol{y}}_p)'(\boldsymbol{y}-\hat{\boldsymbol{y}}_p) = (n-p)^{-1}\boldsymbol{y}'[I-H_1]\boldsymbol{y}$$

where $H_1 = X_1(X_1'X_1)X_1'$ and $\hat{\boldsymbol{y}}_p = X_1\boldsymbol{b}_1 = H_1\boldsymbol{y}$ is the predicted value of $\boldsymbol{y}$ based on the first p independent variables. While $\text{E}(s^2) = \sigma^2$, we need to calculate $\text{E}(s_p^2)$. Since $\text{tr}(I-H_1) = n-p$ and from (11.3) and Theorem B.3, p. 288, we can easily show that $\text{E}(\boldsymbol{y}\boldsymbol{y}') = \sigma^2 I + X\beta\beta'X'$, we get, using various properties of the trace of a matrix (Section A.6, p. 271),

$$\begin{aligned}(n-p)\,\text{E}(s_p^2) &= \text{E}[\boldsymbol{y}'(I-H_1)\boldsymbol{y}] = \text{E}[\,\text{tr}((I-H_1)\boldsymbol{y}\boldsymbol{y}')] \\ &= \text{tr}[(I-H_1)\,\text{E}(\boldsymbol{y}\boldsymbol{y}')] = \text{tr}[(I-H_1)(\sigma^2 I + X\boldsymbol{\beta}\boldsymbol{\beta}'X')] \\ &= (n-p)\sigma^2 + \text{tr}[(I-H_1)X\boldsymbol{\beta}\boldsymbol{\beta}'X'] = (n-p)\sigma^2 + \boldsymbol{\beta}'X'(I-H_1)X\boldsymbol{\beta}\end{aligned}$$

Hence,

$$\text{E}[(n-p)^{-1}\text{RSS}_p] = \text{E}(s_p^2) = \sigma^2 + (n-p)^{-1}\boldsymbol{\beta}'X'(I-H_1)X\boldsymbol{\beta} \quad (11.5)$$

and, since $\text{E}(s^2) = \sigma^2$, it follows that

$$\text{E}(s_p^2 - s^2) = (n-p)^{-1}\boldsymbol{\beta}'X'(I-H_1)X\boldsymbol{\beta} \geq 0.$$

Therefore, s_p^2 is usually a biased estimator of σ^2 and $\text{E}(s_p^2)$ increases when variables are deleted. On the other hand, as shown in the next subsection, the covariance of $\boldsymbol{b}_1$ is less than or equal to the covariance of the estimate of $\boldsymbol{\beta}_1$ based on the full model. Therefore, practical choices involve determination of the trade-offs between improvements in one aspect and deterioration in another, and the reconciliation of these trade-offs with the aims of the analysis.

11.2.3 *Effect on Covariance Matrix of Estimates

Since $\boldsymbol{b}_1 = (X_1'X_1)^{-1}X_1'\boldsymbol{y}$ and $\text{cov}(\boldsymbol{y}) = \sigma^2 I$, we get

$$\text{cov}(\boldsymbol{b}_1) = \sigma^2(X_1'X_1)^{-1}.$$

However, based on the full model

$$\operatorname{cov}(\boldsymbol{b}) = \sigma^2(X'X)^{-1} = \sigma^2 \begin{pmatrix} X_1'X_1 & X_1'X_2 \\ X_2'X_1 & X_2'X_2 \end{pmatrix}^{-1}.$$

Therefore, the covariance of the vector $\boldsymbol{b}^{(1)}$ containing the first p components of $\boldsymbol{b}$ is, using Example A.8, p. 276,

$$\operatorname{cov}(\boldsymbol{b}^{(1)}) = \sigma^2[X_1'X_1 - X_1'X_2(X_2'X_2)^{-1}X_2'X_1]^{-1}. \tag{11.6}$$

Similarly, the vector $\boldsymbol{b}^{(2)}$ of the last $k+1-p$ components of $\boldsymbol{b}$ has covariance

$$\sigma^2[X_2'X_2 - X_2'X_1(X_1'X_1)^{-1}X_1'X_2]^{-1}. \tag{11.7}$$

Since $X_1'X_2(X_2'X_2)^{-1}X_2'X_1$ is positive semi-definite,

$$X_1'X_1 \geq X_1'X_1 - X_1'X_2(X_2'X_2)^{-1}X_2'X_1,$$

where the inequality signs are as defined in Section A.13, p. 279. It follows that

$$(X_1'X_1)^{-1} \leq [X_1'X_1 - X_1'X_2(X_2'X_2)^{-1}X_2'X_1]^{-1}.$$

Hence, $\operatorname{cov}(\boldsymbol{b}_1) \leq \operatorname{cov}(\boldsymbol{b}^{(1)})$; i.e., the covariance matrix of the estimate of first p components of $\boldsymbol{\beta}$ decreases when the remaining components, along with the corresponding columns of X, are deleted.

11.2.4 *Effect on Predicted Values: Mallows' C_p

Since $\mathrm{E}[\hat{\boldsymbol{y}}_p] = H_1 \mathrm{E}[\boldsymbol{y}] = H_1 X\beta$ it follows that $\hat{\boldsymbol{y}}_p$ is biased, unless $X\beta - H_1X\beta = [I - H_1]X\beta = \mathrm{E}[\boldsymbol{e}_p] = 0$, where $\boldsymbol{e}_p$ is the vector of the residuals from the abbreviated model. Define Bias $(\hat{\boldsymbol{y}}_p)$ as $\mathrm{E}[\hat{\boldsymbol{y}}_p - \hat{\boldsymbol{y}}]$. Then

$$\text{Bias}\,(\hat{\boldsymbol{y}}_p) = \mathrm{E}(\hat{\boldsymbol{y}}_p) - \mathrm{E}(\hat{\boldsymbol{y}}) = H_1X\boldsymbol{\beta} - X\boldsymbol{\beta} = -(I - H_1)X\boldsymbol{\beta}$$

and

$$\sum_{i=1}^{n}[\text{Bias}\,(\hat{y}_{pi})]^2 = [\mathrm{E}(\hat{\boldsymbol{y}}_p) - \mathrm{E}(\hat{\boldsymbol{y}})]'[\mathrm{E}(\hat{\boldsymbol{y}}_p) - \mathrm{E}(\hat{\boldsymbol{y}})] = \boldsymbol{\beta}'X'[I - H_1]X\boldsymbol{\beta}$$

where Bias $(\hat{y}_{pi})$ is the ith component of Bias $(\hat{\boldsymbol{y}}_p)$. To make this last expression scale-free, we standardize it by σ^2. Thus a standardized sum of squares of this bias is given by

$$\boldsymbol{\beta}'X'[I - H_1]X\boldsymbol{\beta}/\sigma^2,$$

which, on using (11.5), becomes

$$\mathrm{E}[\mathrm{RSS}_p]/\sigma^2 - (n - p).$$

Hence, an estimate of this standardized bias is

$$\mathrm{RSS}_p/s^2 - (n-p). \tag{11.8}$$

If the bias in $\hat{\boldsymbol{y}}_p$ introduced by dropping variables is negligible, this quantity should be close to zero.

An alternative way to examine bias in $\hat{\boldsymbol{y}}_p$ is to look at the mean square error matrix $\mathrm{MSE}(\hat{\boldsymbol{y}}_p)$, or, more conveniently, at its trace $\mathrm{TMSE}(\hat{\boldsymbol{y}}_p)$ (see equation (B.6), p. 288). Since, $\mathrm{cov}(\hat{\boldsymbol{y}}_p) = H_1 \,\mathrm{cov}(\boldsymbol{y}) H_1 = \sigma^2 H_1$,

$$\begin{aligned}\mathrm{MSE}(\hat{\boldsymbol{y}}_p) &= \mathrm{E}[(\hat{\boldsymbol{y}}_p - X\boldsymbol{\beta})(\hat{\boldsymbol{y}}_p - X\boldsymbol{\beta})'] \\ &= \mathrm{cov}(\hat{\boldsymbol{y}}_p) + \mathrm{Bias}\,(\hat{\boldsymbol{y}}_p)\mathrm{Bias}\,(\hat{\boldsymbol{y}}_p)' = \sigma^2 H_1 + (I - H_1)X\boldsymbol{\beta}\boldsymbol{\beta}'X'(I - H_1).\end{aligned}$$

Therefore, the trace $\mathrm{TMSE}(\hat{\boldsymbol{y}}_p)$ of the MSE matrix, which is the sum of the mean square errors of each component of $\hat{\boldsymbol{y}}_p$, is

$$\mathrm{TMSE}(\hat{\boldsymbol{y}}_p) = \sigma^2 p + \boldsymbol{\beta}'X'(I - H_1)X\boldsymbol{\beta},$$

since $\mathrm{tr}(H_1) = p$.

Let

$$J_p = \mathrm{TMSE}(\hat{\boldsymbol{y}}_p)/\sigma^2 = p + \boldsymbol{\beta}'X'(I - H_1)X\boldsymbol{\beta}/\sigma^2,$$

which, using (11.5), can be estimated by

$$C_p = \frac{\mathrm{RSS}_p}{s^2} - (n-p) + p = \frac{\mathrm{RSS}_p}{s^2} - (n - 2p).$$

This is known as Mallows' C_p statistic (see Mallows, 1973). From the discussion of (11.8) it follows that, if bias is close to zero, C_p should usually be close to p.

11.3 Variable Selection Procedures

The purpose of variable selection procedures is to select or help select from the total number k of candidate variables a smaller subset of, say, $p-1$ variables. There are two types of such procedures: those that *help* choose a subset by presenting several if not all possible combinations of variables with corresponding values of C_p, s_p^2, R_p^2 and possibly other statistics, and those that pretty much *do* the selecting by presenting to the analyst very few (frequently one) subsets of variables for each value of $p-1$. As we have already mentioned, in many situations there is rarely one obviously best equation and the near winners are almost as good. Therefore, we prefer the former approach, which we have called the *search over all possible subsets.* It has also been called the *best subset regression*. However, such methods have voracious appetites for computer time, so that when computer time is at a premium, particularly if there are a large number of variables to select from, other methods might be necessary. Of these, the most popular is the *stepwise procedure* discussed in Section 11.3.2. Draper and Smith (1981) give a fuller discussion of variable search than we do but their ultimate recommendation is somewhat different.

11.3.1 Search Over All Possible Subsets

As the name implies, the search over all possible subsets of independent variables allows us to examine all regression equations constructed out of a given list of variables, along with some measure of fit for each. In our opinion, this procedure is the most useful, particularly if the number of variables is not too large. At the present time, a search over 20 variables is easily feasible on a mainframe computer, although new developments in compiler technology and in supercomputers will soon make it possible to computationally handle much larger numbers of variables. Even now, for large numbers of variables, one may force inclusion of a predetermined list of the variables in all models and search over only the remaining variables. (It is perhaps worth mentioning that computer packages do not actually fit each possible model separately; they use a 'short-cut' method, frequently one based on a procedure given in Furnival and Wilson, 1974, or see Seber, 1977, Chapter 11.)

A difficulty with this procedure stems from the prospect of having to examine huge computer outputs. For example, if even one line is devoted to each combination of variables, 20 variables would necessitate over a million lines. Therefore, several of the packages at least allow the user to use some criterion to eliminate combinations of variables that can be ruled out *a priori*. For example, SAS PROC RSQUARE allows one to choose to be printed for each p only the 'best' (based on R^2) m models and to put bounds on the number of variables p. In BMDP[1] (see Dixon, 1985) the user can choose among R^2, R_a^2 and C_p as the determinant of 'best' and ask that only the 'best' m models of any specified size $p-1$ along with the 'best' model of each size be printed. The Linear Least Squares Curve Fitting Program, which is a companion to Daniel and Wood (1980), uses C_p as the only means for culling and shows a plot of C_p's against p.

The PRESS (acronym for PREdiction Sum of Squares) statistic, first presented by Allen (1971), is another statistic that might be used to compare different models. It is defined as $\sum_{i=1}^n e_{i,-1}^2$ where $e_{i,-1}$ is as in equation (8.12) on p. 161. For each combination of variables, this provides a composite measure of how well it would predict each of the observations had the observation been left out when parameters were estimated. Several other measures also exist — see, for example Amemiya (1980), Judge *et al.* (1985), and Hocking (1976). However, nearly all of them eventually reduce

[1]BMDP Statistical Software Package is a registered trademark of BMDP Statistical Software Inc., Los Angeles, CA

to relatively simple functions of n, p, s_p^2, s^2 and R_p^2 (see SAS, 1985b, p. 715–16) — as, indeed, does C_p.

As will be apparent from Section 11.4, we somewhat favor C_p as a criterion for an initial selection. However, two points should be noted about it. First, when using it, we need to assume that the model with the entire list of independent variables included is unbiased. Second, C_p measures the bias in predicteds $\hat{y}_p$ from the abbreviated model, and these predicteds may not reveal the extent of the bias in estimates of certain future observations (see Example 10.1, p. 220). Whatever criterion statistic is used, in practice one sets bounds in such a way that the subset of models presented includes all those one would seriously consider.

Boyce *et al.* (1974) describe a very flexible program which can be used for the search over all possible subsets, although the primary purpose of the program is to search through all possible combinations of a given number of variables and identify the one with the highest R^2. The program can be obtained by writing to the authors of that monograph.

11.3.2 Stepwise Procedures

Of the stepwise procedures, the only one commonly used in actual applications is the stepwise procedure. Lest this sound silly, we point out that among stepwise procedures, there is one called *the* stepwise procedure. We also discuss the backward elimination procedure and the forward selection procedure, but primarily as an aid to the discussion of the stepwise procedure. Some stepwise procedures are not discussed here. Among the more interesting ones are the MAXR and the MINR procedures given in SAS PROC STEPWISE (see also Myers, 1986).

The Backward Elimination Procedures

Backward elimination procedures start with all variables in the model and eliminate the less important ones one by one. A partially manual version consists of removing one or two variables with low t-values, rerunning, removing some more variables, etc. Such a manual method does not work too badly when the researcher has a good understanding of the underlying relationship. However, it is tedious, and an automated version is available. Mechanically it works the same way but, as with most automated procedures, we pay for the convenience of automation by having to use preset selection criteria. The procedure computes the partial F's corresponding to each variable, given the list of variables included in the model at that step. The partial F statistic (sometimes called 'F to remove') is the square of the t statistic corresponding to each variable. Hence the probabilities obtained and the decisions taken are identical to using the t. If the lowest F value falls below a preset number (the $100 \times \alpha$ per cent point for the F distribution with the appropriate degrees of freedom, where α is either set by the analyst or by the computer package) the corresponding variable is deleted.

After each variable is deleted, partial F's are recomputed and the entire step is repeated with the variables still remaining in the model. The procedure stops when no partial F falls below the appropriate preset number.

While most users pay little attention to these preset numbers and use only the default values supplied by the computer package (for SAS α is .1), it is perhaps appropriate to match the number to the purpose of the analysis. At each step, the minimum value of partial F over all variables still in the model is computed. Hence, if the test is performed at the $100 \times \alpha$ per cent level, the actual probability of including one variable when in fact it has no effect on the dependent variable is much higher than α. Therefore, if one wishes to be particularly careful about not including inappropriate variables, one might wish to set very low α's. On the other hand, if one wishes the model to lean towards inclusivity rather than exclusivity, as one would if prediction was the main purpose for the model, a higher value of α is desirable (see also Forsythe, 1979, p. 855).

Apart from the problem of providing an inadequate list of models for the analyst to choose from, there is one further problem with backward elimination. Suppose we have three independent variables x_1, x_2, x_3, where x_1 is highly correlated with x_2 and x_3 and also with y and we would like to have x_1 in the final model — at least for parsimony. But being highly correlated with both x_2 and x_3, x_1 would have a large standard error and consequently a low t-value and a low partial F-value. As a result, it may get deleted early and we would never see it again.

The Forward Selection Procedure

The forward selection procedure works in the opposite way to the backward elimination procedures. It starts with no variable in the model and first selects that x_j which has the highest correlation with y. Subsequent selections are based on partial correlations, given the variables already selected. The *partial correlation* of y and x_j *given* $x_{j_1}, \ldots, x_{j_s}$, written as $r_{yx_j,x_{j_1}\ldots x_{j_s}}$, is the correlation between

1. the residuals obtained after regressing y on $x_{j_1}, \ldots, x_{j_s}$, and

2. the residuals obtained after regressing x_j on $x_{j_1}, \ldots, x_{j_s}$.

Clearly, the partial correlation measures the relationship between y and x_j after the linear effects of the other variables have been removed.

At every step, the partial F-value is computed for the variable just selected, given that variables previously selected are already in the model (such a partial F is called a 'sequential F' or sometimes 'F to enter'). If this sequential F-value falls below a preset number (e.g., the α-point of the appropriate F distribution — the default value of α in SAS is .5) the variable is deleted and another one is sought. If no suitable variable is found or if all the variables are in the model, the procedure stops. SAS uses a variation in which, instead of the partial correlations, the partial F's are computed for each variable not in the model. If the highest of the F's computed is high enough, the variable is included; otherwise the procedure stops.

A problem with forward selection is the reverse of the one for the backward selection. Suppose that of the highly correlated variables x_1,x_2 and x_3, we want x_2 and x_3 in the model because together they provide a better fit. But x_1 may enter the model first and prevent the others from getting in. Because of such problems, these procedures are now of primarily pedagogical or historical interest, having been replaced in actual use by the stepwise procedure and the all possible subsets search.

The Stepwise Procedure

The stepwise procedure is actually a combination of the two procedures just described. Like the forward selection procedure, it starts with no independent variable and selects variables one by one to enter the model in much the same way. But after each new variable is entered, the stepwise procedure examines every variable already in the model to check if it should be deleted, just as in a backward elimination step. Typically, the significance levels of F for both entry and removal are set differently than for the forward selection and backward elimination methods. It would be counter-productive to have a less stringent criterion for entry and a more stringent criterion for removal, since then we would constantly be picking up variables and then dropping them. SAS uses a default value of .15 for both entry and exit. As for the forward selection procedure, SPSS-X[2] (SPSS, 1986) permits, as an additional criterion, a tolerance level (e.g., $\text{TOL}_j \geq .01$) to be specified which needs to be satisfied for a variable to be considered (see Section 10.3.1, p. 222 for a definition of tolerance).

It is generally accepted that the stepwise procedure is vastly superior to the other stepwise procedures. But if the independent variables are highly correlated, the problems associated with the other stepwise procedures can remain (see Example 11.1 below; also see Boyce *et al.*, 1974). Like the forward selection and backward elimination procedures, usually only one equation is presented at each step. This makes it difficult for the analyst to use his or her intuition, even though most stepwise procedures allow the user to specify a list of variables to be always included.

---

[2]SPSS-X is a trademark of SPSS, Inc., Chicago, IL.

11.3.3 Stagewise and Modified Stagewise Procedures

In stagewise regression the decision to append an additional independent variable is made on the basis of plots of the residuals (from a regression of the dependent variable against all variables already included) against variables which are candidates for inclusion. In the modified stagewise procedure the plot considered is that of

1. the residuals obtained after regressing y on $x_{j_1}, \ldots, x_{j_s}$ against
2. the residuals obtained after regressing x_j on $x_{j_1}, \ldots, x_{j_s}$,

where $x_{j_1}, \ldots, x_{j_s}$ are the variables already in the model. Plots of this latter kind are called *added variable plots*, *partial regression plots*, *partial regression leverage plots* or simply *partial plots*.

In the case of the stagewise procedure, the slope obtained from applying least squares to the residuals is not a least squares estimate in the sense that, if the candidate variable is included in the model and least squares is applied to the resultant multiple regression model, we would get a different estimate for its coefficient. In the case of the modified stagewise procedure (without intercept) the estimates are LS estimates (see Exercise 11.1, also Mosteller and Tukey, 1977, p. 374 *et seq.).*

Modified stagewise least squares might appear to resemble a stepwise technique. But actually they are very different largely because of the way they are practiced. Stagewise and modified stagewise methods are essentially manual techniques — perhaps computer aided but nonetheless manual in essence. At every stage, transformations may be made and outliers dealt with and perhaps even weighting performed. Several examples of what we have called a modified stagewise approach can be found in Mosteller and Tukey (1977, see chapter 12 *et seq.*).

It might be mentioned in passing that some analysts find *partial plots* valuable for the identification of outliers and influential points (see Chatterjee and Hadi, 1986, Cook and Weisberg, 1982, Belsley, Kuh and Welsch, 1980).

11.4 Examples

Example 11.1

The data shown in Exhibit 11.1 are essentially made up by the authors. The independent variable x_1 has values which are the same as an independent variable in a data set in the authors' possession, except that they have been divided by 10 to make them more compatible with the size of x_2, which consists of pseudo-random numbers between 0 and 1. x_3 is $x_1 + x_2$ with an additional random disturbance added and x_4 is $x_1 + x_2$ with a slightly larger disturbance added. Actually, x_4 and x_3 are not too far apart, but as we shall see below, they lead to different behavior on the part of the stepwise procedure. The dependent variable is $.5x_1 + 1.5x_2$ plus a normal pseudo-random number.

x_1	x_2	x_3	x_4	y	x_1	x_2	x_3	x_4	y
0.76	0.05	0.87	0.87	0.14	0.17	0.97	1.39	1.42	1.49
0.47	0.40	1.16	1.20	1.06	0.98	0.82	1.96	1.98	1.92
0.46	0.45	1.03	1.05	1.14	0.23	0.22	0.71	0.74	0.13
0.55	0.26	0.89	0.90	0.50	0.74	0.33	1.09	1.09	1.04
0.55	0.86	1.69	1.72	1.45	0.61	0.73	1.63	1.67	1.09
0.38	0.52	1.19	1.23	1.02	0.62	0.85	1.56	1.57	1.35
0.39	0.31	0.86	0.88	0.78	0.51	0.97	1.56	1.57	1.60
0.46	0.14	0.76	0.78	0.53	0.81	0.16	1.18	1.20	0.47
0.10	0.41	0.52	0.53	0.54	0.03	0.76	1.08	1.12	0.79
0.95	0.26	1.23	1.23	0.69	0.77	0.38	1.22	1.23	1.14

EXHIBIT 11.1: Artificially Created Data for an Example on Variable Search

Step No.	Variable Entered	Variable Removed	$p-1$	Incr. R^2	R^2	C_p	F	p-value
1	x_3		1	0.7072	0.7072	11.7	43.47	0.0001
2	x_2		2	0.1011	0.8083	4.12	8.96	0.0082

EXHIBIT 11.2: Summary Output from Stepwise Procedure for Independent Variables x_1, x_2 and x_3

Exhibit 11.2 shows a summary of actions from a stepwise procedure. Much more detailed action logs are available, and the reader is encouraged to examine one to see exactly how the method proceeds. As is apparent, x_3 was the first variable chosen to enter the model. Then x_2 and x_1 were compared to see which would have the higher partial F if it were introduced into the model (we were using SAS PROC STEPWISE). Since x_2 had the higher value, it was inserted into the model. Then x_2 and x_3, which are now in the model, were examined to see if one of them should be dropped. Both met the partial F criteria for inclusion and were retained. The program then computed the partial F value for the sole variable x_1, still left out of the model, as if it were in the model. The criterion of significance at a .15 level was not met and the procedure stopped. The full output would give the usual least squares printouts associated with each of the models selected (in this case y against x_3 and of y against x_3 and x_2).

Applying the stepwise procedure to x_1, x_2 and x_4 yields more interesting results. The summary lines from an output are presented in Exhibit 11.3.

Step No.	Variable Entered	Variable Removed	$p-1$	Incr. R^2	R^2	C_p	F	p-value
1	x_4		1	0.7070	0.7070	11.66	43.44	.0001
2	x_2		2	0.0977	0.8047	4.43	8.51	.0096
3	x_1		3	0.0258	0.8305	4.00	2.43	.1383
4		x_4	2	0.0001	0.8304	2.01	0.014	.9087

EXHIBIT 11.3: Summary of Stepwise Procedure for Independent Variables x_1, x_2 and x_4

$p-1$	R^2	C_p	b_0	b_1	b_2	b_3
1	0.032	75.45	0.77	0.326		
1	0.705	11.85	0.27		1.361	
1	0.707	11.67	-0.36			1.103
2	0.759	8.79	-0.29	-0.463		1.255
2	0.808	4.12	-0.19		0.781	0.633
2	0.830	2.03	-0.13	0.655	1.487	
3	0.831	4.00	-0.12	0.737	1.589	-0.094

EXHIBIT 11.4: All Possible Subsets Search Over x_1, x_2 and x_3

Now x_4 entered the model first, then x_2 was entered and finally x_1. After all three variables were in the model, x_4 was no longer found to be significant enough to stay and was deleted.

Exhibit 11.4 shows the result of applying an all possible subsets search to x_1, x_2 and x_3. Estimates b_j of the coefficients of x_j for each combination of variables are listed, first in order of the number p of variables and then for each p they are ordered by R^2. The combinations (x_2, x_3) and (x_1, x_2) both appear to be fairly good. But the stepwise procedure chose (x_3, x_2) and gave us little information on (x_1, x_2), although it has indeed the higher R^2, and, as we know from having constructed the data, it is the right model. Although this is a fairly contrived data set, fear of situations like this make us partial to the all possible subsets search. Writing about various stepwise procedures (and perhaps overstating a little), Freund and Littel (1986) point out that:

> In general contradictory results from these procedures are the rule rather than the exception, especially when models contain many variables. Also none of the stepping procedures is clearly superior. ... Of course, PROC RSQUARE does guarantee optimum subsets, but as previously noted, may become prohibitively expensive.

Notice that even for this correct model C_p is not too near 3, as we would have expected. It serves only to remind us that C_p is a random variable. ■

Example 11.2
As another example consider Part 1 of Exercise 2.20, p. 55, on murder rate (M). The model has been included in a number of exercises and the reader perhaps already has ideas about what steps to take in order to get approximate compliance with Gauss-Markov conditions. Several possibilities exist. For example, one could ignore the slight heteroscedasticity and use OLS with Nevada, Utah and perhaps some other points deleted (for fairly obvious but different reasons). Alternatively, one could use weighted least squares.

The approach we chose to pursue is as follows. Although in this case there is no compelling reason to do so, one possible weighting scheme is to use as weights population divided by the expectation of murder rate, i.e., POP/ E[M], which is very close to what one gets using the weight recommended for proportion of counts in Exhibit 6.7, p. 123. When we used these weights, the predicted against residual plot showed little heteroscedasticity, but several other plots did. Since, in particular, variances seemed to be increasing with UR, we decided to also weight with its reciprocal. That is, we used POP/(E[M] × UR) for weights. This achieved approximate homoscedasticity. While we chose the weight by purely empirical experimentation, a partial explanation for its appropriateness follows from the conjecture that rural crime rates and urban crime rates have different variances and our chosen expression is a rough approximation to the expression we would get if we took this into account. Weighting was performed using nonlinear least squares as described in Section 6.4, p. 118.

When we used these weights, the influence of observations was largely due to some of them getting very large weights and there were no obvious influential points with large residuals. Because their populations were small, the OLS outliers got so little weight that there was little reason to bother with them. There also appeared to be no need for transformations.

It is usually a good idea to examine the multicollinearity structure before doing a variable search. Exhibit 11.5 shows eigenvalues and variance components. The three smallest eigenvalues seem to be due to multicollinearities between

1. The intercept, PL (1000 × proportion below poverty level) and HS (1000 × proportion high school graduates);
2. B (birthrate), HT (death rate from heart disease), HS and to a lesser extent CR (crime rate) and PL; and
3. UR (1000 × proportion of population living in urban areas) and CR.

(Before this confirms the worst fears of suburban residents, let us point out that most suburbs are also considered 'urban areas'!)

Exhibit 11.6 shows partial results of applying an all possible subsets search. A number of models are potential candidates for selection. If parsimony is important, we can easily select a five-variable model. Less than five

C. No.	Variance Proportions Corresponding to Int	MA	D	PL	S	B	HT	UR	CR	HS
1.0	.00	.00	.00	.00	.00	.00	.00	.00	.00	.00
8.4	.00	.00	.02	.06	.00	.00	.00	.02	.01	.00
11.1	.00	.00	.14	.01	.01	.00	.04	.00	.02	.00
15.5	.00	.02	.02	.15	.03	.01	.00	.05	.06	.00
19.4	.00	.18	.11	.00	.16	.04	.07	.01	.00	.00
22.1	.00	.03	.46	.03	.31	.01	.07	.03	.03	.00
28.3	.00	.63	.15	.01	.02	.13	.03	.06	.09	.00
37.7	.00	.11	.05	.00	.40	.01	.01	.74	.47	.02
65.3	.08	.00	.05	.19	.05	.77	.59	.03	.25	.19
119	.91	.01	.07	.54	.01	.03	.18	.06	.06	.78

EXHIBIT 11.5: Condition Numbers and Variance Proportions for Murder Rate Model

variables seems to raise the C_p somewhat. But even with five variables, we have some choice. The highest R^2 combination includes two pairs of multicollinear variables: (CR, UR) and (HS, PL). Therefore one might prefer to look upwards in the list for other, less collinear combinations. However, the variables just mentioned are rather persistent in their presence. Moreover, the most obvious replacement variable is B, which, by itself, has the lowest R^2. Assuming that the ultimate model will be used as a means of identifying methods to reduce murders, B is not very helpful (except to someone advocating incentives for abortion). Therefore, we stayed with the model shown in Exhibit 11.7. The slightly higher value of R^2 (than that in Exhibit 11.6) of .863 is due to the fact that the weights were recomputed from estimates of E[M] based on the five independent variables actually included. As policy determinants, the variables have high enough t-values that the presence of multicollinearity is not too unwelcome.

In interpreting the results, one needs to bear in mind that the data are for states. For example, the model does not imply that divorced people are more likely to be murdered or be murderers. All it says is that where there is a high divorce rate (D), murder rate is usually also relatively high — other things being equal, of course. ■

Example 11.3 (Continuation of Example 10.3, Page 228)
Exhibit 11.8 shows part of an output obtained by applying an all possible subsets search procedure to the house price data. The variables, weighting, etc., are as in Example 10.3. Readers who have completed Exercise 10.4 will know that there is not too much multicollinearity left in the model except, perhaps, for a possible relation involving BDR and FLR. As Exhibit 11.8 shows, there is not much hope of deleting either, since they feature in most

k	R^2	C_p	Variables in Model	k	R^2	C_p	Variables in Model
1	.005	254.1	B	5	.842	10.6	CR HS UR PL MA
1	.025	248.0	HT	5	.842	10.3	CR HS UR B MA
1	.031	246.3	S	5	.843	10.2	CR HS UR B HT
1	.045	242.1	MA	5	.847	8.9	CR HS UR PL S
1	.185	199.8	HS	5	.848	8.7	CR HS UR PL HT
1	.196	196.8	D	5	.852	7.5	CR HS UR B S
1	.235	184.9	UR	5	.854	6.8	CR HS UR B D
1	.239	183.6	PL	5	.855	6.6	CR HS UR B PL
1	.258	178.1	CR	5	.860	5.2	CR HS UR PL D
2	.321	161.2	PL D	6	.865	8.5	CR HS UR B PL MA
2	.343	154.5	UR MA	6	.865	8.4	CR HS UR B PL HT
2	.345	153.8	CR D	6	.866	8.2	CR HS UR B D MA
2	.350	152.3	D HS	6	.867	7.9	CR HS UR B D S
2	.419	131.5	UR D	6	.860	7.0	CR HS UR PL D S
2	.622	70.5	CR PL	6	.860	6.9	CR HS UR PL D MA
2	.646	63.4	UR HS	6	.861	6.7	CR HS UR B PL S
2	.733	37.3	CR HS	6	.862	6.5	CR HS UR PL D HT
2	.754	31.0	PL UR	6	.875	5.4	CR HS UR B PL D
3	.755	32.8	PL UR HT	7	.868	9.4	CR HS UR B D S MA
3	.757	32.1	PL UR B	7	.861	8.7	CR HS UR PL D MA S
3	.760	31.1	PL UR S	7	.861	8.6	CR HS UR B PL S HT
3	.771	28.0	CR PL UR	7	.861	8.6	CR HS UR B PL S MA
3	.772	27.6	CR HS B	7	.862	8.2	CR HS UR PL D HT S
3	.778	25.7	PL UR D	7	.864	7.8	CR HS UR PL D HT MA
3	.779	25.5	UR HS D	7	.875	7.4	CR HS UR B PL D HT
3	.780	25.0	CR HS UR	7	.876	7.0	CR HS UR B PL D S
3	.786	23.3	PL UR HS	7	.878	6.4	CR HS UR B PL D MA
4	.800	21.0	CR HS UR MA	8	.797	33.0	CR HS B MA D S PL HT
4	.801	20.7	CR HS UR S	8	.808	29.5	CR PL UR D B MA S HT
4	.806	19.2	PL UR HS HT	8	.844	15.6	PL UR HS D HT S MA B
4	.810	18.0	CR HS UR HT	8	.869	11.4	CR HS UR B D S MA HT
4	.816	16.4	PL UR HS S	8	.862	10.4	CR HS UR B PL S MA HT
4	.825	13.5	CR HS UR D	8	.864	9.7	CR HS UR PL D HT MA S
4	.832	11.4	PL UR HS D	8	.876	9.0	CR HS UR B PL D S HT
4	.840	9.0	CR HS UR PL	8	.879	8.3	CR HS UR B PL D MA HT
4	.842	8.3	CR HS UR B	8	.879	8.1	CR HS UR B PL D MA S
				9	.870	10.0	CR HS UR B PL D MA S HT

EXHIBIT 11.6: All Possible Subsets Search for Murder Rate Model

Variable	b	t-value	TOL
Intercept	73.01	1.29	
D	0.55	2.75	0.660
PL	0.41	3.19	0.315
UR	1.68	5.49	0.332
CR	0.01	2.96	0.281
HS	-3.50	-5.05	0.335

EXHIBIT 11.7: Parameter Estimates, t-Values and Tolerances for Selected Model for Murder Rate

of the 'better' models.

A conservative approach here is to choose between the best ($p - 1 =$) 9-variable model and the two best 8-variable models. The 9-variable model has dropped L1, which did not look important (Exercise 10.4), and it appears that the neighborhood it represents is not too different from the neighborhood described by L1= 0, L2= 0. The two 8-variable models differ in that one contains BTH and no GAR and the other contains GAR but no BTH. The reader is invited to obtain parameter estimates and collinearity diagnostics for both these models. It will be apparent that there is not much to pick and choose between them. GAR has a slightly lower tolerance in the overall model, but BTH is another of the variables which is related to FLR and BDR. But we are clutching at straws! One could just about flip a coin. If the deletion of either BTH or GAR looks unreasonable, as it might to a realtor, we could just choose the 9-variable model. ■

$p-1$	R^2	s	C_p	Variables in Model
3	0.750	4.08	62.6	FLR ST LOT
3	0.773	3.89	55.4	FLR ST COR
3	0.788	3.76	50.7	FLR ST GAR
3	0.788	3.76	50.6	FLR ST BDR
3	0.794	3.71	48.8	FLR ST FP
4	0.826	3.49	40.5	FLR ST FP GAR
4	0.829	3.46	39.7	FLR ST BDR GAR
4	0.838	3.36	36.6	FLR ST FP COR
4	0.839	3.36	36.3	FLR ST BDR LOT
4	0.861	3.11	29.2	FLR ST FP BDR
5	0.874	3.04	27.0	FLR ST FP BDR GAR
5	0.880	2.97	25.3	FLR ST FP BDR BTH
5	0.884	2.92	23.9	FLR ST FP BDR LOT
5	0.884	2.91	23.7	FLR ST BDR LOT COR
5	0.891	2.84	21.9	FLR ST FP BDR COR
6	0.901	2.77	20.6	FLR ST BDR LOT COR BTH
6	0.902	2.76	20.3	FLR ST BDR LOT COR GAR
6	0.909	2.67	18.2	FLR ST FP BDR COR BTH
6	0.918	2.52	15.1	FLR ST BDR LOT GAR L2
6	0.921	2.48	14.2	FLR ST FP BDR COR LOT
7	0.926	2.46	14.0	FLR ST FP BDR COR LOT GAR
7	0.930	2.39	13.2	FLR ST FP BDR COR LOT L2
7	0.932	2.36	12.7	FLR ST BDR LOT COR GAR L2
7	0.933	2.35	12.4	FLR ST FP BDR LOT L2 GAR
7	0.936	2.30	11.6	FLR ST FP BDR COR LOT BTH
8	0.938	2.34	12.9	FLR ST FP BDR COR LOT BTH L1
8	0.938	2.34	12.9	FLR ST BDR LOT COR GAR L2 BTH
8	0.938	2.31	12.5	FLR ST FP BDR LOT BTH L2 GAR
8	0.948	2.14	9.69	FLR ST FP BDR COR LOT BTH L2
8	0.948	2.14	9.65	FLR ST FP BDR COR LOT L2 GAR
9	0.939	2.38	14.3	FLR ST FP BDR COR LOT BTH L1 GAR
9	0.940	2.36	14.1	FLR ST BDR LOT COR GAR L2 BTH L1
9	0.948	2.20	11.5	FLR ST FP BDR COR LOT BTH L2 L1
9	0.948	2.19	11.5	FLR ST FP BDR COR LOT L2 GAR L1
9	0.956	2.03	9.14	FLR ST FP BDR COR LOT BTH L2 GAR
10	0.956	2.09	11.0	FLR ST FP BDR COR LOT BTH L2 GAR L1

EXHIBIT 11.8: All Possible Subsets Search for House Price Data

Problems

Exercise 11.1: Show that the estimate of the regression coefficient obtained when we regress the residuals from the model

$$\boldsymbol{y} = X_{(j)}\boldsymbol{\delta}_{(j)} + \boldsymbol{\psi}$$

against those from the model

$$\boldsymbol{x}_{[j]} = X_{(j)}\boldsymbol{\gamma}_{(j)} + \boldsymbol{\eta},$$

without an intercept term, is the same as that of β_j from the model

$$\boldsymbol{y} = X\boldsymbol{\beta} + \boldsymbol{\epsilon}$$

where $X_{(j)}$ is the matrix X with the jth column $\boldsymbol{x}_{[j]}$ removed, and the other letters have obvious meanings.
[**Hint:** Write the last model above as

$$\boldsymbol{y} = X_{(j)}\boldsymbol{\beta}_{(j)} + \beta_j\boldsymbol{x}_{[j]} + \boldsymbol{\epsilon}$$

and multiply both sides by $M_{(j)} = I - X_{(j)}(X_{(j)}'X_{(j)})^{-1}X_{(j)}'$.]

Exercise 11.2: Interpret the plot of residuals against predicteds for the model of Exercise 4.6, p. 96.

Exercise 11.3: *Let $\hat{\boldsymbol{y}}$ be the predicted value of $\boldsymbol{y}$ based on all k independent variables in the model and let $\hat{\boldsymbol{y}}_p$ be the predicted value of $\boldsymbol{y}$ based only on $p-1$ independent variables. Show that $\text{cov}(\hat{\boldsymbol{y}}) - \text{cov}(\hat{\boldsymbol{y}}_p) \geq 0$.
[**Hint:** Notice that

$$(X_1\ X_2)\begin{pmatrix} \Sigma_{11} & \Sigma_{12} \\ \Sigma_{12}' & \Sigma_{22} \end{pmatrix}^{-1}\begin{pmatrix} X_1' \\ X_2' \end{pmatrix}$$

can be written as

$$X_1\Sigma_{11}^{-1}X_1' + (X_2 - X_1\Sigma_{11}^{-1}\Sigma_{12})\Sigma_{2.1}^{-1}(X_2 - X_1\Sigma_{11}^{-1}\Sigma_{12})',$$

where $\Sigma_{2.1} = \Sigma_{22} - \Sigma_{12}'\Sigma_{11}^{-1}\Sigma_{12}$.]

Exercise 11.4: Is it possible to get a more parsimonious model to predict dial-a-ride patronage? Use the data as we decided in Example 8.3, p. 167, and the weights and transformations as we chose in Example 6.6, p. 124. Now apply your variable search procedures without weighting the model or deleting cases. What do you find?

Exercise 11.5: Fill in the missing details in Example 11.3. Use the weights given in (10.11) and examine both collinearity diagnostics and outputs from variable search procedures. Would you come to different conclusions?

Exercise 11.6: Using the data of Exercise 2.18, p. 53, find a parsimonious model for per capita output using linear and quadratic terms in SI, SP and I.

Exercise 11.7: Using all the variables given in Exhibit 2.2, p. 32, can you produce a better predictive model for house prices? Make sure that you take care of outliers, transformations (if any), etc.

Exercise 11.8: Obtain a simple model to predict percentage savings (PCS) using the data of Exhibit 9.11, p. 203, after making the transformations we decided on in Example 9.9.

Exercise 11.9: Apply both stepwise and all possible subsets search to the model considered in Example 10.2, p. 225. Which combination(s) of variables would you select? Compare the model(s) you chose with the one we selected in Example 10.2 in terms of fit, plausibility of estimates and their standard errors.

Exercise 11.10: After you have completed Exercise 10.5, p. 231, use the data from Exercise 3.14, p. 79, to obtain a parsimonious model for y (oxygen demand).

Exercise 11.11: After deleting whatever cases you think appropriate, use all possible subsets search to obtain more parsimonious models to take the place of each of the two models of Exercise 8.14, p. 178.

Exercise 11.12: Check whether more parsimonious models can be found to take the place of each of the models in Exercise 2.20, p. 55 (excluding part 1). If so, what are they?

Exercise 11.13: Suppose a person wants to buy a 3 bedroom house in the area represented in Exhibit 2.2, p. 32. How would you proceed to get him an idea of the price he would have to pay?

CHAPTER 12

*Biased Estimation

12.1 Introduction

One purpose of variable selection is to reduce multicollinearity, although, as we noted in Section 11.2, reducing the number of independent variables can lead to bias. Obviously, the general principle is that it might be preferable to trade off a small amount of bias in order to substantially reduce the variances of the estimates of $\boldsymbol{\beta}$. There are several other methods of estimation which are also based on trading off bias for variance. This chapter describes three of these: principal component regression, ridge regression and the shrinkage estimator.

The bias that is created is with respect to the model with all the independent variables included. While the amount of the bias can be theoretically computed, the resultant expressions depend on the parameters themselves and hence are not known. Typically for all methods, the sums of squares of residuals get larger (in comparison to OLS) and the usual measures of fit get worse, but the estimates of β have smaller variances.

The use of all the methods described below is mired in controversy, with each having proponents and opponents. There are situations where a certain method might work well and situations where it might not. As a result, it is difficult to make objective and comprehensive recommendations. When carefully applied, the methods have the potential of enhancing our underlying understanding of the situation. In this they are quite frequently useful. However, they should be applied only when the analyst has the time to carefully analyze the results and form a picture of 'what is going on' with the model. They should certainly not be applied in a routine or ritualistic manner.

12.2 Principal Component Regression

Typically, principal component regression is applied either to the centered and scaled version of the model

$$\boldsymbol{y} = X\boldsymbol{\beta} + \boldsymbol{\epsilon} \tag{12.1}$$

if the intercept term is present, or to only the scaled version if the intercept term is absent. That is, we consider the model

$$\boldsymbol{y}_{(0)} = Z_{(s)}\boldsymbol{\delta} + \boldsymbol{\epsilon}_{(0)} \tag{12.2}$$

of Section 2.11, p. 44, if the model (12.1) has an intercept term and the model

$$\boldsymbol{y} = X_{(s)}\boldsymbol{\beta}_{(s)} + \boldsymbol{\epsilon} \tag{12.3}$$

that we used in Chapter 10 (see (10.2)) if it does not. As mentioned in Section 2.11, p. 44, estimating the parameters of the model (12.2) is equivalent to estimating $\boldsymbol{\beta}$ from (12.1). The equivalence of (12.3) and (12.1) is trivial. The notation in the discussion below is with reference to model (12.2).

Let $\Gamma = (\gamma_1, \ldots, \gamma_k)$ be a $k \times k$ orthogonal matrix such that

$$\Gamma'(Z'_{(s)}Z_{(s)})\Gamma = D_\lambda$$

where D_λ is the diagonal matrix diag$(\lambda_1, \ldots, \lambda_k)$. Obviously, $\lambda_1 \geq \ldots \geq \lambda_k \geq 0$ are the eigenvalues of $Z'_{(s)}Z_{(s)}$ and $\boldsymbol{\gamma}_1, \ldots, \boldsymbol{\gamma}_k$ are the corresponding eigenvectors. Since Γ is orthogonal, and therefore $\Gamma\Gamma' = I$, rewrite the model (12.2) as

$$\boldsymbol{y}_{(0)} = Z_{(s)}\Gamma\Gamma'\boldsymbol{\delta} + \boldsymbol{\epsilon}_{(0)} = U\boldsymbol{\eta} + \boldsymbol{\epsilon}_{(0)} \tag{12.4}$$

where $U = (\boldsymbol{u}_1, \ldots, \boldsymbol{u}_k) = Z_{(s)}\Gamma$ and $\boldsymbol{\eta} = (\eta_1, \ldots, \eta_k)' = \Gamma'\boldsymbol{\delta}$. The k new independent variables $\boldsymbol{u}_1, \ldots, \boldsymbol{u}_k$ are called principal components.

Since $U'U = \Gamma' Z'_{(s)}Z_{(s)}\Gamma = D_\lambda$, the least squares estimate of $\boldsymbol{\eta}$ is

$$\hat{\boldsymbol{\eta}} = (U'U)^{-1}U'\boldsymbol{y}_{(0)} = D_\lambda^{-1}U'\boldsymbol{y}_{(0)},$$

and hence the estimate of an individual η_j is $\hat{\eta}_j = \lambda_j^{-1}\boldsymbol{u}'_j\boldsymbol{y}_{(0)}$. More importantly, the covariance matrix of $\hat{\boldsymbol{\eta}}$ is

$$\text{cov}(\hat{\boldsymbol{\eta}}) = \sigma^2 D_\lambda^{-1}$$

where cov$(\boldsymbol{\epsilon})$ is $\sigma^2 I$. Therefore, var$(\hat{\eta}_j) = \sigma^2/\lambda_j$.

If a λ_j is small, its variance is large and typically signals the presence of multicollinearity in the model (12.1) (see Chapter 10). In such cases, we could alleviate this multicollinearity by deleting the $\boldsymbol{u}_j$'s corresponding to the small λ_j's. The usual criteria for variable search, along with the size of var$(\hat{\eta}_j)$, can be used to decide which $\boldsymbol{u}_j$'s to remove. But now, because UU' is diagonal, it follows from the discussion at the end of Section 11.2.1, p. 235, that the $\hat{\eta}_j$'s corresponding to the retained variables remain unchanged after the deletion of the other variables and that $\hat{\boldsymbol{\eta}}_{(r)}$ is an unbiased estimate of $\boldsymbol{\eta}_{(r)}$, i.e.,

$$\text{E}[\hat{\boldsymbol{\eta}}_{(r)}] = \boldsymbol{\eta}_{(r)} \tag{12.5}$$

where $\boldsymbol{\eta}_{(r)}$ is the vector of η_j's corresponding to the retained variables and $\hat{\boldsymbol{\eta}}_{(r)}$ is the OLS estimator of $\boldsymbol{\eta}_{(r)}$.

After renumbering the columns of U, if necessary, write the matrix U as a partitioned matrix $U = (U_{(r)}\ U_{(k-r)})$ where $U_{(r)} = (\boldsymbol{u}_1, \ldots, \boldsymbol{u}_r)$ contains the columns of U we decide to retain and $U_{(k-r)}$ contains the columns we choose to delete. Renumber the η_j's and the columns of Γ to correspond and write $\boldsymbol{\eta}$ and Γ in partitioned matrix form as $\boldsymbol{\eta} = (\boldsymbol{\eta}'_{(r)}\ \boldsymbol{\eta}'_{(k-r)})'$ and $\Gamma = (\Gamma_{(r)}\ \Gamma_{(k-r)})$ where $\boldsymbol{\eta}_{(r)} = (\eta_1, \ldots, \eta_r)'$ and $\Gamma_{(r)} = (\boldsymbol{\gamma}_1, \ldots, \boldsymbol{\gamma}_r)$. From (12.4) we get $U = Z_{(s)}\Gamma$. Therefore, $U_{(r)} = Z_{(s)}\Gamma_{(r)}$, and we can write the abbreviated model as

$$\boldsymbol{y}_{(0)} = U_{(r)}\boldsymbol{\eta}_{(r)} + \tilde{\boldsymbol{\epsilon}}, \tag{12.6}$$

or, writing u_{ij} as the (i,j)th element of U, as

$$\begin{aligned} y_{(0)i} &= u_{i1}\eta_1 + \ldots + u_{ir}\eta_r + \tilde{\epsilon}_i \\ &= z_{i1}(\gamma_{11}\eta_1 + \ldots + \gamma_{1r}\eta_r) + \ldots + z_{ik}(\gamma_{k1}\eta_1 + \ldots \gamma_{kr}\eta_r) + \tilde{\epsilon}_i \end{aligned} \tag{12.7}$$

where $i = 1, \ldots, n$ and γ_{ij} is the ijth element of Γ and $\tilde{\boldsymbol{\epsilon}}$ is the resultant new error term.

Therefore, the deletion of some $\boldsymbol{u}_j$'s need not, and usually does not, result in the removal of the columns of $Z_{(s)}$. What we have done in effect is imposed the constraints $U_{(k-r)} = Z_{(s)}(\boldsymbol{\gamma}_{r+1}, \ldots, \boldsymbol{\gamma}_k) = \mathbf{o}$.

Example 12.1 (Continuation of Example 10.1, Page 220)
In Example 10.1, we, in fact, effectively performed a principal components regression although we did not scale the independent variables. This was not necessary since their sample variances were equal. The rows of the matrix Γ are given by (10.4) and $\boldsymbol{z}_1$ and $\boldsymbol{z}_2$ are the two principal components.

As we saw in that example, elimination of $\boldsymbol{z}_2$ did not yield good predictions when the values of $\boldsymbol{z}_2$ were not close to zero. ■

Principal component regression is quite effective when the variables represented by the deleted principal components must always remain small; i.e., the relationship represented by them is always approximately 'true'. When this is not so, and particularly when it is possible to get additional observations, it might be preferable to obtain independent variable values in the direction of the components we would have deleted. If these design points are well chosen, it frequently does not require too many points to substantially reduce multicollinearity and make multicollinearity-reducing procedures unnecessary.

12.2.1 Bias and Variance of Estimates

The model (12.7) can also be written as

$$\boldsymbol{y}_{(0)} = Z_{(s)}\boldsymbol{\delta}_{(r)} + \tilde{\boldsymbol{\epsilon}} \tag{12.8}$$

where $\boldsymbol{\delta}_{(r)} = \Gamma_{(r)}\boldsymbol{\eta}_{(r)}$. Let $\boldsymbol{d}_{(r)}$ be the estimate $\Gamma_{(r)}\hat{\boldsymbol{\eta}}_{(r)}$ of $\boldsymbol{\delta}_{(r)}$. Since both $\hat{\boldsymbol{\eta}}_{(r)}$ and $\boldsymbol{d}_{(r)}$ minimize the same sum of squares, $\boldsymbol{d}_{(r)}$ is also the OLS estimator of $\boldsymbol{\delta}_{(r)}$. While $\hat{\boldsymbol{\eta}}_{(r)}$ is an unbiased estimator of $\boldsymbol{\eta}_{(r)}$, $\boldsymbol{d}_{(r)}$ is not necessarily an unbiased estimator of $\boldsymbol{\delta}_{(r)}$, as we see below.

Since Γ is orthogonal,

$$\Gamma_{(r)}'\Gamma_{(r)} = I_r, \quad \Gamma_{(k-r)}'\Gamma_{(k-r)} = I_{k-r} \text{ and } \Gamma_{(r)}'\Gamma_{(k-r)} = 0.$$

Hence, $\boldsymbol{\eta}_{(r)} = \Gamma_{(r)}'\boldsymbol{\delta}_{(r)}$ and it follows from (12.5) that

$$\mathrm{E}[\boldsymbol{d}_{(r)}] = \mathrm{E}[\Gamma_{(r)}\hat{\boldsymbol{\eta}}_r] = \Gamma_{(r)}\boldsymbol{\eta}_{(r)} = \Gamma_{(r)}\Gamma_{(r)}'\boldsymbol{\delta}_{(r)}.$$

And because $I = \Gamma\Gamma' = \Gamma_{(r)}\Gamma_{(r)}' + \Gamma_{(k-r)}\Gamma_{(k-r)}'$, it follows that

$$\begin{aligned}\mathrm{E}[\boldsymbol{d}_{(r)}] &= [I - \Gamma_{(k-r)}\Gamma_{(k-r)}']\boldsymbol{\delta}_{(r)} \\ &= \boldsymbol{\delta}_{(r)} - \Gamma_{(k-r)}\Gamma_{(k-r)}'\boldsymbol{\delta}_{(r)} = \boldsymbol{\delta}_{(r)} - \Gamma_{(k-r)}\boldsymbol{\eta}_{(k-r)}.\end{aligned}$$

Therefore, the bias induced by deletion of the $k-r$ variables is $\Gamma_{(k-r)}\boldsymbol{\eta}_{(k-r)}$.

Let $\boldsymbol{d}$ be the estimate of $\boldsymbol{\delta}$ when no $\boldsymbol{u}_j$'s have been deleted. Then

$$\mathrm{cov}(\boldsymbol{d}) = \sigma^2(Z_{(s)}'Z_{(s)})^{-1} = \sigma^2\Gamma D_\lambda^{-1}\Gamma'.$$

Since D_λ^{-1} is diagonal it can be written as

$$\begin{pmatrix} D_{(r)}^{-1} & 0 \\ 0 & D_{(k-r)}^{-1} \end{pmatrix}$$

where $D_{(r)}$ is the diagonal matrix of the latent roots corresponding to the eigenvectors which are the columns of $\Gamma_{(r)}$ and $D_{(k-r)}$ is that of the latent roots corresponding to the eigenvectors which are the columns of $\Gamma_{(k-r)}$. Therefore,

$$\begin{aligned}\mathrm{cov}(\boldsymbol{d}) &= \sigma^2 \begin{pmatrix} \Gamma_{(r)} & \Gamma_{(k-r)} \end{pmatrix} \begin{pmatrix} D_{(r)}^{-1} & 0 \\ 0 & D_{(k-r)}^{-1} \end{pmatrix} \begin{pmatrix} \Gamma_{(r)}' \\ \Gamma_{(k-r)}' \end{pmatrix} \\ &= \sigma^2[\Gamma_{(r)}D_{(r)}^{-1}\Gamma_{(r)}' + \Gamma_{(k-r)}D_{(k-r)}^{-1}\Gamma_{(k-r)}'].\end{aligned}$$

Since it can be shown fairly easily that $\mathrm{cov}(\boldsymbol{d}_{(r)}) = \sigma^2\Gamma_{(r)}D_{(r)}^{-1}\Gamma_{(r)}'$, the reduction in the covariance matrix due to the elimination of variables is $\Gamma_{(k-r)}D_{(k-r)}^{-1}\Gamma_{(k-r)}'$.

12.3 Ridge Regression

Hoerl and Kennard (1970a, 1970b) have suggested a method of combating multicollinearity called *ridge regression*. Usually, though not always, ridge regression is applied to the centered and scaled model (12.2). Then, the ridge estimate of $\boldsymbol{\delta}$ for some $c \geq 0$ is given by

$$\boldsymbol{d}_c = [Z_{(s)}'Z_{(s)} + cI]^{-1}Z_{(s)}'\boldsymbol{y}_{(0)} \tag{12.9}$$

where c is a small nonnegative number, often called the *shrinkage parameter*. The choice of an appropriate value of c is discussed in Section 12.3.2. Obviously, if we do not center or scale, we would replace $Z_{(s)}$ and $\boldsymbol{y}_{(0)}$ by X and $\boldsymbol{y}$.

It is easily seen that (12.9) is equivalent to

$$\boldsymbol{d}_c = [I + c(Z_{(s)}{}'Z_{(s)})^{-1}]^{-1}\boldsymbol{d}$$

where, as before, $\boldsymbol{d}$ is the OLS estimate of $\boldsymbol{\delta}$. Since $Z_{(s)}{}'\mathbf{1} = \mathbf{o}$, (12.9) is also equivalent to

$$\boldsymbol{d}_c = [Z_{(s)}{}'Z_{(s)} + cI]^{-1}Z_{(s)}{}'\boldsymbol{y}.$$

There is also a generalization of ridge regression available. The estimate is

$$\boldsymbol{d}_C = [Z_{(s)}{}'Z_{(s)} + \Gamma C \Gamma']^{-1}Z_{(s)}{}'\boldsymbol{y}$$

where, as in Section 12.2, Γ is the matrix of eigenvectors of $Z_{(s)}{}'Z_{(s)}$ and C is a diagonal matrix with nonnegative elements. We do not consider this further in this book. See Judge *et al.* (1985, p. 913, *et seq.*) for a treatment of this generalized ridge estimator.

12.3.1 Physical Interpretations of Ridge Regression

We can always find (at least in theory) an $n \times k$ matrix V, the columns of which are orthogonal to those of the matrix $Z_{(s)}$ and also to $\boldsymbol{y}_{(0)}$. Let $W' = c^{1/2}(V'V)^{-1/2}V'$. Then $W'W = cI$, $W'Z_{(s)} = 0$ and $W'\boldsymbol{y}_{(0)} = \mathbf{o}$. Suppose we now perturb the matrix of independent variables $Z_{(s)}$ by the amount W and obtain the least squares estimate of $\boldsymbol{\delta}$ in the model (12.2) with the matrix of independent variables as $Z_{(s)} + W$. This estimate is given by

$$\begin{aligned}[(Z_{(s)} + W)'(Z_{(s)} + W)]^{-1}(Z_{(s)} + W)'\boldsymbol{y}_{(0)} \\ = [Z_{(s)}{}'Z_{(s)} + cI]^{-1}Z_{(s)}{}'\boldsymbol{y}_{(0)} = \boldsymbol{d}_c,\end{aligned}$$

i.e., the ridge estimates can be found by suitably disturbing the independent variable values by a small amount. In Exhibit 10.2, p. 221, the manifestation of multicollinearity was the narrowness of the strip within which the data lay. According to this interpretation of ridge regression, we would presumably be effectively widening this strip by adding some small numbers to the independent variables.

Another similar interpretation consists of adding additional 'cases' to the data set by augmenting the design matrix by a matrix W such that $W'W = cI$ and the dependent variable values by the appropriate number of zeros. This augmented model can be written as

$$\mathrm{E}\left[\begin{pmatrix} \boldsymbol{y}_{(0)} \\ \mathbf{o} \end{pmatrix}\right] = \begin{pmatrix} Z_{(s)} \\ W \end{pmatrix}.$$

It can easily be shown that the OLS estimator for this model is $\boldsymbol{d}_c$. An example of such a W is $\sqrt{c}I$, which provides an alternative way to compute ridge estimates using simply an OLS package.

Yet another interpretation can be given in terms of constrained least squares. Suppose we impose the constraint that $\boldsymbol{\delta}'\boldsymbol{\delta} = \kappa^2$ and minimize

$$(\boldsymbol{y}_{(0)} - Z_{(s)}\boldsymbol{\delta})'(\boldsymbol{y}_{(0)} - Z_{(s)}\boldsymbol{\delta})$$

subject to this constraint. Then setting $\partial \mathcal{S}/\partial \boldsymbol{\delta} = 0$ where

$$\mathcal{S} = (\boldsymbol{y}_{(0)} - Z_{(s)}\boldsymbol{\delta})'(\boldsymbol{y}_{(0)} - Z_{(s)}\boldsymbol{\delta}) + c(\boldsymbol{\delta}'\boldsymbol{\delta} - \kappa^2),$$

we get

$$(Z_{(s)}{}'Z_{(s)} + cI)\boldsymbol{\delta} = Z_{(s)}{}'\boldsymbol{y}_{(0)}$$

which yields the ridge estimate (12.9). Substituting (12.9) into the constraint $\boldsymbol{\delta}'\boldsymbol{\delta} = \kappa^2$, we get a relationship between c and κ.

Thus, a ridge estimate can be viewed as a constrained least squares estimate where the parameter $\boldsymbol{\delta}$ has been required to lie on the surface of a sphere. In the presence of acute multicollinearity, the estimates of the components δ_j of $\boldsymbol{\delta}$ often start compensating for each other and can get numerically very large. In such cases, ridge regression is often effective.

12.3.2 Bias and Variance of Estimates

Since

$$\begin{aligned} \mathrm{E}(\boldsymbol{d}_c) &= [Z_{(s)}{}'Z_{(s)} + cI]^{-1}Z_{(s)}{}'Z_{(s)}\boldsymbol{\delta} \\ &= [Z_{(s)}{}'Z_{(s)} + cI]^{-1}[Z_{(s)}{}'Z_{(s)} + cI - cI]\boldsymbol{\delta} = \boldsymbol{\delta} - c(Z_{(s)}{}'Z_{(s)} + cI)^{-1}\boldsymbol{\delta}, \end{aligned}$$

the bias of the ridge estimate is

$$\text{Bias}\,(\boldsymbol{d}_c) = c(Z_{(s)}{}'Z_{(s)} - cI)^{-1}\boldsymbol{\delta}. \tag{12.10}$$

The covariance matrix of $\boldsymbol{d}_c$ is given by

$$\text{cov}(\boldsymbol{d}_c) = (Z_{(s)}{}'Z_{(s)} + cI)^{-1}Z_{(s)}{}'Z_{(s)}(Z_{(s)}{}'Z_{(s)} + cI)^{-1}\sigma^2. \tag{12.11}$$

Therefore, since the trace of a matrix is the sum of eigenvalues and the eigenvalues of the inverse of a matrix A are the reciprocals of the eigenvalues of A (see Appendix A), the sum of variances of the components d_{cj} of $\boldsymbol{d}_c$ is

$$\begin{aligned} &\text{tr}[\,\text{cov}(\boldsymbol{d}_c)] \\ &\quad = \sigma^2\,\text{tr}[(Z_{(s)}{}'Z_{(s)} + cI)^{-1}Z_{(s)}{}'Z_{(s)}(Z_{(s)}{}'Z_{(s)} + cI)^{-1}] \\ &\quad = \text{tr}[(Z_{(s)}{}'Z_{(s)} + cI)^{-2}Z_{(s)}{}'Z_{(s)}] = \sigma^2\sum_{j=1}^{k}\lambda_j^{-1}(1 + c\lambda_j^{-1})^{-2}, \end{aligned} \tag{12.12}$$

where $\lambda_1 \geq \cdots \geq \lambda_k$ are the k characteristic roots of the matrix $Z_{(s)}{}'Z_{(s)}$. The sum of the variances of the components of the ordinary least squares estimator $\boldsymbol{d}$ of $\boldsymbol{\delta}$ is

$$\begin{aligned} \operatorname{tr}[\operatorname{cov}(\boldsymbol{d})] &= \sigma^2 \operatorname{tr}[(Z_{(s)}{}'Z_{(s)})^{-1}] = \sigma^2 \sum_{j=1}^{k} \lambda_j^{-1} \\ &\geq \sigma^2 \sum_{j=1}^{k} \lambda_j^{-1}(1 + c\lambda_j^{-1})^{-2}, \end{aligned} \tag{12.13}$$

i.e., the total of the variances of the components of a ridge estimator is lower than the corresponding total for the OLS estimator. The price we pay is, of course, the bias.

The mean square error matrix of the biased estimator $\boldsymbol{d}_c$ is given by (see Appendix B, near the end of Section B.2, p. 286)

$$\operatorname{MSE}(\boldsymbol{d}_c) = \operatorname{E}[(\boldsymbol{d}_c - \boldsymbol{\delta})(\boldsymbol{d}_c - \boldsymbol{\delta})'] = \operatorname{cov}(\boldsymbol{d}_c) + \operatorname{Bias}(\boldsymbol{d}_c)\operatorname{Bias}(\boldsymbol{d}_c)'$$

where $\operatorname{cov}(\boldsymbol{d}_c)$ and $\operatorname{Bias}(\boldsymbol{d}_c)$ are given in (12.11) and (12.10) respectively. The total mean square error (Section B.2) is the trace of $\operatorname{MSE}(\boldsymbol{d}_c)$ and is, therefore, (from (12.10) and (12.11))

$$\begin{aligned} \operatorname{TMSE}(\boldsymbol{d}_c) &= \operatorname{tr}[\operatorname{cov}(\boldsymbol{d}_c)] + \operatorname{tr}[\operatorname{Bias}(\boldsymbol{d}_c)\operatorname{Bias}(\boldsymbol{d}_c)] \\ &= \sigma^2 \sum_{j=1}^{k} \lambda_j^{-1}(1 + c\lambda_j^{-1})^{-2} + c^2 \boldsymbol{\delta}'(Z_{(s)}{}'Z_{(s)} + cI)^{-2}\boldsymbol{\delta} \\ &= \sigma^2 \sum_{j=1}^{k} \lambda_j(\lambda_j + c)^{-2} + c^2 \boldsymbol{\delta}'(Z_{(s)}{}'Z_{(s)} + cI)^{-2}\boldsymbol{\delta}. \end{aligned}$$

Hoerl and Kennard (1970a) have shown that there exists a value of c, say c_0, such that for $0 < c < c_0$, $\operatorname{TMSE}(\boldsymbol{d}_c) \leq \operatorname{TMSE}(\boldsymbol{d})$. However, c_0 depends on the unknown parameters $\boldsymbol{\delta}$ and σ^2 and as such cannot be determined (see Thistead, 1978).

Perhaps the most popular method for choosing c is through the use of what is called a ridge trace. Estimates $\boldsymbol{d}_c$ are obtained for different values of c over some interval (say, from 0 to 1) and then components d_{cj} of $\boldsymbol{d}_c$ are plotted against c. The resultant curves, called ridge traces, are then examined to find the minimum value of c after which values of d_{cj} are moderately stable. The $\boldsymbol{d}_c$ for this value of c is then adopted as the chosen ridge estimator. We could also examine the diagonal elements of $\sigma^{-2}\operatorname{cov}[\hat{\boldsymbol{b}}_c]$, which are actually akin to the variance inflation factors, in order to see the effect of c on the variances of estimates.

The subject of proper choice of c has received so much attention in the literature that we have but touched on the subject in this section. A fuller review is given in Draper and Van Nostrand (1979), Hocking (1976) and Judge *et al.* (1985).

Example 12.2 (Continuation of Example 12.1, Page 255)
Exhibit 12.1 shows plots of $\hat{d}_{c1}$, $\hat{d}_{c2}$ and $\sigma^{-2}\text{var}(\hat{d}_{c1})$ ($\sigma^{-2}\text{var}(\hat{d}_{c2})$ had a very similar plot) for the data of Exhibit 10.1, p. 220, with x_1 and x_2 as independent variables and $y^{(1)}$ as the dependent variable. Values of c shown are in the range $0 < c \leq .1$ which was chosen after a previous run over the interval $0 < c \leq 1$, which is usually quite adequate. While typically one uses scaled variables, we did not do so here since both independent variables had the same sample standard deviations.

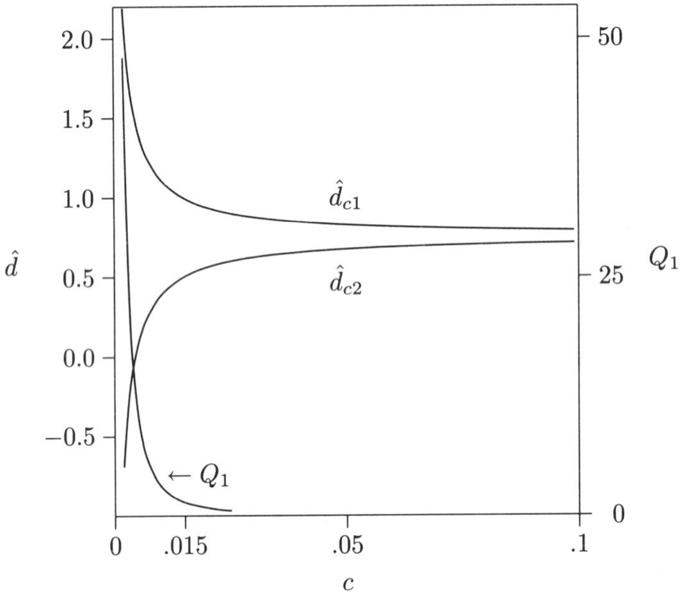

EXHIBIT 12.1: Ridge Traces and $Q_1 = \sigma^{-2}\text{var}(\hat{d}_{c1})$ for Ridge Regression of $y^{(1)}$ Against x_1 and x_2

Although several programs are available for doing ridge regression (e.g., PROC RIDGEREG documented in SAS, 1986), we wrote our own program using SAS PROC MATRIX. Only a few lines are required and no problems need occur unless one computes estimates for $c = 0$ in an extremely multicollinear situation, like the one being considered here. The values .997 and .503 of $\hat{d}_{c1}$ and $\hat{d}_{c2}$ for $c = .015$ are actually very close to the 'true values' (see Example 10.1, p. 220) $\beta_1 = 1$ and $\beta_2 = .5$ (this does not always happen — see Exercise 12.7). However, had we not already known these true values, it would have been difficult to select them from the ridge traces shown. Perhaps we would have selected something like (.79, .71), which is similar to what we get from principal components regression. With increasing c the reduction in variance is dramatic, as can also be seen from Exhibit 12.1.

While we have illustrated only the methods of choosing c that we have discussed in the text, we did apply a number of other methods suggested in the literature to this problem. The results were not satisfactory. ■

12.4 Shrinkage Estimator

There is another class of biased estimators introduced by Stein (1956) and James and Stein (1961). While the original thrust of ridge regression was to alleviate effects of multicollinearity, that of the shrinkage estimator was aimed solely at reducing the MSE. For the model $\boldsymbol{y} = X\boldsymbol{\beta} + \boldsymbol{\epsilon}$, where X is an $n \times p$ matrix and $\boldsymbol{\beta}$ is a vector of dimension p, it is given by

$$\hat{\boldsymbol{\beta}} = \left[1 - \frac{\kappa(n-p)s^2}{\boldsymbol{b}'X'X\boldsymbol{b}}\right]\boldsymbol{b}, \tag{12.14}$$

where κ is a number discussed below, $\boldsymbol{b}$ is the least squares estimator of $\boldsymbol{\beta}$ and, as usual, $s^2 = (n-p)^{-1}(\boldsymbol{y} - X\boldsymbol{b})'(\boldsymbol{y} - X\boldsymbol{b})$. The estimator $\hat{\boldsymbol{\beta}}$ is biased. Thus, in order to compare it with the least squares estimator $\boldsymbol{b}$, we need to consider the mean squared error matrix $\text{MSE}(\hat{\boldsymbol{\beta}}) = \text{E}[(\hat{\boldsymbol{\beta}} - \boldsymbol{\beta})(\hat{\boldsymbol{\beta}} - \boldsymbol{\beta})']$ and compare it with $\text{E}[(\boldsymbol{b} - \boldsymbol{\beta})(\boldsymbol{b} - \boldsymbol{\beta})']$, the covariance of $\boldsymbol{b}$. The latter is, of course, $\sigma^2(X'X)^{-1}$. However, it is more convenient to compare the scalar quantities $\text{tr}[(X'X)\text{MSE}(\hat{\boldsymbol{\beta}})]$ and $\text{tr}[(X'X)\,\text{cov}(\boldsymbol{b})]$. (Note that

$$\text{tr}[(X'X)\text{MSE}(\hat{\boldsymbol{\beta}})] = \text{E}[(\hat{\boldsymbol{\beta}} - \boldsymbol{\beta})'X'X(\hat{\boldsymbol{\beta}} - \boldsymbol{\beta})]$$

and

$$\text{tr}[(X'X)\,\text{cov}(\boldsymbol{b})] = \text{E}[(\boldsymbol{b} - \boldsymbol{\beta})'X'X(\boldsymbol{b} - \boldsymbol{\beta})].)$$

The latter is easily seen to equal $\sigma^2 p$. It has been shown by Stein that, if $0 < \kappa < 2(p-2)(n-p+2)^{-1}$ and $p \geq 3$, the former quantity — i.e., $\text{tr}[(X'X)\text{MSE}(\hat{\boldsymbol{\beta}})]$ — is always less than $\sigma^2 p$ and attains its minimum value at $\kappa = (p-2)/(n-p+2)$.

It may be shown (see Bilodeau and Srivastava, 1989, Carter *et al.*, 1990) that a uniformly minimum variance (UMV) unbiased estimator of the bias vector of $\hat{\boldsymbol{\beta}}$ is

$$\hat{\boldsymbol{\beta}} - \boldsymbol{b} = -\kappa\frac{(n-p)s^2}{\boldsymbol{b}'X'X\boldsymbol{b}}\boldsymbol{b}$$

and a UMV unbiased estimator of the MSE matrix of $\hat{\boldsymbol{\beta}}$ is

$$s^2\left[1 - \frac{2\kappa(n-p)^2 s^2}{((n-p)+2)\boldsymbol{b}'X'X\boldsymbol{b}}\right](X'X)^{-1}$$
$$+\kappa\left(\kappa + \frac{4}{(n-p)+2}\right)\left(\frac{(n-p)s^2}{\boldsymbol{b}'X'X\boldsymbol{b}}\right)^2\boldsymbol{b}\boldsymbol{b}'.$$

A further improvement in the estimator $\hat{\boldsymbol{\beta}}$ in (12.14) obtains if we modify it as follows:

$$\hat{\boldsymbol{\beta}}_{(+)} = \left[\max\left(0, 1 - \frac{\kappa(n-p)s^2}{\boldsymbol{b}'X'X\boldsymbol{b}}\right)\right]\boldsymbol{b}. \tag{12.15}$$

Problems

Exercise 12.1: Consider the model given in (12.2), where $\mathrm{E}(\boldsymbol{\epsilon}) = \mathbf{o}$, and $\mathrm{cov}(\boldsymbol{\epsilon}) = \sigma^2 I$. Show that

1. $\mathrm{cov}(\boldsymbol{\epsilon}_{(0)}) = \sigma^2 A$ where $A = (I - n^{-1}\mathbf{1}\mathbf{1}')$,
2. the generalized inverse of A is A, and
3. the generalized least squares estimate of $\boldsymbol{\delta}$ is

$$\boldsymbol{d} = (Z_{(s)}'Z_{(s)})^{-1}Z_{(s)}'\boldsymbol{y}_{(0)} = (Z_{(s)}'Z_{(s)})^{-1}Z_{(s)}'\boldsymbol{y}.$$

Exercise 12.2: Let $\boldsymbol{z}$ be a p dimensional random vector and let $\boldsymbol{z} \sim N(\boldsymbol{\theta}, \sigma^2 I_p)$. Show that

$$\mathrm{E}[(\boldsymbol{z} - \boldsymbol{\theta})'(\boldsymbol{z} - \boldsymbol{\theta})] = \sigma^2 p.$$

Exercise 12.3: Suppose $\boldsymbol{y} = X\boldsymbol{\beta} + \boldsymbol{\epsilon}$, where $\boldsymbol{\epsilon} \sim N(\mathbf{o}, \sigma^2 I)$ and X is an $n \times p$ matrix of rank $p \leq n$. Let $\boldsymbol{b} = (X'X)^{-1}X'\boldsymbol{y}$. Show that

$$\mathrm{E}[(\boldsymbol{b} - \boldsymbol{\beta})'X'X(\boldsymbol{b} - \boldsymbol{\beta})] = \sigma^2 p.$$

[**Hint:** Let $\boldsymbol{z} = (X'X)^{1/2}\boldsymbol{b}$ and $\boldsymbol{\theta} = (X'X)^{1/2}\boldsymbol{\beta}$ where $(X'X) = (X'X)^{1/2} \cdot (X'X)^{1/2}$. Now, show that $\boldsymbol{z} \sim N(\boldsymbol{\theta}, \sigma^2 I_p)$ and use the result from the last exercise.]

Exercise 12.4: Let $\boldsymbol{z}$ be a p dimensional random vector and let $\boldsymbol{z} \sim N(\boldsymbol{\theta}, \sigma^2 I_p)$. Show that

$$\mathrm{E}\left[\frac{\boldsymbol{z}'(\boldsymbol{z} - \boldsymbol{\theta})}{\boldsymbol{z}'\boldsymbol{z}}\right] = (p-2)\,\mathrm{E}\left(\frac{1}{\boldsymbol{z}'\boldsymbol{z}}\right).$$

[**Hint:** Let $\Phi(\boldsymbol{\theta}, I; \boldsymbol{z})$ denote the pdf of $\boldsymbol{z} = (z_1, \ldots, z_p)'$ and let $\boldsymbol{\theta} = (\theta_1, \ldots, \theta_p)'$. Then the left side is

$$\mathrm{E}\left[\frac{\sum_{j=1}^p z_j(z_j - \theta_j)}{\boldsymbol{z}'\boldsymbol{z}}\right] = -\int_{R^p} [\boldsymbol{z}'\boldsymbol{z}]^{-1}\left(\sum_{j=1}^p z_j \frac{\partial}{\partial z_j}\Phi(\boldsymbol{\theta}, I; \boldsymbol{z})\right) d\boldsymbol{z},$$

where $R^p = \{(z_1, \ldots, z_p) : -\infty < z_i < \infty\}$ and $d\boldsymbol{z} = dz_1 dz_2 \ldots dz_p$. It may be shown that

$$\int_{-\infty}^{\infty} \frac{z_i}{\boldsymbol{z}'\boldsymbol{z}} \frac{\partial}{\partial z_i}\Phi(\boldsymbol{\theta}, I; \boldsymbol{z})\, dz_i = -\int_{-\infty}^{\infty} \Phi(\boldsymbol{\theta}, I; \boldsymbol{z})\left[\frac{1}{\boldsymbol{z}'\boldsymbol{z}} - \frac{2z_i^2}{(\boldsymbol{z}'\boldsymbol{z})^2}\right] dz_i$$

using the integration by parts formula $\int u\,dv = uv - \int v\,du$.]

Exercise 12.5: Using the results of the last problem, show that when $p \geq 3$ the trace of the mean square error matrix of the Stein estimator given in (12.14) attains its minimum value at $\kappa = (p-2)/(n-p+2)$.

Exercise 12.6: Apply principal components regression to the supervisor rating data set given in Exhibit 10.3. Are your results very different from those we obtained in Example 10.2, p. 225?

Exercise 12.7: Using each of the variables $y^{(2)}$ and $y^{(3)}$ of Exhibit 10.1, p. 220, as dependent variables, obtain ridge traces and plots of $\sigma^{-2}\text{var}(\hat{d}_{c1})$ and $\sigma^{-2}\text{var}(\hat{d}_{c2})$. Comment on what you observe and select a suitable ridge estimator.

Exercise 12.8: Try both ridge regression and principle components regression on the data of Exercise 3.14, p. 79. How does removing outliers affect your results?

Exercise 12.9: How would you do weighted principal components regression and weighted ridge regression? Apply weighted principal components regression to the house price example (with outliers deleted). How does your chosen model compare with the one found in Example 11.3, p. 247? Apply weighted ridge regression to the same data and discuss your results with particular reference to the sign of BDR. If you were to choose a single model to give to give to a client, which one would you choose and why?

Exercise 12.10: Compute the shrinkage estimator for the data for Exercise 12.7. Also compute the estimate of the mean square error matrix.

Exercise 12.11: Using the methods considered in this chapter, can you find better models to take the place of those in Exercise 2.20, p. 55?

Appendices

APPENDIX A

Matrices

A rectangular array

$$A = \begin{pmatrix} a_{11} & a_{12} & a_{13} & \dots & a_{1c} \\ a_{21} & a_{22} & a_{23} & \dots & a_{2c} \\ \dots & \dots & \dots & \dots & \dots \\ \dots & \dots & \dots & \dots & \dots \\ a_{r1} & a_{r2} & a_{r3} & \dots & a_{rc} \end{pmatrix}$$

of rc numbers a_{ij}, when it is subject to certain rules of addition and multiplication given in Section A.1, is called a matrix of dimension $r \times c$ with elements a_{ij}. We shall use capital letters to denote matrices; when we wish to make explicit mention of a dimension we place it after a colon, e.g., $A : r \times c$, $B : 3 \times 5$. etc. When r=c the matrix $A : r \times c$ is called square and in that case a single number will suffice to give its dimension; i.e., $A : r \times r = A : r$. When we wish to make specific mention of the elements a_{ij} of a matrix we shall denote the matrix as (a_{ij}) or if necessary as $(a_{ij}):r \times c$ or as $(a_{ij}):r$ when $r = c$. Sometimes, when no ambiguity arises, we shall employ the notation $A_{r\times c}$, $(a_{ij})_{r\times c}$, A_r, $(a_{ij})_r$ to denote $A:r \times c$, $(a_{ij}):r \times c$, $A:r$ and $(a_{ij}):r$ respectively. A matrix of dimension $r \times c$ will also be called an $r \times c$ matrix; a square matrix of size r will also be called an r matrix.

Two matrices $(a_{ij}):r \times c$ and $(b_{ij}):r' \times c'$ are equal if and only if $r = r'$, $c = c'$ and $a_{ij} = b_{ij}$ for all i and j.

A.1 Addition and Multiplication

Two matrices $A = (a_{ij}):r \times c$ and $B = (b_{ij}):r \times c$, of the same dimension, can be added and their sum is $(a_{ij} + b_{ij}) : r \times c$. Matrices of unequal dimensions cannot be added.

The product of a number α and a matrix $(a_{ij}) : r \times c$ is the matrix $(\alpha a_{ij}) : r \times c$ — i.e., each element gets multiplied by the number. The product $C = AB$ of two matrices $A = (a_{ij}):r \times s$ and $B = (b_{ij}):p \times q$ is defined if and only if $s = p$, and then $C = (c_{ij}):r\times q$ with $c_{ij} = \sum_{k=1}^{p} a_{ik}b_{kj}$. Note that the existence of the product AB does not assure the existence of BA and even if the latter exists, AB will not necessarily equal BA. In the sequel, if sums of products of matrices are written, the assumption is made that their dimensions are appropriate. It is easy to show that $A(B_1 + B_2) = AB_1 + AB_2$ and $\sum_{k=1}^{K} AB_k = A(\sum_{k=1}^{K} B_k)$.

A.2 The Transpose of a Matrix

If the rows and columns of a matrix A are interchanged, the resulting matrix is called the *transpose* of A and is denoted by A', i.e., $(a_{ij})' = (a_{ji})$. A square matrix A is said to be *symmetric* when it is its own transpose, i.e., $A = A'$.

The following identities are easily verified:

$$(A')' = A, \quad (A+B)' = A' + B', \quad (\sum_{k=1}^{K} A_k)' = \sum_{k=1}^{K} A'_k,$$
$$(AB)' = B'A', \ (ABC)' = C'B'A'.$$

A matrix A of the form XX' is always symmetric since $(XX')' = (X')'X' = XX'$.

Example A.1
Let

$$A = \begin{pmatrix} 2 & 1 & 3 \\ 4 & 6 & 1 \end{pmatrix}, \ B = \begin{pmatrix} 6 & 8 & 1 \\ 7 & 1 & 2 \end{pmatrix} \text{ and } C = \begin{pmatrix} 2 & 1 \\ -1 & 3 \end{pmatrix}.$$

Then

$$A + B = \begin{pmatrix} 8 & 9 & 4 \\ 11 & 7 & 3 \end{pmatrix} \quad \text{and} \quad CA = \begin{pmatrix} 8 & 8 & 7 \\ 10 & 17 & 0 \end{pmatrix}.$$

However, A and C cannot be added and AC does not exist.

$$A' = \begin{pmatrix} 2 & 4 \\ 1 & 6 \\ 3 & 1 \end{pmatrix} \text{ and } AA' = \begin{pmatrix} 14 & 17 \\ 17 & 53 \end{pmatrix}$$

is symmetric. ■

A.3 Null and Identity Matrices

If all elements of A are zeros, then A is said to be a *null matrix* or a zero matrix and is denoted by 0 or, if we need to make the dimension explicit, by $0{:}r \times c$, $0{:}r$, $0_{r\times c}$ or 0_r.

The elements along the north-west to south-east diagonal of a square matrix A comprise the *diagonal* (sometimes also called the principal diagonal) of A. When the only non-zero elements of a matrix are in its diagonal, it is called a diagonal matrix; e.g.,

$$\begin{pmatrix} 1 & 0 & 0 \\ 0 & 2 & 0 \\ 0 & 0 & 3 \end{pmatrix} \tag{A.1}$$

is a diagonal matrix. If the diagonal entries of the diagonal matrix A are $a_1, \ldots, a_r$, then A is sometimes written as $\operatorname{diag}(a_1, \ldots, a_r)$. Therefore, (A.1) is the same as $\operatorname{diag}(1, 2, 3)$.

The matrix $\operatorname{diag}(1, 1, \ldots, 1)$ is called an *identity matrix* and is denoted by I or $I{:}r$ or I_r, e.g.,

$$I_1 = (1),\ I_2 = \begin{pmatrix} 1 & 0 \\ 0 & 1 \end{pmatrix}, \text{ and } I_3 = \begin{pmatrix} 1 & 0 & 0 \\ 0 & 1 & 0 \\ 0 & 0 & 1 \end{pmatrix}.$$

It is easy to see that if B is any matrix of suitable dimension, $IB = BI = B$.

A.4 Vectors

A matrix A consisting of a single column is often called a *column vector* and a matrix consisting of only one row is sometimes called a *row vector*. In this book, a *vector* always means a column vector, and, when its dimension is r, it could also be called an r-vector. Vectors will usually be denoted by a boldface lower case Latin or Greek letter, e.g., $\boldsymbol{b}$, $\boldsymbol{\beta}$, $\boldsymbol{x}_i$. Note that when this notation is used the subscript does *not* indicate dimension.

The columns of an $r \times c$ dimensional matrix A may be viewed as column vectors $\boldsymbol{\alpha}_1, \boldsymbol{\alpha}_2, \cdots, \boldsymbol{\alpha}_c$, where $\boldsymbol{\alpha}_i$'s are r-vectors, and it will sometimes be more convenient to write A as $A = (\boldsymbol{\alpha}_1, \cdots, \boldsymbol{\alpha}_c)$. Rows can be treated in the same way; e.g.,

$$A = \begin{pmatrix} \boldsymbol{\beta}_1' \\ \vdots \\ \boldsymbol{\beta}_r' \end{pmatrix}$$

where each $\boldsymbol{\beta}_i'$ is a row vector which is the transpose of a c-vector.

Example A.2

Let

$$\boldsymbol{\alpha}_1 = \begin{pmatrix} 1 \\ 2 \\ 3 \end{pmatrix},\ \boldsymbol{\alpha}_2 = \begin{pmatrix} 1 \\ 0 \\ 1 \end{pmatrix} \text{ and } \boldsymbol{\alpha}_3 = \begin{pmatrix} -1 \\ 0 \\ 1 \end{pmatrix}.$$

Then

$$A = \begin{pmatrix} \boldsymbol{\alpha}_1 & \boldsymbol{\alpha}_2 & \boldsymbol{\alpha}_3 \end{pmatrix} = \begin{pmatrix} 1 & 1 & -1 \\ 2 & 0 & 0 \\ 3 & 1 & 1 \end{pmatrix}.$$

Also,

$$B = \begin{pmatrix} \boldsymbol{\alpha}_1' \\ \boldsymbol{\alpha}_2' \\ \boldsymbol{\alpha}_3' \end{pmatrix} = \begin{pmatrix} 1 & 2 & 3 \\ 1 & 0 & 1 \\ -1 & 0 & 1 \end{pmatrix} = A',$$

since $\boldsymbol{\alpha}_1' = (1\ 2\ 3)$, $\boldsymbol{\alpha}_2' = (1\ 0\ 1)$ and $\boldsymbol{\alpha}_3' = (-1\ 0\ 1)$. ■

A vector $\boldsymbol{a} = (a_1, \ldots, a_n)'$ is said to be a *non-null vector* if at least one a_i is different from zero. If all the a_i's are zero, it is called a *null vector* and is denoted by $\mathbf{o}$.

We designate a vector consisting entirely of 1's as $\mathbf{1} = (1, 1, 1, \ldots, 1)'$. Let $\boldsymbol{x} = (x_1, \ldots, x_n)'$. Then $\bar{x} = \sum_{i=1}^{n} x_i/n = n^{-1}\mathbf{1}'\boldsymbol{x}$ and the vector $(x_1 - \bar{x}, \ldots, x_n - \bar{x})'$ may be written as

$$\boldsymbol{x} - n^{-1}\mathbf{1}\mathbf{1}'\boldsymbol{x} = (I_n - n^{-1}\mathbf{1}\mathbf{1}')\boldsymbol{x}. \tag{A.2}$$

The *length* of a vector $\boldsymbol{x}$ is $(\boldsymbol{x}'\boldsymbol{x})^{1/2}$ and is often written as $||\boldsymbol{x}||$. For two vectors $\boldsymbol{x}$ and $\boldsymbol{y}$, we have the well-known Cauchy-Schwartz inequality,

$$|\boldsymbol{x}'\boldsymbol{y}| \leq ||\boldsymbol{x}||\,||\boldsymbol{y}||.$$

The quantity $\boldsymbol{x}'\boldsymbol{y}/[||x||\,||y||]$ is the cosine of the angle between the two vectors. Therefore, $\boldsymbol{x}'\boldsymbol{y} = 0$ implies that $\boldsymbol{x}$ and $\boldsymbol{y}$ are at right angles; $\boldsymbol{x}$ and $\boldsymbol{y}$ are then said to be *orthogonal*. If, in addition, $||\boldsymbol{x}|| = 1$ and $||\boldsymbol{y}|| = 1$, $\boldsymbol{x}$ and $\boldsymbol{y}$ are called *orthonormal*.

Two matrices A and B are said to be *orthogonal* if $AB = 0$. A single matrix Γ is said to be *orthogonal* if it is a square matrix such that $\Gamma\Gamma' = I$. Note that this definition is not exactly the analog of the definition of orthogonality for vectors. In fact, $\Gamma'\Gamma = I$ requires the columns of Γ to be ortho*normal*. Clearly, then, $\Gamma' = \Gamma^{-1}$ and for any vector $\boldsymbol{\alpha}$,

$$||\Gamma\boldsymbol{\alpha}|| = (\boldsymbol{\alpha}'\Gamma'\Gamma\boldsymbol{\alpha})^{-1/2} = ||\boldsymbol{\alpha}||, \tag{A.3}$$

i.e., multiplication by an orthogonal matrix leaves length unchanged.

A.5 Rank of a Matrix

The set $\boldsymbol{\alpha}_1, \ldots, \boldsymbol{\alpha}_n$ is said to be *linearly independent* if no $\boldsymbol{\alpha}_i$ can be expressed as a linear combination of the others. This is equivalent to saying that there is no non-null vector $\boldsymbol{c} = (c_1, \ldots, c_n)'$ such that $\sum_{i=1}^{n} c_i\boldsymbol{\alpha}_i = \mathbf{o}$. If $\boldsymbol{\alpha}_1, \cdots, \boldsymbol{\alpha}_n$ are not linearly independent they are said to be *linearly dependent*.

It may be proved that the number of independent columns of a matrix A is equal to the number of independent rows (see Rao, 1973, Section 1a.3 for a proof). The number of linearly independent columns of a matrix is called its *rank*, $\rho(A)$. If the rank of a square matrix A_r is the same as its dimension r, then it is said to be *non-singular*; otherwise it is called *singular*. The following properties of $\rho(A)$ are readily verified:

1. $\rho(A) = 0$ if and only if $A = 0$,
2. if $A\!: r \times c \neq 0$, then $1 \leq \rho(A) \leq \min(r, c)$,
3. $\rho(A) = \rho(A')$,

4. $\rho(AB) \leq \min(\rho(A), \rho(B))$,

5. $\rho(AB) = \rho(A) = \rho(B'A')$ if $\rho(B\!:\!r \times c) = r \leq c$,

6. $\rho(A) = \rho(AA') = \rho(A'A)$.

A.6 Trace of a Matrix

For a square matrix $A = (a_{ij})\!:\!r$, the sum of its diagonal elements is called its trace, $\text{tr}(A)$; i.e., $\text{tr}(A) = \sum_{i=1}^{r} a_{ii}$. It is easy to verify that

1. $\text{tr}(A) = \text{tr}(A')$,

2. $\text{tr}(AB) = \text{tr}(BA)$,

3. $\text{tr}(ABC) = \text{tr}(BCA) = \text{tr}(CAB)$,

4. $\text{tr}(A + B) = \text{tr}(B + A) = \text{tr}(A) + \text{tr}(B)$,

5. $\text{tr}(\sum_{\alpha=1}^{k} A_\alpha) = \sum_{\alpha=1}^{k} \text{tr}(A_\alpha)$,

6. $\text{tr}(kA) = k\,\text{tr}(A)$ where k is a real number.

Example A.3
If $\boldsymbol{x}$ is a vector, $\text{tr}(\boldsymbol{x}\boldsymbol{x}') = \text{tr}(\boldsymbol{x}'\boldsymbol{x}) = \boldsymbol{x}'\boldsymbol{x}$ since $\boldsymbol{x}'\boldsymbol{x}$ is a scalar. ■

Example A.4
Let Γ be an orthogonal matrix. Then

$$\text{tr}[\Gamma \boldsymbol{x}\boldsymbol{x}'\Gamma'] = \text{tr}[\boldsymbol{x}\boldsymbol{x}'\Gamma'\Gamma] = \text{tr}[\boldsymbol{x}\boldsymbol{x}'],$$

which as we have just seen is $\boldsymbol{x}'\boldsymbol{x}$. ■

A.7 Partitioned Matrices

A partitioned matrix is a matrix of matrices. For example, we can write the matrix

$$\begin{pmatrix} 1 & 2 & 3 & 4 \\ 5 & 6 & 7 & 8 \\ 9 & 10 & 11 & 12 \\ 13 & 14 & 15 & 16 \\ 17 & 18 & 19 & 20 \end{pmatrix} \text{ as } \begin{pmatrix} A_{11} & A_{12} \\ A_{21} & A_{22} \end{pmatrix} \text{ or as } \begin{pmatrix} B_{11} & B_{12} \\ B_{21} & B_{22} \end{pmatrix},$$

where

$$A_{11} = \begin{pmatrix} 1 & 2 \\ 5 & 6 \end{pmatrix}, \; A_{12} = \begin{pmatrix} 3 & 4 \\ 7 & 8 \end{pmatrix},$$

$$A_{21} = \begin{pmatrix} 9 & 10 \\ 13 & 14 \\ 17 & 18 \end{pmatrix}, \; A_{22} = \begin{pmatrix} 11 & 12 \\ 15 & 16 \\ 19 & 20 \end{pmatrix},$$

$$B_{11} = \begin{pmatrix} 1 & 2 & 3 \\ 5 & 6 & 7 \end{pmatrix}, \; B_{12} = \begin{pmatrix} 4 \\ 8 \end{pmatrix},$$

$$B_{21} = \begin{pmatrix} 9 & 10 & 11 \\ 13 & 14 & 15 \\ 17 & 18 & 19 \end{pmatrix}, \text{ and } B_{22} = \begin{pmatrix} 12 \\ 16 \\ 20 \end{pmatrix}.$$

The rules for multiplying two partitioned matrices are the same as for multiplying two ordinary matrices, only now the submatrices are the 'elements'.

A.8 Determinants

A *permutation* $\pi(n)$ of the set of numbers $1, 2, \ldots, n$ is any reordering of the set; e.g., $1, 3, 2, 4$ is a permutation of $1, 2, 3, 4$, as is $4, 2, 3, 1$. The ith element of $\pi(n)$ will be referred to as $\pi(n)_i$ or as π_i. The number $\#(\pi(n))$ of *inversions* of a permutation $\pi(n)$ is the number of exchanges of pairs of numbers in $\pi(n)$ to bring them to their natural order. For example, exchanging 4 and 1 and then 2 and 3 brings $4, 3, 2, 1$ into the order $1, 2, 3, 4$; therefore, the number of inversions in this case is 2. The number of inversions of a permutation is not unique, but it might be shown that $(-1)^{\#(\pi(n))}$ is unique. The *determinant* of a square matrix $A = (a_{ij}) : n$ is defined as

$$\det(A) = \sum_{\pi(n)} (-1)^{\#(\pi(n))} \prod_{i=1}^{n} a_{\pi_i, i}.$$

For example,

$$\det \begin{pmatrix} a_{11} & a_{12} \\ a_{21} & a_{22} \end{pmatrix} = a_{11} \times a_{22} - a_{21} \times a_{12}$$

and

$$\det \begin{pmatrix} 3 & 1 \\ 1 & 2 \end{pmatrix} = 3 \times 2 - 1 \times 1 = 5.$$

The following properties are easily verified:

1. $\det(A') = \det(A)$.

2. If the ith row (or column) of a matrix is multiplied by a number c, its determinant is multiplied by c. Therefore, $\det(cA) = c^n \det(A)$ if A is an $n \times n$ matrix.

3. If two rows (or columns) of a matrix are interchanged, the sign of the determinant changes. It follows that if two rows or columns are identical the value of the determinant is zero.

4. The value of a determinant is unchanged if to the ith row (column) is added c times the jth row (column). Hence, if a matrix has linearly dependent rows (or columns), its determinant is zero.

5. $\det[\operatorname{diag}(a_1, \ldots, a_n)] = \prod_{i=1}^n a_i$ and, in particular, $\det(I) = 1$.

6. $\det(AB) = \det(A)\det(B)$.

7. $\det(AA') \geq 0$.

8. If A and B are square matrices,

$$\det\begin{pmatrix} I & 0 \\ C & A \end{pmatrix} = \det(A), \quad \det\begin{pmatrix} A & C \\ 0 & I \end{pmatrix} = \det(A),$$
$$\det\begin{pmatrix} A & C \\ 0 & B \end{pmatrix} = \det\begin{pmatrix} A & 0 \\ C & B \end{pmatrix} = \det(A)\det(B).$$

9. $\det(I_p + AB) = \det(I_q + BA)$, where A and B are matrices of dimension $p \times q$ and $q \times p$ respectively.

A.9 Inverses

If, for a square matrix A, there exists a matrix B such that

$$AB = BA = I,$$

then B is called the *inverse* of A and is denoted as A^{-1}. If A is non-singular such a matrix always exists and is unique. For a singular matrix there is no inverse.

Inverses may be computed as follows. Consider the three operations:

O1 Multiplying a row (or column) of a matrix A by a number c.

O2 Replacing a row (column) $\boldsymbol{a}_i$ of A by $\boldsymbol{a}_i + \lambda \boldsymbol{a}_j$ where λ is a number and $\boldsymbol{a}_j$ is another row(column).

O3 Interchanging two rows or columns.

The application of a sequence of such operations to any nonsingular matrix will reduce it to an identity matrix. The application of the same sequence of operations (in the same order) to an identity matrix will yield the inverse. This is easily seen as follows: Notice that each of the row operations O1, O2 and O3 is equivalent to the multiplication of A from the left by a matrix, say O_i. Therefore, a sequence of such operations is equivalent to multiplication by a matrix O which is the product of a sequence of these individual matrices O_i. Since $AA^{-1} = I$ and $OA = I$, it follows, on multiplying both sides of the former by O, that $A^{-1} = OI$.

However, in order to reduce accumulated round-off errors, particularly in near singular situations (Chapter 10), actual computer procedures are somewhat more complex. A description of many of the procedures in common use is given in Chapters 11 and 12 of Seber (1977). While most users tend to ignore the exact computational procedure used by packages, and nowadays most of those given in *serious statistical* packages are quite reliable, a paper by Longley (1967) provides a sobering thought. He examined a number of the packages of the day with a highly multicollinear data set and found that estimates for one of parameters varied from -41 to 27 for the different packaged programs (the correct estimate found by careful hand calculation was about 15). A portion of his data is given in Exhibit 10.11, p. 232.

Example A.5
The following sequence of operations:

1. divide Row 1 by 5,
2. subtract Row 2 from Row 1 and replace Row 1 by the result,
3. divide Row 1 by -2,
4. subtract 3 times Row 1 from Row 2 and place the result in Row 2,
5. divide Row 2 by 2,

reduces

$$A = \begin{pmatrix} 5 & 10 \\ 3 & 2 \end{pmatrix} \text{ to } I_2.$$

The same sequence of operations changes

$$I_2 \text{ to } \begin{pmatrix} -.1 & .5 \\ .15 & -.25 \end{pmatrix} = B.$$

It is easily verified that $AB = BA = I$, i.e., A and B are inverses of each other. ■

Let P and Q be non-singular matrices. Since $(Q^{-1}P^{-1})PQ = I$, it follows that $(PQ)^{-1} = Q^{-1}P^{-1}$.

Theorem A.1 *Let P and Q be n-dimensional non-singular matrices such that $Q = P + UV$, where U has dimension $n \times q$ and V has dimension $q \times n$. Then*

$$Q^{-1} = (P + UV)^{-1} = P^{-1} - P^{-1}U(I_q + VP^{-1}U)^{-1}VP^{-1}. \tag{A.4}$$

PROOF: Since $Q^{-1}P + Q^{-1}UV = I$,

$$Q^{-1} + Q^{-1}UVP^{-1} = P^{-1}. \tag{A.5}$$

Therefore,

$$P^{-1}U = (Q^{-1}U) + (Q^{-1}U)VP^{-1}U = Q^{-1}U(I + VP^{-1}U)$$

and it follows that

$$Q^{-1}U = P^{-1}U(I + VP^{-1}U)^{-1}. \tag{A.6}$$

On substituting for $Q^{-1}U$ from (A.6) into (A.5) we get (A.4). □

Example A.6
Let $X_n = (\boldsymbol{x}_1, \ldots, \boldsymbol{x}_n)'$ be a $k \times n$ dimensional matrix where the $\boldsymbol{x}_i$'s are k-vectors. Let $X_{n+1} = (\boldsymbol{x}_1, \ldots, \boldsymbol{x}_n, \boldsymbol{x}_{n+1})'$, $Q = X'_{n+1}X_{n+1}$ and $P = X'_nX_n$. Then

$$Q = X'_{n+1}X_{n+1} = \begin{pmatrix} X'_n & \boldsymbol{x}_{n+1} \end{pmatrix} \begin{pmatrix} X_n \\ \boldsymbol{x}'_{n+1} \end{pmatrix} = P + \boldsymbol{x}_{n+1}\boldsymbol{x}'_{n+1}.$$

It follows that

$$\begin{aligned} Q^{-1} &= P^{-1} - P^{-1}\boldsymbol{x}_{n+1}[I_1 + \boldsymbol{x}'_{n+1}P^{-1}\boldsymbol{x}_{n+1}]^{-1}\boldsymbol{x}'_{n+1}P^{-1} \\ &= P^{-1} - \frac{P^{-1}\boldsymbol{x}_{n+1}\boldsymbol{x}'_{n+1}P^{-1}}{1 + \boldsymbol{x}'_{n+1}P^{-1}\boldsymbol{x}_{n+1}} \\ &= (X'_nX_n)^{-1} - \frac{(X'_nX_n)^{-1}\boldsymbol{x}_{n+1}\boldsymbol{x}'_{n+1}(X'_nX_n)^{-1}}{1 + \boldsymbol{x}'_{n+1}(X'_nX_n)^{-1}\boldsymbol{x}_{n+1}}, \end{aligned}$$

from Theorem A.1. ■

Example A.7
Let

$$Q_n = \begin{pmatrix} 2 & 1 & \ldots & 1 \\ 1 & 2 & \ldots & 1 \\ \vdots & \vdots & \ddots & \vdots \\ 1 & 1 & \ldots & 2 \end{pmatrix}.$$

Then,

$$Q^{-1} = I_n - \frac{\mathbf{1}\mathbf{1}'}{1 + \mathbf{1}'\mathbf{1}} = I_n - \frac{\mathbf{1}\mathbf{1}'}{n+1},$$

since $Q = I_n + \mathbf{1}\mathbf{1}'$. ∎

Example A.8
Let

$$A = \begin{pmatrix} A_{11} & A_{12} \\ A_{21} & A_{22} \end{pmatrix} \text{ and } B = \begin{pmatrix} B_{11} & B_{12} \\ B_{21} & B_{22} \end{pmatrix}$$

be two partitioned matrices. Then if $B = A^{-1}$, we get from $BA = I$,

$$B_{11}A_{11} + B_{12}A_{21} = I, \quad B_{11}A_{12} + B_{12}A_{22} = 0,$$

$$B_{21}A_{11} + B_{22}A_{21} = 0 \text{ and } B_{21}A_{12} + B_{22}A_{22} = I,$$

whence solving the first pair of simultaneous equations and then the second pair, we get

$$\begin{aligned} B_{11} &= (A_{11} - A_{12}A_{22}^{-1}A_{21})^{-1} \\ B_{12} &= -A_{11}^{-1}A_{12}(A_{22} - A_{21}A_{11}^{-1}A_{12})^{-1} \\ B_{21} &= -(A_{22} - A_{21}A_{11}^{-1}A_{12})^{-1}A_{21}A_{11}^{-1} \\ B_{22} &= (A_{22} - A_{21}A_{11}^{-1}A_{12})^{-1}. \end{aligned}$$

In particular, if

$$A = \begin{pmatrix} A_{11} & 0 \\ 0 & A_{22} \end{pmatrix}, \text{ then } A^{-1} = \begin{pmatrix} A_{11}^{-1} & 0 \\ 0 & A_{22}^{-1} \end{pmatrix},$$

as can be verified from the formulæ above or directly by multiplication. ∎

A.10 Characteristic Roots and Vectors

If for a square matrix A, we can write $A\boldsymbol{x} = c\boldsymbol{x}$ for some non-null vector $\boldsymbol{x}$, then c is called a *characteristic root* (or *eigenvalue* or *latent root*) of A and $\boldsymbol{x}$ is called the corresponding *characteristic vector*. A symmetric matrix $A : r$ has r real valued characteristic roots, its rank equals the number of its non-zero characteristic roots and its trace is the sum of its characteristic roots (see Srivastava and Khatri, 1979, Chapter 1). The characteristic roots of a diagonal matrix are its diagonal elements; hence, obviously, all characteristic roots of an identity matrix are one.

If an r-dimensional matrix A is symmetric, then there exists an orthogonal matrix Γ and a diagonal matrix $D = \text{diag}\,(d_1, \ldots, d_r)$ such that

$$A = \Gamma D \Gamma'. \tag{A.7}$$

Moreover, the d_i's are the characteristic roots of A, and the columns of Γ are the corresponding eigenvectors (see Srivastava and Khatri, 1979, p. 18).

Example A.9
Let $X\boldsymbol{c} = \sum_{j=1}^{k} c_j \boldsymbol{x}_j$ be any linear combination of the columns $\boldsymbol{x}_j$ of $X : n \times k$. Since $X'X$ is symmetric (see Section A.2), it can be written as $\Gamma D \Gamma'$ where Γ is orthogonal and D is the diagonal matrix diag $(d_1, \ldots, d_k)$ of the eigenvalues. Let $\boldsymbol{\gamma} = (\gamma_1, \ldots, \gamma_k)' = \Gamma' \boldsymbol{c}$. Then $||\boldsymbol{c}|| = ||\boldsymbol{\gamma}||$ and

$$\min_{||\boldsymbol{c}||=1} ||X\boldsymbol{c}|| = \min_{||\boldsymbol{\gamma}||=1} \boldsymbol{\gamma}' D \boldsymbol{\gamma} = \min_{||\boldsymbol{\gamma}||=1} \sum_{j=1}^{k} d_j \gamma_j^2 = d_{min}, \tag{A.8}$$

the smallest eigenvalue (call it d_j) of $X'X$, and the minimum occurs when $\boldsymbol{\gamma}$ has a 1 in the row corresponding to d_j and has zeros in other positions. For such a $\boldsymbol{\gamma}$, $\boldsymbol{c}$ is the jth column of Γ; i.e., $\boldsymbol{c}$ is the eigenvector corresponding to the smallest eigenvalue. ■

A.11 Idempotent Matrices

A square matrix A will be called *idempotent* if $A^2 = AA = A$. In this book *all idempotent matrices are assumed to be symmetric.* For such matrices $\text{tr}(A) = \rho(A)$. Also it may be shown that $A{:}r$ is idempotent if and only if $\rho(A) + \rho(I_r - A) = r$ (see Srivastava and Khatri, 1979, p. 14). Yet another useful property is that the characteristic roots of an idempotent matrix A are either 0 or 1 and $\rho(A)$ is equal to the number of roots that are 1 (see Exercise A.9).

Example A.10
Let $M^{(0)} = I_n - n^{-1}\mathbf{1}\mathbf{1}'$. Since $\mathbf{1}'\mathbf{1} = n$,

$$(M^{(0)})^2 = I_n - n^{-1}\mathbf{1}\mathbf{1}' - n^{-1}\mathbf{1}\mathbf{1}' + n^{-2}\mathbf{1}\mathbf{1}'\mathbf{1}\mathbf{1}' = M^{(0)}.$$

Hence, $M^{(0)}$ is idempotent. The matrix $\mathbf{1}\mathbf{1}'$ is an $n \times n$ dimensional matrix of 1's; hence $\text{tr}(n^{-1}\mathbf{1}\mathbf{1}') = 1$. Obviously $\text{tr}(I) = n$. Consequently, the trace of $M^{(0)}$ is $n-1$, which is also its rank. Moreover, it has $n-1$ characteristic roots which are 1's and one zero root. ■

Example A.11
Let X be an $n \times (k+1)$ dimensional matrix with $n > k+1$ and let $H = X(X'X)^{-1}X'$. Then

$$HH = X(X'X)^{-1}X'X(X'X)^{-1}X' = X(X'X)^{-1}X' = H.$$

Hence, H is idempotent. From Property 2 of traces (Section A.6)

$$\mathrm{tr}[X(X'X)^{-1}X'] = \mathrm{tr}[X'X(X'X)^{-1}] = \mathrm{tr}(I_{k+1}) = k+1. \quad \text{(A.9)}$$

Hence the rank of H is $k+1$. If $M = I - H$,

$$MM = (I-H)(I-H) = I - H - H + HH = I - H = M,$$

showing that M is also idempotent. Moreover, since

$$\mathrm{tr}(M) = \mathrm{tr}(I) - \mathrm{tr}(H) = n - k - 1, \quad \text{(A.10)}$$

M has rank $n-k-1$. ∎

A.12 The Generalized Inverse

A matrix B is said to be a generalized inverse of A if it satisfies $ABA = A$. A generalized inverse B of A is denoted as A^-. While a singular matrix has no inverse, all matrices have generalized inverses. Notice that if A is non-singular, $AA^{-1}A = A$. Hence an inverse is a generalized inverse. But when A is singular A^- is not unique.

Example A.12
Let X be an $n \times p$ dimensional matrix of rank $p \le n$. Then a generalized inverse of X is given by

$$X^- = (X'X)^{-1}X',$$

since $XX^-X = X(X'X)^{-1}X'X = X$. ∎

Example A.13
Let A be an idempotent matrix. Then $A^- =$ A, since $AAA = AA = A$. If A^- is a generalized inverse for an arbitrary matrix A, then $A = AA^-A$, and hence $A^-A = A^-AA^-A$. Hence the matrix A^-A is an idempotent matrix. Similarly, since $AA^- = AA^-AA^-$, it follows that AA^- is an idempotent matrix. Also, since $\rho(A) = \rho(AA^-A) \le \rho(A^-A) \le \rho(A)$, it follows that $\rho(A) = \rho(A^-A)$. ∎

In fact, it can be shown that if there exists a matrix B such that BA is idempotent and $\rho(BA) = \rho(A)$, then B is a generalized inverse of A. This, along with some other results on generalized inverses, are presented in the following theorem.

Theorem A.2 *Let A be an $m \times n$ dimensional matrix. Then B is a generalized inverse of A if and only if*

1. $\rho(I_n - BA) = n - \rho(A)$, *or*

2. $\rho(BA) = \rho(A)$ *and* BA *is idempotent, or*

3. $\rho(AB) = \rho(A)$ *and* AB *is idempotent.*

Corollary A.1

1. $(A'A)^- A'$ *is a generalized inverse of* A *and* $A(A'A)^-$ *is a generalized inverse of* A',

2. $A(A'A)^- A'A = A$ *and* $(A'A)(A'A)^- A' = A'$, *and*

3. $A(A'A)^- A'$ *is symmetric, idempotent, of rank* $\rho(A)$, *and unique.*

For proofs of these results, see Srivastava and Khatri (1979, p. 12–13) or Rao and Mitra (1971, p. 22–23).

A.13 Quadratic Forms

Let A be a symmetric matrix and $\boldsymbol{x} = (x_1, \ldots, x_r)'$ be a vector. Then $\boldsymbol{x}'A\boldsymbol{x}$, which is a second degree polynomial in the x_i's, is called a quadratic form in $\boldsymbol{x}$. The matrix A will be said to be positive definite (semi-definite) if and only if $\boldsymbol{x}'A\boldsymbol{x} > 0$ (≥ 0) for all $\boldsymbol{x} \neq \mathbf{o}$. The fact that a matrix A is positive definite (semi-definite) is sometimes indicated as $A > 0 (A \geq 0)$. In this book any matrix that is positive definite or positive semi-definite is assumed to be symmetric.

Since

$$\boldsymbol{z}'X'X\boldsymbol{z} = (X\boldsymbol{z})'(X\boldsymbol{z})$$

is a sum of squares and is hence non-negative, $X'X$ is positive semi-definite for all matrices X. Using (A.7), it may be shown that the characteristic roots of a positive definite (semi-definite) matrix are all positive (non-negative). Also, from (A.7) it follows that a matrix A is positive semi-definite if and only if it can be written in the form $A = X'X$, since we can write $A = \Gamma D \Gamma' = \Gamma D^{1/2} D^{1/2} \Gamma' = X'X$, where D is a diagonal matrix with non-negative diagonal elements.

Lemma A.1 *Let* $S : p$ *be a positive definite matrix and* $B : p \times m$ *and* $C : p \times (p - m)$ *be two matrices of ranks* m *and* $(p - m)$, *respectively, such that* $C'B = 0$. *Then*

$$S^{-1} = S^{-1}B(B'S^{-1}B)^{-1}B'S^{-1} + C(C'SC)^{-1}C'.$$

For a proof, see Srivastava and Khatri (1979).

Corollary A.2 *For matrices* B *and* C *as defined above*

$$I = B(B'B)^{-1}B' + C(C'C)^{-1}C'.$$

A.14 Vector Spaces

The material of this section has not been directly used in the text. However, the concepts in it provide an alternative, geometrical discussion of least squares.

A vector of the form $\sum_{i=1}^{n} c_i \boldsymbol{\alpha}_i$ is said to be a linear combination of vectors $\boldsymbol{\alpha}_1, \ldots, \boldsymbol{\alpha}_n$. The set of all linear combinations of a set of vectors constitutes a *vector space*. Obviously a vector space is closed under addition and multiplication by a scalar. Let $\boldsymbol{x}_1, \ldots, \boldsymbol{x}_p$ be $p \leq n$ members of a vector space $\mathcal{S}$. Then all linear combinations of them form a vector (sub)-space which is called a *linear manifold* $\mathcal{M}(X)$ spanned by $X = (\boldsymbol{x}_1, \ldots, \boldsymbol{x}_p)$. All vectors in $\mathcal{S}$ which are orthogonal to each vector in $\mathcal{M}(X)$ also form a vector subspace which is called the *orthogonal complement of* $\mathcal{M}(X)$ and is denoted by $\mathcal{M}^{\perp}(X)$. Needless to say, $\boldsymbol{x} \in \mathcal{M}(X)$ and $\boldsymbol{y} \in \mathcal{M}^{\perp}(X)$ implies $\boldsymbol{x}'\boldsymbol{y} = 0$. It can be shown that any vector $\boldsymbol{z} \in \mathcal{S}$ can be written as $\boldsymbol{z} = \boldsymbol{x} + \boldsymbol{y}$ where $\boldsymbol{x} \in \mathcal{M}(X)$ and $\boldsymbol{y} \in \mathcal{M}^{\perp}(X)$ (Rao, 1973, p. 11).

Example A.14
Linear combinations of the vectors

$$\boldsymbol{\alpha}_1 = \begin{pmatrix} 1 \\ 0 \\ 0 \\ 0 \end{pmatrix}, \ \boldsymbol{\alpha}_2 = \begin{pmatrix} 0 \\ 1 \\ 0 \\ 0 \end{pmatrix}, \ \boldsymbol{\alpha}_3 = \begin{pmatrix} 0 \\ 0 \\ 1 \\ 0 \end{pmatrix}, \ \boldsymbol{\alpha}_4 = \begin{pmatrix} 0 \\ 0 \\ 0 \\ 1 \end{pmatrix}$$

constitute a vector space. Any vector of the form

$$(c_1, c_2, c_3, c_4)' = \sum_{i=1}^{4} c_i \boldsymbol{\alpha}_i$$

is in it. All vectors of the form $(c_1, c_2, 0, 0)'$ are included in the manifold $\mathcal{M}(\boldsymbol{\alpha}_1, \boldsymbol{\alpha}_2)$ and are orthogonal to any vector of the form $(0, 0, c_3, c_4)' \in \mathcal{M}^{\perp}(\boldsymbol{\alpha}_3, \boldsymbol{\alpha}_4)$. ■

Let $\mathcal{S}$ be a vector space and $\mathcal{M}(X)$ be a manifold contained in it and let $\mathcal{M}^{\perp}(X)$ be the orthogonal complement of $\mathcal{M}(X)$. Let $\boldsymbol{z} = \boldsymbol{x} + \boldsymbol{y}$ where $\boldsymbol{z} \in \mathcal{S}$, $\boldsymbol{x} \in \mathcal{M}(X)$ and $\boldsymbol{y} \in \mathcal{M}^{\perp}(X)$. Then a matrix $\mathcal{P}$ such that $\mathcal{P}\boldsymbol{z} = \boldsymbol{x}$ is called a *projection matrix* and is said to *project* $\boldsymbol{z}$ on $\mathcal{M}(X)$ *along* $\mathcal{M}^{\perp}(X)$. It is unique for a given $\mathcal{S}$ and $\mathcal{M}(X)$, and $I - \mathcal{P}$ projects $\boldsymbol{z}$ along $\mathcal{M}(X)$ on $\mathcal{M}^{\perp}(X)$. Moreover, for $\boldsymbol{x} \in \mathcal{M}(X)$ and any $\boldsymbol{y} \in \mathcal{S}$, $\|\boldsymbol{y} - \boldsymbol{x}\|$ assumes its minimum value for $\boldsymbol{x} = \mathcal{P}\boldsymbol{y}$. In order for a matrix to be a projection, it is necessary and sufficient that it be (symmetric and) idempotent (Rao, 1973, pp. 46-48).

Example A.15
H, as defined in Example A.11, is an idempotent matrix. Hence, it is a projection and, since $H\boldsymbol{y} = X\boldsymbol{b}$ if $\boldsymbol{b} = (X'X)^{-1}X\boldsymbol{y}$, H projects a vector

on $\mathcal{M}(X)$. Therefore $M = I - H$ projects a vector on the orthogonal complement of $\mathcal{M}(X)$. Let $\boldsymbol{e} = M\boldsymbol{y}$. Then $\boldsymbol{e}$ and $\hat{\boldsymbol{y}} = H\boldsymbol{y}$ are orthogonal. This can also be verified directly, since $HM = H(I - H) = H - H = 0$ and, therefore, $\hat{\boldsymbol{y}}'\boldsymbol{e} = (H\boldsymbol{y})'M\boldsymbol{y} = \boldsymbol{y}'HM\boldsymbol{y} = 0$.

With the usual 'least squares' meanings (see Chapter 2) given to $\boldsymbol{y}$, $\boldsymbol{e}$, X and $\hat{\boldsymbol{y}}$, this example provides a geometric interpretation of linear least squares which several analysts, including the authors, have found useful. ■

Problems

Exercise A.1: Let

$$A = \begin{pmatrix} 1 & 2 \\ 2 & 1 \end{pmatrix}, \quad B = \begin{pmatrix} 2 & -1 \\ 1 & 2 \end{pmatrix}, \quad C = \begin{pmatrix} 1 & 0 & 1 \\ 0 & 1 & 0 \\ 0 & 0 & 1 \end{pmatrix},$$

$$D = \begin{pmatrix} 1 & 0 & 1 \\ 1 & 1 & 1 \\ 2 & 2 & 3 \end{pmatrix}, \quad \text{and } E = \begin{pmatrix} 1 & 2 & 3 & 4 \end{pmatrix}.$$

1. Obtain $A+B$, $A'+B'$, $C+D$, $C'+D'$, $(C+D)'$.
2. Compute AB, $B'A'$, $A'B'$, $(AB)'$, CD, $C'D'$, $D'C'$, $(CD)'$.
3. Compute EE' and $E'E$.

Exercise A.2: Let $B = [(1-\rho)I_n + \rho \mathbf{1}\mathbf{1}']$.

1. Find the determinant of B and give conditions under which it is positive.
2. Find the inverse of B.

Exercise A.3: Find the eigenvalues of

$$\begin{pmatrix} 1 & \rho \\ \rho & 1 \end{pmatrix}.$$

Exercise A.4: Let

$$X' = \begin{pmatrix} 1 & 1 & 1 \\ 1 & 2 & 3 \end{pmatrix}.$$

Find the eigenvalues of $X'X$, $(X'X)^2$ and $(X'X)^{-1}$.

Exercise A.5: Find the inverse of each of the following three matrices:

$$\begin{pmatrix} 5 & 0 \\ 0 & 4 \end{pmatrix}, \quad \begin{pmatrix} 5 & 0 \\ 3 & 4 \end{pmatrix} \quad \text{and} \quad \begin{pmatrix} 5 & 2 \\ 3 & 4 \end{pmatrix}.$$

Exercise A.6: Find the determinant of the matrix Q in Example A.7.

Exercise A.7: Show that the matrix $\Omega^{-1/2}X(X'\Omega^{-1}X)^{-1}X'\Omega^{-1/2}$, where X is an $n \times p$ matrix and $\Omega^{-1} = \Omega^{-1/2}\Omega^{-1/2}$, is idempotent. What is its rank?

Exercise A.8: Show that the matrix

$$A = \begin{pmatrix} \frac{1}{2} & -\frac{1}{2} \\ -\frac{1}{2} & \frac{1}{2} \end{pmatrix}$$

is an idempotent matrix of rank 1. Find its eigenvalues.

Exercise A.9: Let A be a symmetric, idempotent matrix of rank r. Show that the characteristic roots of A are either zero or 1. Also show that $\rho(A) = r$.

APPENDIX B

Random Variables and Random Vectors

B.1 Random Variables

B.1.1 INDEPENDENT RANDOM VARIABLES

We assume the reader is familiar with the notions of random variables (r.v.'s) and the independence and expectation of r.v.'s. The expectation $E(u)$ of an r.v. u is sometimes called the mean of u and we denote the variance $E(u - E(u))^2$ of the r.v. u as $\text{var}(u)$. The square root of $\text{var}(u)$ is called the standard deviation of u. Given constants a and b, and an r.v. u, the following relationships are true and may be verified from a definition of $E(u)$:

1. $E(au + b) = a E(u) + b$.
2. $\text{var}(u) = \text{var}(u \pm b)$.
3. $\text{var}(au \pm b) = a^2 \text{var}(u)$.
4. $E(u^2) = \text{var}(u) + [E(u)]^2$.

We also have the well known Chebyshev's inequality:

$$P\{|u - E(u)| \geq \tau\} \leq \tau^{-2} \text{var}(u).$$

A function $t(\boldsymbol{x})$ of n random variables $\boldsymbol{x} = (x_1, \ldots, x_n)'$ is called a *statistic*. Now, suppose we wish to estimate a parameter θ by $t(\boldsymbol{x})$. If $E[t(\boldsymbol{x})] = \theta$, then the statistic $t(\boldsymbol{x})$ is said to be an *unbiased* estimator of θ and a measure of precision of this unbiased estimator is its variance $E[t(\boldsymbol{x}) - \theta]^2$. If, on the other hand, $E[t(\boldsymbol{x})] \neq \theta$, then $t(\boldsymbol{x})$ is called a *biased* estimator of θ. A measure of its precision is also $E[t(\boldsymbol{x}) - \theta]^2$, but now since $E[t(\boldsymbol{x})] \neq \theta$, this quantity is not the variance. It is called the *mean square error*. Let $E[t(\boldsymbol{x})] = \eta$. Then

$$\begin{aligned} E[t(\boldsymbol{x}) - \theta]^2 &= E[t(\boldsymbol{x}) - \eta + \eta - \theta]^2 \\ &= E[t(\boldsymbol{x}) - \eta]^2 + (\eta - \theta)^2 + 2(\eta - \theta) E[t(\boldsymbol{x}) - \eta] \\ &= E[t(\boldsymbol{x}) - \eta]^2 + (\eta - \theta)^2 = \text{var}[t(\boldsymbol{x})] + (\eta - \theta)^2. \end{aligned}$$

The quantity $\eta - \theta$ is called the bias, $\text{Bias}[t(\boldsymbol{x})]$, of the estimator $t(\boldsymbol{x})$. Hence,

5. $\text{MSE}(t) = \text{var}(t) + [\text{Bias}(t)]^2$.

B.1.2 Correlated Random Variables

Let u and v be two r.v.'s with means ξ and η respectively. Then the covariance between u and v is defined by

$$\text{cov}(u,v) = \text{E}[(u-\xi)(v-\eta)] = \text{E}[(u-\xi)v] = \text{E}[u(v-\eta)]. \qquad \text{(B.1)}$$

If the random variables u and v are independent, then $\text{E}(uv) = \text{E}(u)\,\text{E}(v)$. Therefore, if u and v are *independent* random variables, then

$$\text{cov}(u,v) = \text{E}[u(v-\eta)] = \text{E}(u)\,\text{E}(v-\eta) = 0.$$

However, $\text{cov}(u,v) = 0$ does not imply independence, as the following example shows. Let the random variable x have a symmetric distribution with mean zero. Then $\text{cov}(x,x^2) = \text{E}(x^3) = 0$, since the odd moments of a symmetric distribution are zero, but x and x^2 are not independent.

The covariance between two random variables depends on measurement units since

$$\text{cov}(au+b, cv+d) = ac\,\text{cov}(u,v).$$

In order to make it free of units, we define the correlation between the two r.v.'s u and v as

$$\rho = \frac{\text{cov}(u,v)}{[\text{var}\,(u)\text{var}\,(v)]^{1/2}}. \qquad \text{(B.2)}$$

B.1.3 Sample Statistics

For a random variable x with observations $x_1,\ldots,x_n$, the sample mean is defined as $\bar{x} = n^{-1}\sum_{i=1}^{n} x_i$ and the sample variance as $(n-1)^{-1}\sum_{i=1}^{n}(x_i - \bar{x})^2$. The square root of the sample variance is called the sample standard deviation.

Let $x_{11},\ldots,x_{n1}$ and $x_{12},\ldots,x_{n2}$ be observations on the variables x_1 and x_2 respectively. Then, the sample correlation coefficient between x_1 and x_2 is defined by

$$r_{x_1,x_2} = r = \frac{\sum_{i=1}^{n}(x_{i1}-\bar{x}_1)(x_{i2}-\bar{x}_2)}{[\sum_{i=1}^{n}(x_{i1}-\bar{x}_1)^2\sum_{i=1}^{n}(x_{i2}-\bar{x}_2)^2]^{1/2}}.$$

The sample correlation coefficient r lies between -1 and 1 and is a measure of the linear relationship between the two variables x_1 and x_2.

The word 'sample' is sometimes dropped when it is obvious from the context that observations from a sample, rather than random variables, are being referred to.

B.1.4 Linear Combinations of Random Variables

Let u be a linear combination of n random variables, i.e.,

$$u = a_1u_1 + \cdots + a_nu_n = \sum_{i=1}^{n} a_iu_i;$$

where $\mathrm{E}(u_i) = \theta_i$, $\mathrm{var}\,(u_i) = \sigma_i^2$ and $\mathrm{cov}(u_i, u_j) = \sigma_{ij}$ when $i \neq j$. Then

1. $\mathrm{E}(u) = \sum_{i=1}^n a_i\theta_i$,
2. $\mathrm{var}\,(u) = \sum_{i=1}^n a_i^2\sigma_i^2 + \sum_{\substack{i,j=1 \\ i\neq j}}^n a_i a_j \sigma_{ij}$,
3. $\mathrm{var}\,(u) = \sum_{i=1}^n a_i^2\sigma_i^2$, if $\sigma_{ij} = 0$ for all $i \neq j$.

In addition, if u_i's are normally distributed, then u is also normal with mean $(\sum_{i=1}^n a_i\theta_i)$ and variance $\sum_{i=1}^n a_i^2\sigma_i^2 + \sum_{\substack{i,j=1 \\ i\neq j}}^n a_i a_j \sigma_{ij}$.

Example B.1
Let $u_1, \ldots, u_n$ be random variables with means $\theta_1, \ldots, \theta_n$ and common variance σ^2. Further, let $\mathrm{cov}(u_i, u_j) = 0$. Then,

$$\mathrm{var}\,(\bar{u}) = n^{-2}\sum_{i=1}^n \mathrm{var}\,(u_i) = \sigma^2/n,$$

where $\bar{u} = n^{-1}\sum_{i=1}^n u_i$. ■

B.2 Random Vectors

When we have several random variables, it is often convenient to write them as vectors or matrices. For example, we may write $\boldsymbol{u} = (u_1, \ldots, u_n)'$ to denote the n r.v.'s, $u_1, \ldots, u_n$. Then $\boldsymbol{u}$ is called a vector of n r.v.'s or simply a vector r.v. or a random n-vector. Similarly, we can have a matrix $U = (u_{ij})$ of r.v.'s u_{ij} where $i = 1, \ldots, I$ and $j = 1, \ldots, J$. We define

$$\mathrm{E}(\boldsymbol{u}') = (\mathrm{E}(u_1), \mathrm{E}(u_2), \ldots, \mathrm{E}(u_n)), \text{ and } \mathrm{E}(U) = (\mathrm{E}(u_{ij})).$$

If A, B and C are constant matrices (of appropriate dimensions), it is easily verified that

1. $\mathrm{E}(\boldsymbol{u} + \boldsymbol{v}) = \mathrm{E}(\boldsymbol{u}) + \mathrm{E}(\boldsymbol{v})$,
2. $\mathrm{E}(A\boldsymbol{u}) = A\,\mathrm{E}(\boldsymbol{u})$,
3. $\mathrm{E}(AU) = A\,\mathrm{E}(U)$,
4. $\mathrm{E}(AUB + C) = A\,\mathrm{E}(U)B + C$.

Thus, if $u = \boldsymbol{a}'\boldsymbol{u} = a_1u_1 + \ldots + a_nu_n$, where $\boldsymbol{a}' = (a_1, \ldots, a_n)$ and $\boldsymbol{u} = (u_1, \ldots, u_n)'$, then $\mathrm{E}(u) = \boldsymbol{a}'\,\mathrm{E}(\boldsymbol{u}) = \boldsymbol{a}'\boldsymbol{\theta}$, where $\boldsymbol{\theta} = (\theta_1, \ldots, \theta_n)'$ and $\theta_i = \mathrm{E}(u_i)$.

We define the covariance matrix of a random vector $\boldsymbol{u}$ with mean vector $\boldsymbol{\theta}$ as

$$\mathrm{cov}(\boldsymbol{u}) = \mathrm{E}[(\boldsymbol{u} - \boldsymbol{\theta})(\boldsymbol{u} - \boldsymbol{\theta})'],$$

where $\boldsymbol{\theta} = \mathrm{E}(\boldsymbol{u})$. It can readily be seen that $\mathrm{cov}(\boldsymbol{u})$ is the expectation of the matrix

$$\begin{pmatrix} (u_1-\theta_1)^2 & (u_1-\theta_1)(u_2-\theta_2) & \dots & (u_1-\theta_1)(u_n-\theta_n) \\ (u_1-\theta_1)(u_2-\theta_2) & (u_2-\theta_2)^2 & \dots & (u_2-\theta_2)(u_n-\theta_n) \\ \dots & \dots & \dots & \dots \\ \dots & \dots & \dots & \dots \\ (u_1-\theta_1)(u_n-\theta_n) & (u_2-\theta_2)(u_n-\theta_n) & \dots & (u_n-\theta_n)^2 \end{pmatrix},$$

i.e., the diagonal elements of $\mathrm{cov}(\boldsymbol{u})$ are the variances of the individual components of $\boldsymbol{u}$ and the other elements are the covariances between pairs of components of $\boldsymbol{u}$. Since $(\boldsymbol{u}-\theta)(\boldsymbol{u}-\theta)'$ is clearly a symmetric matrix and the covariance matrix is simply its expectation, the covariance matrix is always a symmetric matrix.

The variance of $u = \boldsymbol{a}'\boldsymbol{u}$ is

$$\mathrm{var}\,(u) = \mathrm{E}[\boldsymbol{a}'\boldsymbol{u} - \boldsymbol{a}'\boldsymbol{\theta}]^2 = \mathrm{E}[\boldsymbol{a}'(\boldsymbol{u}-\boldsymbol{\theta})]^2 = \mathrm{E}[\boldsymbol{a}'(\boldsymbol{u}-\boldsymbol{\theta})(\boldsymbol{u}-\boldsymbol{\theta})'\boldsymbol{a}],$$

since $\boldsymbol{a}'(\boldsymbol{u}-\boldsymbol{\theta})$ is scalar and is equal to $(\boldsymbol{u}-\boldsymbol{\theta})'\boldsymbol{a}$. Hence

$$\mathrm{var}\,(u) = \boldsymbol{a}'[\,\mathrm{cov}(\boldsymbol{u})]\boldsymbol{a}. \tag{B.3}$$

Theorem B.1 *Let $\boldsymbol{u}$ be a random n-vector, with mean vector $\boldsymbol{\theta}$ and covariance matrix Σ. Then for any $r \times n$ matrix A of constants,*

$$\mathrm{E}(A\boldsymbol{u}) = A\boldsymbol{\theta} \tag{B.4}$$

and

$$\mathrm{cov}(A\boldsymbol{u}) = A\Sigma A'. \tag{B.5}$$

PROOF: The identity (B.4) is trivial and since

$$\begin{aligned} \mathrm{cov}(A\boldsymbol{u}) &= \mathrm{E}[(A\boldsymbol{u} - A\boldsymbol{\theta})(A\boldsymbol{u} - A\theta)'] \\ &= \mathrm{E}[A(\boldsymbol{u}-\boldsymbol{\theta})(\boldsymbol{u}-\boldsymbol{\theta})'A'] = A\,\mathrm{cov}(\boldsymbol{u})A' \end{aligned}$$

we get (B.5). □

Theorem B.2 *The matrix $\Sigma = \mathrm{cov}(\boldsymbol{u})$ is symmetric and at least positive semi-definite.*

PROOF: We have already seen that the covariance matrix is always symmetric. For any linear combination of the vector $\boldsymbol{u}$, say $\boldsymbol{a}'\boldsymbol{u}$, we get from (B.3)

$$\mathrm{var}\,(\boldsymbol{a}'\boldsymbol{u}) = \boldsymbol{a}'\,\mathrm{cov}(\boldsymbol{u})\boldsymbol{a} \geq 0$$

since the variance of any random variable is always non-negative. The theorem follows. □

For a random p-vector $\boldsymbol{u}$, $\mathrm{tr}[\,\mathrm{cov}(\boldsymbol{u})]$ is called the *total variance* of $\boldsymbol{u}$. The total variance is, obviously, the sum of the variances of the individual components of $\boldsymbol{u}$.

Theorem B.3 *Let $\boldsymbol{u}$ be a vector of dimension p with mean vector $\boldsymbol{\theta}$ and covariance matrix Σ. Then*

1. $\mathrm{E}(\boldsymbol{u}\boldsymbol{u}') = \Sigma + \boldsymbol{\theta}\boldsymbol{\theta}'$,

2. $\mathrm{cov}(\boldsymbol{u} + \mathbf{d}) = \mathrm{cov}(\boldsymbol{u})$, *for any constant p-vector* $\mathbf{d}$,

3. $\mathrm{cov}(A\boldsymbol{u} + \mathbf{c}) = A\Sigma A'$, *for any $r \times p$ matrix of constants A and any r-vector $\mathbf{c}$ of constants.*

The proof is simple.

Let $\boldsymbol{x}_1, \ldots, \boldsymbol{x}_n$ random p-vectors, and suppose $\boldsymbol{t}(\boldsymbol{x}_1, \ldots, \boldsymbol{x}_n)$ is a vector statistic (i.e., a vector-valued function of $\boldsymbol{x}_1, \ldots, \boldsymbol{x}_n$) to estimate the parameter vector $\boldsymbol{\theta}$. Then $\boldsymbol{t}$ is said to be an unbiased estimator of $\boldsymbol{\theta}$ if $\mathrm{E}[\boldsymbol{t}(\boldsymbol{x}_1, \ldots, \boldsymbol{x}_n)] = \boldsymbol{\theta}$; otherwise it is called a biased estimator of $\boldsymbol{\theta}$. Let $\mathrm{E}[\boldsymbol{t}] = \boldsymbol{\eta}$. Then $\boldsymbol{\eta} - \boldsymbol{\theta}$ is called the bias of $\boldsymbol{t}$ and is denoted by Bias $[\boldsymbol{t}]$. The *mean square error matrix* of $\boldsymbol{t}$ is defined as $\mathrm{MSE}(\boldsymbol{t}) = \mathrm{E}[(\boldsymbol{t} - \boldsymbol{\theta})(\boldsymbol{t} - \boldsymbol{\theta})']$. It is easy to see that

$$\begin{aligned} \mathrm{MSE}(\boldsymbol{t}) &= \mathrm{E}[(\boldsymbol{t} - \boldsymbol{\eta})(\boldsymbol{t} - \boldsymbol{\eta})'] + (\boldsymbol{\theta} - \boldsymbol{\eta})(\boldsymbol{\theta} - \boldsymbol{\eta})' \\ &= \mathrm{cov}(\boldsymbol{t}) + [\mathrm{Bias}\,(\boldsymbol{t})][\mathrm{Bias}\,(\boldsymbol{t})]'. \end{aligned} \tag{B.6}$$

The trace of the MSE matrix is called the *total mean square, TMSE*. That is,

$$\begin{aligned} \mathrm{TMSE}(\boldsymbol{t}) &= \mathrm{tr}[\mathrm{MSE}(\boldsymbol{t})] = \mathrm{E}[(\boldsymbol{t} - \boldsymbol{\theta})'(\boldsymbol{t} - \boldsymbol{\theta})] \\ &= \mathrm{E}[(\boldsymbol{t} - \boldsymbol{\eta})'(\boldsymbol{t} - \boldsymbol{\eta})] + (\boldsymbol{\theta} - \boldsymbol{\eta})'(\boldsymbol{\theta} - \boldsymbol{\eta}). \end{aligned}$$

B.3 The Multivariate Normal Distribution

A standard normal (i.e., normal with mean 0 and variance 1) r.v. is defined by its probability density function (pdf)

$$(2\pi)^{-1/2} \exp(-\tfrac{1}{2} z^2). \tag{B.7}$$

That z is normal with mean 0 and variance 1 will be denoted by $z \sim N(0, 1)$. A variable u has a normal distribution with mean μ and variance $\sigma^2 > 0$ if u has the same distribution as $\mu + \sigma z$. Then we shall write $u \sim N(\mu, \sigma^2)$.

The joint pdf of n independent standard normal r.v.'s $z_1, \ldots, z_n$ is easily obtained from (B.7); it is

$$\prod_{i=1}^{n} (2\pi)^{-1/2} \exp(-\tfrac{1}{2} z_i^2) = (2\pi)^{-n/2} \exp[-\tfrac{1}{2}(z_1^2 + \cdots + z_n^2)]. \tag{B.8}$$

Writing $\boldsymbol{z} = (z_1, \cdots, z_n)'$, we can verify that $\mathrm{E}(\boldsymbol{z}) = \mathbf{o}$ and $\mathrm{cov}(\boldsymbol{z}) = \mathrm{E}(\boldsymbol{z}\boldsymbol{z}') = I$. We shall say $\boldsymbol{z}$ has a multivariate normal distribution with mean $\mathbf{o}$ and covariance matrix I and denote this fact as $\boldsymbol{z} \sim N(\mathbf{o}, I)$.

An n-dimensional r.v. $\boldsymbol{u}$ is defined as having the multivariate normal distribution with mean $\boldsymbol{\theta}$ and covariance matrix Σ (denoted by $\boldsymbol{u} \sim N(\boldsymbol{\theta}, \Sigma)$) when it has the same distribution as $\boldsymbol{\theta} + A\boldsymbol{z}$ where $AA' = \Sigma$ and $\boldsymbol{z} \sim N(\mathbf{o}, I)$. We shall sometimes write simply normal instead of multivariate normal when context makes the meaning obvious.

If $\boldsymbol{x} \sim N(\boldsymbol{\theta}, \Sigma)$, then $\boldsymbol{a}'\boldsymbol{x} \sim N(\boldsymbol{a}'\boldsymbol{\theta}, \boldsymbol{a}'\Sigma\boldsymbol{a})$, a univariate normal distribution. The converse of this statement is also true. That is, if $\boldsymbol{a}'\boldsymbol{x} \sim N(\boldsymbol{a}'\boldsymbol{\theta}, \boldsymbol{a}'\Sigma\boldsymbol{a})$ for every vector $\boldsymbol{a}$, then $\boldsymbol{x} \sim N(\boldsymbol{\theta}, \Sigma)$. This fact is sometimes used to define the multivariate normal distribution.

Lemma B.1 *Let $\boldsymbol{u} \sim N(\boldsymbol{\theta}, \Sigma)$ and let A be an $r \times n$ matrix of constants; then $A\boldsymbol{u} \sim N(A\boldsymbol{\theta}, A\Sigma A')$.*

The proof can be found in Srivastava and Khatri (1979).

Letting $A = (0, 0, \ldots, 0, 1, 0, \ldots 0)$ (i.e., A consists of a single row with a 1 in the ith position and zeros elsewhere), we see from Lemma B.1 that if $\boldsymbol{u} \sim N(\boldsymbol{\theta}, \boldsymbol{\Sigma})$ and $\Sigma = (\sigma_{ij})$ then $u_i \sim N(\theta_i, \sigma_{ii})$. That is, the marginal distributions of components of normally distributed vectors are also normal. In essentially a similar way it may be shown that if

$$\boldsymbol{u} = \begin{pmatrix} \boldsymbol{u}_1 \\ \boldsymbol{u}_2 \end{pmatrix} \sim N\left(\begin{pmatrix} \boldsymbol{\theta}_1 \\ \boldsymbol{\theta}_2 \end{pmatrix}, \begin{pmatrix} \Sigma_{11} & \Sigma_{12} \\ \Sigma_{12}' & \Sigma_{22} \end{pmatrix} \right), \tag{B.9}$$

then $\boldsymbol{u}_1 \sim N(\boldsymbol{\theta}_1, \Sigma_{11})$.

In (B.9) the vectors $\boldsymbol{u}_1$ and $\boldsymbol{u}_2$ are independently distributed if and only if $\Sigma_{12} = 0$. Note that since $\Sigma_{12} = \operatorname{cov}(\boldsymbol{u}_1, \boldsymbol{u}_2) = \mathrm{E}[(\boldsymbol{u}_1 - \boldsymbol{\theta}_1)(\boldsymbol{u}_2 - \boldsymbol{\theta}_2)']$, which is a matrix consisting of the covariances between components of $\boldsymbol{u}_1$ and those of $\boldsymbol{u}_2$, independence of $\boldsymbol{u}_1$ and $\boldsymbol{u}_2$ will always imply that $\Sigma_{12} = 0$ regardless of the distributions of $\boldsymbol{u}_1$ and $\boldsymbol{u}_2$. While $\Sigma_{12} = 0$ does *not* always imply the independence of $\boldsymbol{u}_1$ and $\boldsymbol{u}_2$, under normality it does.

Example B.2
Let $\boldsymbol{y} \sim N(\boldsymbol{\theta}, \sigma^2 I)$ and $u = \boldsymbol{a}'\boldsymbol{y} = \sum a_i y_i$. Then $u \sim N(\boldsymbol{a}'\boldsymbol{\theta}, \sigma^2 \boldsymbol{a}'\boldsymbol{a})$. As a special case with $a_1 = \cdots = a_n = n^{-1}$ and $\theta_1 = \cdots = \theta_n = \theta$, we get $\bar{y} \sim N(\theta, \sigma^2/n)$. ■

Example B.3
Let $v = \boldsymbol{b}'\boldsymbol{y}$ and $u = \boldsymbol{a}'\boldsymbol{y}$ where $\boldsymbol{y} \sim N(\boldsymbol{\theta}, \sigma^2 I)$. Then

$$\begin{pmatrix} u \\ v \end{pmatrix} \sim N(\delta, \Sigma)$$

where

$$\boldsymbol{\delta} = \begin{pmatrix} \boldsymbol{a}'\boldsymbol{\theta} \\ \boldsymbol{b}'\boldsymbol{\theta} \end{pmatrix} \text{ and } \Sigma = \sigma^2 \begin{pmatrix} \boldsymbol{a}'\boldsymbol{a} & \boldsymbol{a}'\boldsymbol{b} \\ \boldsymbol{a}'\boldsymbol{b} & \boldsymbol{b}'\boldsymbol{b} \end{pmatrix}.$$

This follows from the fact that

$$\begin{pmatrix} u \\ v \end{pmatrix} = \begin{pmatrix} \boldsymbol{a}'\boldsymbol{y} \\ \boldsymbol{b}'\boldsymbol{y} \end{pmatrix} = \begin{pmatrix} \boldsymbol{a}' \\ \boldsymbol{b}' \end{pmatrix} \boldsymbol{y} \equiv C\boldsymbol{y}, \text{ say,}$$

which has a normal distribution with mean $C\boldsymbol{\theta}$ and covariance

$$\sigma^2 CC' = \sigma^2 \begin{pmatrix} \boldsymbol{a}' \\ \boldsymbol{b}' \end{pmatrix} \begin{pmatrix} \boldsymbol{a} & \boldsymbol{b} \end{pmatrix} = \sigma^2 \begin{pmatrix} \boldsymbol{a}'\boldsymbol{a} & \boldsymbol{a}'\boldsymbol{b} \\ \boldsymbol{b}'\boldsymbol{a} & \boldsymbol{b}'\boldsymbol{b} \end{pmatrix}$$

(since $\boldsymbol{a}'\boldsymbol{b} = \sum a_i b_i = \boldsymbol{b}'\boldsymbol{a}$).

Therefore, u and v are independently normally distributed if and only if $\boldsymbol{a}'\boldsymbol{b} = \sum a_i b_i = 0$.

In a similar way it may be shown that if $\boldsymbol{u} = A\boldsymbol{y}$ and $\boldsymbol{v} = B\boldsymbol{y}$, then $\boldsymbol{u}$ and $\boldsymbol{v}$ are independently normally distributed if and only if $AB' = 0$. ■

Example B.4

Suppose $\boldsymbol{e} = [I - X(X'X)^{-1}X']\boldsymbol{y} \equiv M\boldsymbol{y}$ and $\boldsymbol{b} = (X'X)^{-1}X'\boldsymbol{y} = B\boldsymbol{y}$. Since

$$\begin{aligned} MB' &= [I - X(X'X)^{-1}X']X(X'X)^{-1} \\ &= [X - X(X'X)^{-1}X'X](X'X)^{-1} = [X - X](X'X)^{-1} = 0 \end{aligned}$$

it follows that $\boldsymbol{e}$ and $\boldsymbol{b}$ are independently normally distributed if y_i's are normally distributed. ■

B.4 Quadratic Forms of Normal Variables: The Chi-Square Distributions

Let $u_1, \ldots, u_n$ be independent normal r.v.'s and let $\boldsymbol{u}' = (u_1, \ldots, u_n)$. Then we can define the central and noncentral chi-square distributions in the following way:

1. If for each i, u_i has a normal distribution with mean 0 and variance 1, then $\sum_{i=1}^n u_i^2 = \boldsymbol{u}'\boldsymbol{u}$ has a (central) chi-square distribution with n degrees of freedom. We denote this fact as $\sum_{i=1}^n u_i^2 \sim \chi_n^2$ or simply as $\sum_{i=1}^n u_i^2 \sim \chi^2$.

2. If for each i, $u_i \sim N(\theta_i, 1)$, then $\sum_{i=1}^n u_i^2 = \boldsymbol{u}'\boldsymbol{u}$ has a noncentral chi-square distribution with n degrees of freedom and noncentrality parameter $\sum_{i=1}^n \theta_i^2 = \boldsymbol{\theta}'\boldsymbol{\theta}$, where $\boldsymbol{\theta} = (\theta_1, \ldots, \theta_n)'$. We denote this as $\sum_{i=1}^n u_i^2 \sim \chi_n^2(\boldsymbol{\theta}'\boldsymbol{\theta})$.

Clearly, if the noncentrality parameter is zero, we have a central chi-square distribution. It also follows from the definition above that if $u_i \sim$

$N(\theta_i, \sigma_i)$, then $\sum_{i=1}^n (u_i - \theta_i)^2/\sigma_i^2$ has a central chi-square distribution and $\sum_{i=1}^n \sigma_i^{-2} u_i^2$ has a noncentral chi-square distribution. In the latter case the noncentrality parameter is $\sum_{i=1}^n \theta_i^2/\sigma_i^2$, and in either case the number of degrees of freedom is n.

Example B.5
Let $\boldsymbol{y} \sim N(X\boldsymbol{\beta}, \sigma^2 I)$, where $\boldsymbol{y} = (y_1, \ldots, y_n)'$, X is an $n \times p$ matrix and $\boldsymbol{\beta}$ is a p-vector. Further, let $\boldsymbol{b} = A\boldsymbol{y}$ where A is an $r \times n$ matrix. Then $\boldsymbol{b} \sim N(AX\boldsymbol{\beta}, \sigma^2 AA')$. If we choose $A = (X'X)^{-1}X'$, then $AX\boldsymbol{\beta} = \boldsymbol{\beta}$, $AA' = (X'X)^{-1}$,

$$\boldsymbol{b} \sim N(\boldsymbol{\beta}, \sigma^2(X'X)^{-1}) \tag{B.10}$$

and

$$\sigma^{-2}(\boldsymbol{b} - \boldsymbol{\beta})'(X'X)(\boldsymbol{b} - \boldsymbol{\beta}) \sim \chi_p^2. \tag{B.11}$$

■

Example B.6
Let $\boldsymbol{e} \sim N(\mathbf{o}, M)$, where M is an idempotent matrix of rank r. Then $\boldsymbol{e}'M\boldsymbol{e}$ has a central chi-square distribution with r degrees of freedom. This follows from the fact that there exists an orthogonal matrix Γ such that $M = \Gamma D_\lambda \Gamma'$, where $D_\lambda = \operatorname{diag}(1, \ldots, 1, 0, \ldots, 0)$ with the number of ones being equal to the rank of the matrix M. Thus $\boldsymbol{e}'M\boldsymbol{e} = \boldsymbol{e}'\Gamma D_\lambda \Gamma'\boldsymbol{e} = \boldsymbol{z}'D_\lambda \boldsymbol{z} = \sum_{i=1}^r z_i^2$, where $\boldsymbol{z} = (z_1, \ldots, z_n)' = \Gamma'\boldsymbol{e} \sim N(0, \Gamma M \Gamma') = N(0, D_\lambda)$. ■

In fact, there is a stronger result than the above example, which, because of its importance, is stated in the following theorem; the proof can be found in Srivastava and Khatri (1979).

Theorem B.4 *Let $u_1, \ldots, u_n$ be independently normally distributed with means $\theta_1, \ldots, \theta_n$ and common variance σ^2. Then $\sigma^{-2}\mathbf{u}'A\mathbf{u}$, for any symmetric matrix A, has a noncentral chi-square distribution with $r = \operatorname{tr}(A)$ degrees of freedom if and only if A is an idempotent matrix. The noncentrality parameter is given by $\boldsymbol{\theta}'A\boldsymbol{\theta}/\sigma^2$ with $\boldsymbol{\theta}' = (\theta_1, \ldots, \theta_n)$.*

Example B.7
Let $y_1, \ldots, y_n$ be independently normally distributed with common mean θ and common variance σ^2. Then

$$\sigma^{-2} \sum_{i=1}^n (y_i - \bar{y}^2) \sim \chi_{n-1}^2.$$

Since $\bar{y} = n^{-1}\mathbf{1}'\boldsymbol{y}$, where, as in Section A.4, $\mathbf{1} = (1, \ldots, 1)'$, it follows that $n\bar{y}^2 = n^{-1}\boldsymbol{y}'\mathbf{1}\mathbf{1}'\boldsymbol{y}$ and

$$\sum_{i=1}^n (y_i - \bar{y})^2 = \sum_{i=1}^n y_i^2 - n\bar{y}^2 = \boldsymbol{y}'[I - n^{-1}\mathbf{1}\mathbf{1}']\boldsymbol{y} = \boldsymbol{y}'A\boldsymbol{y}.$$

We have seen in Example A.10 that $A^2 = A$ and $\text{tr}(A) = n-1$. Hence, the result follows from Theorem B.4, since $\text{E}[\boldsymbol{y}] = \theta\mathbf{1}$ and $\theta^2\mathbf{1}'A\mathbf{1} = 0$. ∎

The proofs of the following two theorems can also be found in Srivastava and Khatri (1979).

Theorem B.5 (Cochran's Theorem) *Let $\boldsymbol{z} = (z_1, \ldots, z_n)'$ and assume that its components z_i have independent normal distributions with common variance 1 and means $\theta_1, \ldots, \theta_n$. Let $q_1, \ldots, q_p$ be the p quadratic forms $q_j = \boldsymbol{z}'A_j\boldsymbol{z}$ with $\rho(A_j) = n_j$, such that*

$$\boldsymbol{z}'\boldsymbol{z} = q_1 + \cdots + q_p.$$

Then $q_1, \ldots, q_p$ are independently distributed as noncentral chi-squares, $q_j \sim \chi^2_{n_j}(\lambda_j)$ with $\lambda_j = \boldsymbol{\theta}'A_j\boldsymbol{\theta}$, if and only if $n = \sum_{j=1}^p n_j$ and $\boldsymbol{\theta}'\boldsymbol{\theta} = \sum_{j=1}^p \lambda_j$ where $\boldsymbol{\theta} = (\theta_1, \ldots, \theta_p)'$.

Theorem B.6 *Let $z_1, \ldots, z_n$ be n independent normally distributed r.v.'s with means $\theta_1, \ldots, \theta_n$ and common variance σ^2, and let $\boldsymbol{z} = (z_1, \ldots, z_n)'$. Furthermore, let $q_1 = \boldsymbol{z}'A_1\boldsymbol{z}$ and $q_2 = \boldsymbol{z}'A_2\boldsymbol{z}$, where A_1 and A_2 are $n \times n$ symmetric matrices. Then q_1 and q_2 are independently distributed if and only if $A_1A_2 = 0$.*

Example B.8
Suppose $\boldsymbol{y} \sim N(\boldsymbol{\theta}, I)$, and let

$$q = \boldsymbol{y}'\boldsymbol{y} = \boldsymbol{y}'[I - X(X'X)^{-1}X']\boldsymbol{y} + \boldsymbol{y}'[X(X'X)^{-1}X']\boldsymbol{y} = q_1 + q_2,$$

where X is an $n \times p$ matrix of rank p and $\boldsymbol{y}$ is an n-vector. From Example A.11, $H = X(X'X)^{-1}X'$ and $M = [I - X(X'X)^{-1}X']$ are matrices of rank p and $n-p$ respectively. Since $(n-p) + p = n$, q_1 and q_2 are independently distributed as $\chi^2_{n-p}(\delta_1)$ and $\chi^2_p(\delta_2)$ respectively, where

$$\delta_1 = \boldsymbol{\theta}'[I - X(X'X)^{-1}X']\boldsymbol{\theta} \text{ and } \delta_2 = \boldsymbol{\theta}'X(X'X)^{-1}X'\boldsymbol{\theta}.$$

If $\boldsymbol{\theta} = X\boldsymbol{\beta}$, then $\delta_1 = 0$ and $\delta_2 = \boldsymbol{\beta}'(X'X)\boldsymbol{\beta}$.

Independence could also have been demonstrated using Theorem B.6 and Example A.15. ∎

B.5 The F and t Distributions

The ratio of two independent chi-square r.v.'s each divided by its degrees of freedom has an F distribution; i.e.,

$$F_{m,n} = (u/m)/(v/n),$$

where $u \sim \chi^2_m$, $v \sim \chi^2_n$ and u and v are independent, has the F distribution with m and n degrees of freedom. The quantities u/m and v/n are often referred to as averaged chi-square variables. It follows immediately from the definition that $F_{m,n} = 1/F_{n,m}$. Tables for F distribution are given on p. 322 and the three pages following it.

If $z \sim N(0,1)$ and $v \sim \chi^2_n$ and z and v are independent,

$$t_n = z/\sqrt{v/n}$$

has the Student's t distribution with n degrees of freedom. Thus it follows immediately that $(t_n)^2 = z^2/(v/n)$ has the F distribution with 1 and n degrees of freedom. However, some caution should be exercised in the use of this identification since t_n can be used for one-sided hypothesis testing while $F_{1,n}$ can be used only for two-sided tests. A table for the t distribution is given on p. 320.

Since in the above discussion we assumed that u and v are central chi-squares and $z \sim N(0,1)$, we could have called the F and t distributions as defined above the central F and t distributions. When u has a noncentral chi-square distribution with m degrees of freedom and noncentrality parameter δ, then $F_{m,n}$ is said to have a *noncentral F distribution* with (m,n) degrees of freedom and noncentrality parameter δ. Similarly, if $z \sim N(\theta,1)$, then the t_n is said to have a noncentral t distribution with n degrees of freedom and noncentrality parameter θ.

Example B.9

Let $\boldsymbol{y} = (y_1,\ldots,y_n)' \sim N(\mathbf{o},\sigma^2 I)$, $\bar{y} = n^{-1}\sum_{i=1}^n y_i$ and $s^2 = (n-1)^{-1}\sum_{i=1}^n (y_i-\bar{y})^2$. From

$$\sum_{i=1}^n y_i^2 = [\sum_{i=1}^n y_i^2 - n\bar{y}^2] + n\bar{y}^2 = [\sum_{i=1}^n (y_i-\bar{y})^2] + n\bar{y}^2,$$

it follows that

$$q = \sigma^{-2}\boldsymbol{y}'\boldsymbol{y} = \sigma^{-2}\boldsymbol{y}'[I - n^{-1}\mathbf{1}\mathbf{1}']\boldsymbol{y} + \sigma^{-2}n^{-1}\boldsymbol{y}'\mathbf{1}\mathbf{1}'\boldsymbol{y} = q_1 + q_2.$$

Hence, from Example A.10 and Cochran's theorem, q_1 and q_2 are independently distributed as χ^2_{n-1} and χ^2_1 respectively. Therefore, $n\bar{y}^2/s^2 = q_2/[(n-1)^{-1}q_1] \sim F_{1,n-1}$. ■

B.6 Jacobian of Transformations

Let $\boldsymbol{x} = (x_1,\ldots,x_n)'$ be a random vector of n variables and let the pdf of $\boldsymbol{x}$ be $f(\boldsymbol{x})$ when $\boldsymbol{x} \in A$, where A is a subset of the n-dimensional Euclidean space R^n. Let $\boldsymbol{y} = \boldsymbol{g}(\boldsymbol{x})$ be a one-to-one continuously differentiable transformation of $\boldsymbol{x} \in A$, where $\boldsymbol{y} = (y_1, \ldots, y_n)'$ and $\boldsymbol{g} = (g_1, \ldots, g_n)'$, and let the inverse transformation be $\boldsymbol{x} = \boldsymbol{h}(\boldsymbol{y})$. Then the Jacobian of the transformation from $\boldsymbol{x}$ to $\boldsymbol{y}$, denoted by $J(\boldsymbol{x} \to \boldsymbol{y})$, is given by

$$J(\boldsymbol{x} \to \boldsymbol{y}) = \det{}_+ \begin{pmatrix} \frac{\partial x_1}{\partial y_1} & \cdots & \frac{\partial x_1}{\partial y_n} \\ \cdots & \cdots & \cdots \\ \cdots & \cdots & \cdots \\ \frac{\partial x_n}{\partial y_1} & \cdots & \frac{\partial x_n}{\partial y_n} \end{pmatrix},$$

where $\det_+$ stands for the absolute value of the determinant. It may be shown that

$$J(\boldsymbol{x} \to \boldsymbol{y}) = 1/J(\boldsymbol{y} \to \boldsymbol{x})$$

and that the pdf of $\boldsymbol{y}$ is

$$f(\boldsymbol{h}(\boldsymbol{y}))J(\boldsymbol{x} \to \boldsymbol{y})$$

for $\boldsymbol{y} = \{\boldsymbol{g}(\boldsymbol{x}) : \boldsymbol{x} \in A\}$. These results follow from those concerning change of variables in multiple integration (see any advanced calculus book, e.g., Apostol, 1957, p. 271 *et seq.*).

Example B.10

Let x_1 and x_2 be independently normally distributed each with mean 0 and variance 1. Then the pdf of $\boldsymbol{x} = (x_1, x_2)'$ is

$$f(\boldsymbol{x}) = (2\pi)^{-1} \exp[-\tfrac{1}{2}\boldsymbol{x}'\boldsymbol{x}].$$

Let

$$x_1 = r\cos(\theta) \text{ and } x_2 = r\sin(\theta),$$

where $r > 0$ and $0 \leq \theta < 2\pi$. Then, of course, $r^2 = x_1^2 + x_2^2 = \boldsymbol{x}'\boldsymbol{x}$ and

$$\partial x_1/\partial r = \cos(\theta), \quad \partial x_1/\partial\theta = -r\sin(\theta)$$
$$\partial x_2/\partial r = \sin(\theta) \text{ and } \partial x_2/\partial\theta = r\cos(\theta).$$

Hence,

$$J(\boldsymbol{x} \to (r, \theta)') = \det{}_+ \begin{pmatrix} \cos(\theta) & -r\sin(\theta) \\ \sin(\theta) & r\cos(\theta) \end{pmatrix} = r.$$

Thus, the pdf of $(r, \theta)'$ is

$$(2\pi)^{-1} r \exp[-\tfrac{1}{2}r^2],$$

where $r > 0$ and $0 \leq \theta < 2\pi$. ■

Example B.11

Consider the following transformations:

$$y_i^{(\lambda)} = (y_i^\lambda - 1)/\lambda$$

where $\lambda \neq 0$, $y_i > 0$ and $i = 1, \ldots, n$. Let $\boldsymbol{y} = (y_1, \ldots, y_n)'$ and $\boldsymbol{y}^{(\lambda)} = (y_1{}^{(\lambda)}, \ldots, y_n{}^{(\lambda)})'$. Here the Jacobian $J(\boldsymbol{y} \rightarrow \boldsymbol{y}^{(\lambda)})$ is

$$1/J(\boldsymbol{y}^{(\lambda)} \rightarrow \boldsymbol{y}) = 1/\prod_{i=1}^{n}[\partial y_i{}^{(\lambda)}/\partial y_i] = 1/\prod_{i=1}^{n} y_i^{\lambda-1}.$$

This example is used in Section 9.4.3. ■

B.7 Multiple Correlation

Suppose we have k variables $x_1, \ldots, x_k$. Then, the maximum correlation between x_1 and all possible (non-null) linear combinations of $x_2, \ldots, x_k$ is called the multiple correlation coefficient between x_1 and $x_2, \ldots, x_k$. If we write the covariance matrix of $\boldsymbol{x} = (x_1, \ldots, x_n)'$ as

$$\operatorname{cov}(\boldsymbol{x}) = \begin{pmatrix} \sigma_{11} & \boldsymbol{\sigma}_{12}' \\ \boldsymbol{\sigma}_{12} & \Sigma_{22} \end{pmatrix},$$

then it may be shown that the multiple correlation is given by

$$R = (\boldsymbol{\sigma}_{12}'\Sigma_{22}^{-1}\boldsymbol{\sigma}_{12})^{1/2}/\sigma_{11}^{1/2}$$

and its value lies between 0 and 1 (see Srivastava and Khatri, 1979, Chapter 2).

Let

$$Z = \begin{pmatrix} x_{11} - \bar{x}_1 & \ldots & x_{1k} - \bar{x}_k \\ \cdots\cdots & \cdots\cdots & \cdots\cdots \\ \cdots\cdots & \cdots\cdots & \cdots\cdots \\ x_{n1} - \bar{x}_1 & \ldots & x_{nk} - \bar{x}_k \end{pmatrix} = \begin{pmatrix} \boldsymbol{z}_1 & Z_2 \end{pmatrix},$$

where $x_{1j}, \ldots, x_{nj}$ are n observations on each x_j, $\bar{x}_j = n^{-1}\sum_{i=1}^{n} x_{ij}$ and $j = 1, \ldots, k$. Then, if

$$W = Z'Z = \begin{pmatrix} w_{11} & \boldsymbol{w}_{12}' \\ \boldsymbol{w}_{12} & W_{22} \end{pmatrix} = (w_{ij}),$$

we may write $w_{11} = \boldsymbol{z}_1'\boldsymbol{z}_1$, $\boldsymbol{w}_{12}' = \boldsymbol{z}_1'Z_2$, $W_{22} = Z_2'Z_2$ and

$$w_{ij} = \sum_{\ell=1}^{n}(x_{\ell i} - \bar{x}_i)(x_{\ell j} - \bar{x}_j), \text{ and } i, j = 1, \ldots, k.$$

The sample version of the multiple correlation coefficient is defined as follows:

Definition B.1 *The sample multiple correlation $R_{1(2,\ldots,k)}$, between x_1 and $x_2, \ldots, x_k$, is defined by*

$$R_{1(2,\ldots,k)}^2 = \boldsymbol{w}_{12}'W_{22}^{-1}\boldsymbol{w}_{12}/w_{11}.$$

It follows that

$$R^2_{1(2,\ldots,k)} = \boldsymbol{z}_1' Z_2 (Z_2' Z_2)^{-1} Z_2' \boldsymbol{z}_1 / (\boldsymbol{z}_1' \boldsymbol{z}_1) = \boldsymbol{x}_1' Z_2 (Z_2' Z_2)^{-1} Z_2' \boldsymbol{x}_1 / (\boldsymbol{z}_1' \boldsymbol{z}_1)$$

where $\boldsymbol{x}_1 = (x_{11}, \ldots, x_{n1})'$. The last expression in the equation is obtained using the fact that the variables in Z_2 are centered and therefore $\mathbf{1}' Z_2 = \mathbf{o}$.

From the definition it follows, using the results of Example A.8 p. 276, that

$$1 - R^2_{1(2,\ldots,k)} = (w_{11} - \boldsymbol{w}_{12}' W_{22}^{-1} \boldsymbol{w}_{12}) / w_{11} = 1/(w^{11} w_{11}),$$

where w^{11} is the $(1,1)$st element of W^{-1}. If $w_{11} = 1$, then

$$w^{11} = (1 - R^2_{1(2,\ldots,k)})^{-1}.$$

Problems

Exercise B.1: Let $x \sim N(\mu, \sigma^2)$. Show that

$$\mathrm{E}(e^{tx}) = e^{\mu t + \frac{1}{2} t^2 \sigma^2}.$$

Exercise B.2: Let $x \sim N(0, 2)$. Show that $\mathrm{E}|x| = (14/11)^{1/2} = 1.128$.

Exercise B.3: Let $x_1, \ldots, x_n$ be independently and identically distributed with mean 0 and variance σ^2. Let $\bar{x} = n^{-1} \sum_{i=1}^{n} x_i$, and let $y_i = x_i - \bar{x}$ for all $i = 1, \ldots, n$. Find the covariance matrix of $\boldsymbol{y} = (y_1, \ldots, y_n)'$. What can you say about the distribution of $\boldsymbol{y}$ when x_i's are normally distributed?

Exercise B.4: In the above exercise, show that $\bar{x}$ and $\boldsymbol{y}$ are independently distributed under the normality assumption.

Exercise B.5: Let $\boldsymbol{b} = (X'X)^{-1}X'\boldsymbol{y}$ where $\boldsymbol{y} \sim N_n(X\boldsymbol{\beta}, \sigma^2 I_n)$, X is an $n \times p$ matrix of rank $p (\leq n)$ and $\boldsymbol{\beta}$ is a p-vector. Are

$$\boldsymbol{b}'G'[G(X'X)^{-1}G']^{-1}G\boldsymbol{b} \text{ and } \boldsymbol{y}'[I - X(X'X)^{-1}X']\boldsymbol{y}$$

independently distributed?

APPENDIX C

Nonlinear Least Squares

We have made a number of forays into nonlinear least squares (NLS), and on one occasion (Section 6.4, p. 118) we used NLS to fit a linear function because we wished to perform iterative weighting. In this portion of the appendix we present an overview of nonlinear least squares with an emphasis on parameter estimation techniques.

It is usual to use different notation for nonlinear least squares and linear least squares. However, we shall not do this with one exception. We shall continue to use $\boldsymbol{y} = (y_1, \ldots, y_n)'$ to stand for the vector of n dependent variable values and $\boldsymbol{x}_i = (x_{i1}, \ldots, x_{ik})'$ to stand for the vector of values of the k independent variables corresponding to the ith observation. As for linear least squares, we shall let $X = (\boldsymbol{x}_1, \ldots, \boldsymbol{x}_n)'$ stand for the matrix of independent variable values and set $\boldsymbol{\epsilon} = (\epsilon_1, \ldots, \epsilon_n)'$ for the error vector. However, the parameter vector $\boldsymbol{\beta}$ will be replaced by $\boldsymbol{\theta}^* = (\theta_1^*, \ldots, \theta_k^*)'$. This is the true value we are going to attempt to estimate, while $\boldsymbol{\theta}$ will refer to a general point in the parameter space.

Our model is

$$y_i = f(\boldsymbol{x}_i, \boldsymbol{\theta}^*) + \epsilon_i, \tag{C.1}$$

where $i = 1, \ldots, n$. The form of the function f needs to be specified by the user and is usually done on the basis of theoretical considerations in the substantive field of application. The ϵ_i's are assumed to have mean zero and also be homoscedastic and uncorrelated, i.e., $\mathrm{E}[\boldsymbol{\epsilon}] = \mathbf{o}$ and $\mathrm{E}[\boldsymbol{\epsilon}\boldsymbol{\epsilon}'] = \sigma^2 I$. The NLS estimate of $\boldsymbol{\theta}^*$ is the value of $\boldsymbol{\theta}$ which minimizes

$$S^2(\boldsymbol{\theta}) = \sum_{i=1}^{n} (y_i - f(\boldsymbol{x}_i, \boldsymbol{\theta}))^2. \tag{C.2}$$

Assuming adequate differentiability of f, and setting partial derivatives of (C.2) with respect to each θ_j equal to zero, we get

$$\sum_{i=1}^{n} (y_i - f(\boldsymbol{x}_i, \boldsymbol{\theta}))v_{ij} = 0, \tag{C.3}$$

for all $j = 1, \ldots, k$, where $v_{ij} = \partial f(\boldsymbol{x}_i, \boldsymbol{\theta})/\partial\theta_j$. While on occasion the system of k equations can be easily solved, frequently this is not possible and some iterative scheme needs to be used. Statistical packages provide such algorithms and we shall describe some in this part of the appendix.

Example C.1
Let the form of f be $f(\boldsymbol{x}_i, \boldsymbol{\theta}) = \exp[\theta_1 x_{i1}] + \exp[\theta_2 \boldsymbol{x}_{i2}]$. Then, in order to minimize

$$S^2(\boldsymbol{\theta}) = S^2(\theta_1, \theta_2) = \sum_{i=1}^{n} [y_i - \exp(\theta_1 x_{i1}) - \exp(\theta_2 \boldsymbol{x}_{i2})]^2,$$

we set its partial derivatives with respect to θ_1 and θ_2 equal to zero and get the two equations

$$\sum_{i=1}^{n} [y_i - \exp(\theta_1 x_{i1}) - \exp(\theta_2 \boldsymbol{x}_{i2})] x_{i1} \exp(\theta_1 x_{i1}) = 0$$

$$\text{and} \quad \sum_{i=1}^{n} [y_i - \exp(\theta_1 x_{i1}) - \exp(\theta_2 \boldsymbol{x}_{i2})] x_{i2} \exp(\theta_2 x_{i2}) = 0,$$

which are not easy to solve for θ_1 and θ_2. ■

C.1 Gauss-Newton Type Algorithms

Several methods have been given in the literature for the minimization of (C.2). Those based on the Gauss-Newton procedure are perhaps the most popular.

C.1.1 The Gauss-Newton Procedure

Assuming adequate differentiability of f, we can approximate it using linear terms of a Taylor expansion around a value $\boldsymbol{\theta}^{(0)}$ of $\boldsymbol{\theta}$ as follows:

$$f(\boldsymbol{x}_i, \boldsymbol{\theta}) \approx f(\boldsymbol{x}_i, \boldsymbol{\theta}^{(0)}) + \sum_{j=1}^{k} v_{ij}{}^{(0)} (\theta_j - \theta_j{}^{(0)}) \tag{C.4}$$

for $i = 1, \ldots, n$ where $v_{ij}{}^{(0)}$ is the partial derivative

$$v_{ij} = \partial f(\boldsymbol{x}_i, \boldsymbol{\theta}) / \partial \theta_j$$

evaluated at $\boldsymbol{\theta}^{(0)}$. Hence,

$$y_i - f(\boldsymbol{x}_i, \boldsymbol{\theta}) \approx e_i{}^{(0)} - \sum_{j=1}^{k} v_{ij}{}^{(0)} \tau_j{}^{(0)} \tag{C.5}$$

where $\boldsymbol{e}^{(0)} = (e_1{}^{(0)}, \ldots, e_n{}^{(0)})'$, with $e_i{}^{(0)} = y_i - f(\boldsymbol{x}_i, \boldsymbol{\theta}^{(0)})$ and $\boldsymbol{\tau}^{(0)} = (\tau_1{}^{(0)}, \ldots, \tau_k{}^{(0)})' = \boldsymbol{\theta} - \boldsymbol{\theta}^{(0)}$. Therefore, if the approximation (C.4) is close, minimizing $S^2(\boldsymbol{\theta})$ is approximately equivalent to minimizing

$$\sum_{i=1}^{n}(e_i{}^{(0)} - \sum_{j=1}^{k} v_{ij}{}^{(0)}\tau_j{}^{(0)})^2 = (\boldsymbol{e}^{(0)} - V^{(0)}\boldsymbol{\tau}^{(0)})'(\boldsymbol{e}^{(0)} - V^{(0)}\boldsymbol{\tau}^{(0)}) \quad \text{(C.6)}$$

with respect to $\boldsymbol{\tau}^{(0)}$ where $V^{(0)}$ is the matrix $(v_{ij}{}^{(0)})$ of $v_{ij}{}^{(0)}$'s. Since $V^{(0)}\boldsymbol{\tau}^{(0)}$ is linear in $\boldsymbol{\tau}^{(0)}$, (C.6) is obviously minimized by the linear least squares estimate $\boldsymbol{t}^{(0)}$ of $\boldsymbol{\tau}^{(0)}$; i.e., by

$$\boldsymbol{t}^{(0)} = [(V^{(0)})'V^{(0)}]^{-1}(V^{(0)})'\boldsymbol{e}^{(0)}.$$

Therefore, if the approximation (C.4) is reasonably good, one should expect that

$$S^2(\boldsymbol{\theta}^{(0)} + \boldsymbol{t}^{(0)}) < S^2(\boldsymbol{\theta}^{(0)}). \quad \text{(C.7)}$$

Unfortunately, the approximation (C.4) is not always very good and (C.7) does not always hold true, but we shall postpone discussion of remedies for this till later.

Whether (C.7) holds or not, we can mechanically continue with the procedure. At the next (i.e., second) iteration of the procedure, we would start with $\boldsymbol{\theta}^{(1)} = \boldsymbol{\theta}^{(0)} + \boldsymbol{t}^{(0)}$ and do exactly what we did earlier, only with $\boldsymbol{\theta}^{(1)}$ instead of $\boldsymbol{\theta}^{(0)}$. At the $(r+1)$th step, we would start with $\boldsymbol{\theta}^{(r)} = \boldsymbol{\theta}^{(r-1)} + \boldsymbol{t}^{(r-1)}$ and then apply least squares to the model $\boldsymbol{e}^{(r)} = V^{(r)}\boldsymbol{\tau}^{(r)} + \boldsymbol{\epsilon}$ to get the estimate

$$\boldsymbol{t}^{(r)} = [(V^{(r)})'V^{(r)}]^{-1}(V^{(r)})'\boldsymbol{e}^{(r)}$$

where $\boldsymbol{e}^{(r)} = (e_1{}^{(r)}, \ldots, e_n{}^{(r)})'$ with $e_i{}^{(r)} = y_i - f(\boldsymbol{x}_i, \boldsymbol{\theta}^{(r)})$, $V^{(r)}$ is the matrix $(v_{ij}{}^{(r)})$ of the partial derivatives $v_{ij}{}^{(r)} = v_{ij} = \partial f(\boldsymbol{x}_i, \boldsymbol{\theta})/\partial \theta_j$ evaluated at $\boldsymbol{\theta}^{(r)}$ and $\boldsymbol{\tau}^{(r)} = \boldsymbol{\theta} - \boldsymbol{\theta}^{(r)}$.

These iterations are continued until a suitable criterion is met. Typically, iterations are stopped when either

1. $[S^2(\boldsymbol{\theta}^{(r)}) - S^2(\boldsymbol{\theta}^{(r-1)})]/[S^2(\boldsymbol{\theta}^{(r)}) + \delta^*] < \delta$ where δ^* and δ are small numbers (e.g., 10^{-6}), or
2. a specified number of iterations or allotted computer time is exceeded.

Obviously, one can think of possible replacements for 1; e.g., close enough consecutive $\boldsymbol{\theta}^{(r)}$'s.

C.1.2 Step Halving

If at any step $S^2(\boldsymbol{\theta}^{(r+1)}) = S^2(\boldsymbol{\theta}^{(r)} + \boldsymbol{t}^{(r)}) \geq S^2(\boldsymbol{\theta}^{(r)})$, a process called step halving can be undertaken. Instead of setting $\boldsymbol{\theta}^{(r+1)} = \boldsymbol{\theta}^{(r)} + \boldsymbol{t}^{(r)}$ we compute values of $S^2(\boldsymbol{\theta}^{(r)} + \nu\boldsymbol{t}^{(r)})$ for $\nu = .5, .25, .125, \ldots$, and choose a ν for which $S^2(\boldsymbol{\theta}^{(r)} + \nu\boldsymbol{t}^{(r)}) < S^2(\boldsymbol{\theta}^{(r)})$. Then we set $\boldsymbol{\theta}^{(r+1)} = \boldsymbol{\theta}^{(r)} + \nu_0\boldsymbol{t}^{(r)}$ where ν_0 is the value of ν chosen and continue with the Gauss-Newton iterations.

Other methods could have been used to find a suitable ν. These include methods such as golden-section search, which searches over a number of possible values of ν to find one which minimizes $S^2(\boldsymbol{\theta}^{(r)} + \nu \mathbf{t}^{(r)})$.

As can be seen from the argument at the beginning of Section C.2, for sufficiently smooth f, the process of step halving will ultimately always reduce ν to a point where $S^2(\boldsymbol{\theta}^{(r)} + \nu \mathbf{t}^{(r)}) < S^2(\boldsymbol{\theta}^{(r)})$. In fact, the Gauss-Newton iterations with the inclusion of this modification always converge. However, sometimes the speed of convergence is slow enough to render the procedure practically infeasible. Computer packages restrict the number of step halvings and stop when the maximum number is reached. In practical situations we have encountered, a very large number of step halvings usually occur near a local minimum of $S^2(\boldsymbol{\theta})$. However, this need not always be so and then other methods need to be considered. Nevertheless, this modified Gauss-Newton procedure is a default option in SAS PROC NLIN.

C.1.3 Starting Values and Derivatives

To run a Gauss-Newton procedure, the initial estimate $\boldsymbol{\theta}^{(0)}$ of the parameter vector $\boldsymbol{\theta}$ and the partial derivatives of f are needed. Computer programs usually require the user to supply initial estimates, although some (e.g., SAS) will, if requested, compute $S^2(\boldsymbol{\theta})$ for each combination of several possible values of each θ_j to find that combination $\boldsymbol{\theta}^{(0)}$ which minimizes $S^2(\boldsymbol{\theta})$ and will use this $\boldsymbol{\theta}^{(0)}$ as the initial estimate.

Computer programs also frequently require the user to input algebraic expressions for the derivatives. Since this is not always too easy when f is complicated, some programs have the ability to numerically obtain approximate derivatives. One method is to use $[F(\boldsymbol{x}_i, \boldsymbol{\theta} + h\boldsymbol{\Delta}_j) - F(\boldsymbol{x}_i, \boldsymbol{\theta})]/h$, where $\boldsymbol{\Delta}_j$ is a vector of dimension k with its jth element 1 and all other elements zero. Since this is rather time consuming, after the kth iteration derivatives are sometimes estimated from values of $f(\boldsymbol{x}_i, \boldsymbol{\theta}^{(s)})$ obtained from earlier iterations. For example, for $s = r-1, r-2, \ldots, r-k$, $\boldsymbol{\theta}^{(s)}$ can be substituted for $\boldsymbol{\theta}$ in

$$f(\boldsymbol{x}_i, \boldsymbol{\theta}) - f(\boldsymbol{x}_i, \boldsymbol{\theta}^{(r)}) \approx \sum_{j=1}^{k} {v_{ij}}^{(r)} (\theta_j - {\theta_j}^{(r)})$$

and the resultant system of approximate equations solved for each ${v_{ij}}^{(r)}$.

A derivative-free approach has been taken by Ralston and Jennrich (1979; the method is also described in Bates and Watts, 1988, p. 82 *et seq.*) in their DUD (standing for 'Doesn't Use Derivatives') method, which is available in SAS PROC NLIN.

C.1.4 Marquardt Procedure

Since at each step, the Gauss-Newton procedure uses a linear least squares estimate $\boldsymbol{t}^{(r)} = [(V^{(r)})'V^{(r)}]^{-1}(V^{(r)})'\boldsymbol{e}^{(r)}$, it is not unreasonable to expect that if $V^{(r)}$ is ill-conditioned — i.e., if the columns of $V^{(r)}$ are multicollinear — problems would arise. This is indeed what happens. Sometimes in ill-conditioned cases, the Gauss-Newton procedure is excruciatingly slow, requires excessive step halving and becomes practically unusable.

Such situations are handled well by a procedure called Marquardt's procedure, in which at each iteration, $\boldsymbol{\tau}^{(r)} = \boldsymbol{\theta}^{(r+1)} - \boldsymbol{\theta}^{(r)}$ is estimated by

$$\boldsymbol{t}_c{}^{(r)} = [(V^{(r)})'V^{(r)} + cI]^{-1}(V^{(r)})'\boldsymbol{e}^{(r)}, \qquad \text{(C.8)}$$

which is rather akin to the ridge estimate (Section 12.3, p. 256). Marquardt's (1963) paper motivates the method in a slightly different way: as a compromise between the Gauss-Newton procedure, to which Marquardt's procedure is congruent when $c = 0$, and the steepest descent procedure (Section C.2.1), which it resembles for large c. (This procedure was actually suggested by Levenberg, 1947, while Marquardt suggested one in which the cI in (C.8) is replaced by cD where D is the diagonal matrix consisting of the diagonal entries in $(V^{(r)})'V^{(r)}$.)

In actual use, c is usually initially set equal to a very small number (e.g., 10^{-3} in SAS). If, at any iteration, $S^2(\boldsymbol{\theta}^{(r+1)}) > S^2(\boldsymbol{\theta}^{(r)})$ (where $\boldsymbol{\theta}^{(r+1)} = \boldsymbol{\theta}^{(r)} + \boldsymbol{t}_c{}^{(r)}$), then c is scaled up by some factor (10 in SAS) and step (C.8) is repeated. This scaling-up and recomputation of (C.8) may have to be done several times. On the other hand, if $S^2(\boldsymbol{\theta}^{(r+1)}) < S^2(\boldsymbol{\theta}^{(r)})$, then c is scaled down (by .1 in SAS) for the *next iteration*. Thus, if at each iteration $S^2(\boldsymbol{\theta}^{(r)})$ declines, without having to change c within the iteration, then $c \to 0$ as $r \to \infty$ and with increasing r the Marquardt procedure becomes essentially the Gauss-Newton procedure.

C.2 Some Other Algorithms

Another cluster of approaches for obtaining estimates of $\boldsymbol{\theta}$ involve minimizing $S^2(\boldsymbol{\theta})$ — see (C.2) — using one of the standard methods in nonlinear optimization. We present three of these in this section.

The first two require f to be differentiable. Let $\boldsymbol{\nabla}^{(r)}$ be the gradient vector of $S^2(\boldsymbol{\theta})$ at the point $\boldsymbol{\theta} = \boldsymbol{\theta}^{(r)}$, i.e., $\boldsymbol{\nabla}^{(r)} = (\vartheta^{(1)}, \ldots, \vartheta^{(k)})'$ where $\vartheta^{(j)}$'s are the partial derivatives $\vartheta^{(j)} = \partial S^2(\boldsymbol{\theta})/\partial\theta_j$ evaluated at the point $\boldsymbol{\theta}^{(r)}$. From (C.2) and (C.3) it is easy to verify that $\boldsymbol{\nabla}^{(r)} = -2(V^{(r)})'\boldsymbol{e}^{(r)}$ where $V^{(r)}$ and $\boldsymbol{e}^{(r)}$ are as in the last section. Then, taking a Taylor series expansion of $S^2(\boldsymbol{\theta})$ about the point $\boldsymbol{\theta}^{(r)}$, we get, as for (C.4),

$$S^2(\boldsymbol{\theta}) - S^2(\boldsymbol{\theta}^{(r)}) \approx (\boldsymbol{\nabla}^{(r)})'\boldsymbol{\tau}^{(r)}, \qquad \text{(C.9)}$$

where as before, $\boldsymbol{\tau}^{(r)} = \boldsymbol{\theta} - \boldsymbol{\theta}^{(r)}$. By the Cauchy-Schwartz inequality,

$$|(\boldsymbol{\nabla}^{(r)})'\boldsymbol{\tau}^{(r)}| \leq ||\boldsymbol{\nabla}^{(r)}||\,||\boldsymbol{\tau}^{(r)}||,$$

with equality holding if, and only if, for some α, $\boldsymbol{\tau}^{(r)} = \alpha\boldsymbol{\nabla}^{(r)}$. Therefore, the right side of (C.9) is minimized when $\boldsymbol{\tau}^{(r)} = -\alpha^{(r)}\boldsymbol{\nabla}^{(r)}$ with $\alpha^{(r)} > 0$; i.e., when $\boldsymbol{\tau}^{(r)}$ lies along $-\boldsymbol{\nabla}$, or, equivalently, when $\boldsymbol{\theta}$ moves away from $\boldsymbol{\theta}^{(r)}$ in the direction of $-\boldsymbol{\nabla}^{(r)}$. This is called the direction of steepest descent.

Obviously, there are many other directions which also reduce the right side of (C.9). For any k dimensional positive definite matrix P, the direction $-\alpha^{(r)}P\boldsymbol{\nabla}^{(r)}$ $(\alpha^{(r)} > 0)$ will reduce it, since then $(\boldsymbol{\nabla}^{(r)})'\boldsymbol{\tau}^{(r)} = -\alpha^{(r)}(\boldsymbol{\nabla}^{(r)})'P\boldsymbol{\nabla}^{(r)} < 0$. This fact yields a number of alternative methods for minimizing $S^2(\boldsymbol{\theta})$. Actually, since $[(V^{(r)})'V^{(r)}]^{-1}$ and $[(V^{(r)})'V^{(r)} + cI]^{-1}$ are both positive definite, the Gauss-Newton procedure and the Marquardt procedure are, in a sense, applications of this principle.

C.2.1 Steepest Descent Method

As the name implies, after starting at some initial choice $\boldsymbol{\theta}^{(0)}$ of $\boldsymbol{\theta}$, in each iteration of this method, we move in the direction of steepest descent; i.e., at the $(r+1)$th iteration

$$\boldsymbol{\theta}^{(r+1)} = \boldsymbol{\theta}^{(r)} - \alpha^{(r)}\boldsymbol{\nabla}^{(r)} = \boldsymbol{\theta}^{(r)} + 2\alpha^{(r)}V^{(r)\prime}\boldsymbol{e}^{(r)}.$$

The major computational question is essentially the choice of $\alpha^{(r)}$. If $\alpha^{(r)}$ are chosen too small, although the right side of (C.9) will be negative, the entire process will be inefficient. On the other hand, $S^2(\boldsymbol{\theta})$, which initially declines as we move in the $-\boldsymbol{\nabla}^{(r)}$ direction, can start increasing again and for large enough $\alpha^{(r)}$ can exceed $S^2(\boldsymbol{\theta}^{(r)})$.

Computer packages obtain $\alpha^{(r)}$ by computing possible values of $S^2(\boldsymbol{\theta})$ for several possible values of $\alpha^{(r)}$. A very careful choice of $\alpha^{(r)}$ is time consuming, but a cruder choice (which is what is usually made) increases the number of iterations. Frequently, the value of $S^2(\boldsymbol{\theta})$ declines rapidly for the first few iterations and then the rate of decline slows down. Particularly if multicollinearity is present, the procedure can get so slow that it is preferable to use one of the other procedures except under unusual circumstances.

C.2.2 Quasi-Newton Algorithms

Before describing this family of algorithms, we shall describe the Newton or, as it is sometimes called, the Newton-Raphson algorithm. After some initial choice $\boldsymbol{\theta}^{(0)}$, this procedure selects successive $\boldsymbol{\theta}^{(r)}$'s using the formula

$$\boldsymbol{\theta}^{(r+1)} = \boldsymbol{\theta}^{(r)} + \alpha^{(r)}(\mathcal{H}^{(r)})^{-1}\boldsymbol{\nabla}^{(r)}, \qquad \text{(C.10)}$$

where $\alpha^{(r)}$ is a suitable number and $\mathcal{H}^{(r)}$ is the Hessian matrix of $S^2(\boldsymbol{\theta})$ at the point $\boldsymbol{\theta} = \boldsymbol{\theta}^{(r)}$; i.e., $\mathcal{H}^{(r)}$ is the matrix the (j,ℓ)th element of which is the second partial derivative $\partial^2 S^2(\boldsymbol{\theta})/\partial\theta_j\partial\theta_\ell$ evaluated at $\boldsymbol{\theta} = \boldsymbol{\theta}^{(r)}$. The motivation behind the use of (C.10) is the following: Consider a Taylor expansion of $S^2(\boldsymbol{\theta})$ about $\boldsymbol{\theta}^{(r)}$, keeping all linear and quadratic terms —

$$S^2(\boldsymbol{\theta}) - S^2(\boldsymbol{\theta}^{(r)}) \approx (\boldsymbol{\nabla}^{(r)})'\boldsymbol{\tau}^{(r)} + \tfrac{1}{2}(\boldsymbol{\tau}^{(r)})'\mathcal{H}^{(r)}\boldsymbol{\tau}^{(r)}. \tag{C.11}$$

It is easily shown, using matrix differentiation, that the right side of (C.11) is minimized only if

$$\boldsymbol{\nabla}^{(r)} + \mathcal{H}^{(r)}\boldsymbol{\tau}^{(r)} = 0, \tag{C.12}$$

from which, because $(\mathcal{H}^{(r)})^{-1}$ is positive definite, we see that the choice (C.10) minimizes the right side of (C.11) for small enough $\alpha^{(r)}$.

Notice that (C.12) is also the necessary condition for maxima and saddle points, which makes the choice of an initial point slightly more critical compared with other procedures. But possibly the most serious problem is that this procedure requires the computation of the Hessian. Since computer packages are intended to be user friendly, and a requirement of computing and keyboarding expressions for several second derivatives is not likely to win friends, statistical packages provide quasi-Newton methods rather than the Newton method.

In quasi-Newton methods an approximate equivalent $P^{(r)}$ of the inverse of the Hessian is found numerically using a relationship of the form $P^{(r+1)} = P^{(r)} + Q^{(r)}$. These $P^{(r)}$'s are used in place of the $(\mathcal{H}^{(r)})^{-1}$'s in (C.10). Taking a linear Taylor series approximation for the gradient vector, we get

$$\boldsymbol{\nabla}^{(r+1)} - \boldsymbol{\nabla}^{(r)} \approx \mathcal{H}^{(r+1)}(\boldsymbol{\theta}^{(r+1)} - \boldsymbol{\theta}^{(r)}),$$

which yields, if $\mathcal{H}^{(r+1)}$ is nonsingular,

$$(\mathcal{H}^{(r+1)})^{-1}(\boldsymbol{\nabla}^{(r+1)} - \boldsymbol{\nabla}^{(r)}) \approx (\boldsymbol{\theta}^{(r+1)} - \boldsymbol{\theta}^{(r)}).$$

If we now replace $(\mathcal{H}^{(r+1)})^{-1}$ by $P^{(r+1)} = P^{(r)} + Q^{(r)}$ we get

$$Q^{(r)}(\boldsymbol{\nabla}^{(r+1)} - \boldsymbol{\nabla}^{(r)}) \approx (\boldsymbol{\theta}^{(r+1)} - \boldsymbol{\theta}^{(r)}) - P^{(r)}(\boldsymbol{\nabla}^{(r+1)} - \boldsymbol{\nabla}^{(r)}) = \boldsymbol{\xi}. \tag{C.13}$$

A number of choices of $Q^{(r)}$ will satisfy (C.13). One suggestion, made by Fletcher and Powell (1963), is

$$\begin{aligned} Q^{(r)} = &\frac{(\boldsymbol{\theta}^{(r+1)} - \boldsymbol{\theta}^{(r)})(\boldsymbol{\theta}^{(r+1)} - \boldsymbol{\theta}^{(r)})'}{(\boldsymbol{\theta}^{(r+1)} - \boldsymbol{\theta}^{(r)})'(\boldsymbol{\nabla}^{(r+1)} - \boldsymbol{\nabla}^{(r)})} \\ &- \frac{P^{(r)}(\boldsymbol{\nabla}^{(r+1)} - \boldsymbol{\nabla}^{(r)})(\boldsymbol{\nabla}^{(r+1)} - \boldsymbol{\nabla}^{(r)})'P^{(r)}}{(\boldsymbol{\nabla}^{(r+1)} - \boldsymbol{\nabla}^{(r)})'P^{(r)}(\boldsymbol{\nabla}^{(r+1)} - \boldsymbol{\nabla}^{(r)})}. \end{aligned} \tag{C.14}$$

The initial $P^{(0)}$ can be chosen to be I. If $\alpha^{(r)}$ is chosen to yield a minimum of $S^2(\boldsymbol{\theta})$, then $P^{(r+1)}$ chosen in this way will be positive definite (Fletcher and Powell, 1963).

C.2.3 The Simplex Method

The simplex method is a derivative-free method which is only very distantly related to the simplex method used in linear programming. Roughly, the procedure is the following. Around any given value $\boldsymbol{\theta}^{(r)}$ of $\boldsymbol{\theta}$ consider the point $\boldsymbol{\theta}^{(r)} + h\mathbf{1}$ and the k points $\boldsymbol{\theta}^{(r)} + h\boldsymbol{\Delta}_j$ for $j = 1, \ldots, k$ where, as in Section C.1.3, $\boldsymbol{\Delta}_j$ is a vector with all components zero except the jth, which is one, h is a positive number and, as before, $\mathbf{1} = (1, \ldots, 1)'$. These $k+1$ points are the vertices of a simplex in k dimensional space. (Notice that in two dimensions a simplex is a triangle and in three it is a tetrahedron.) These vertices are examined to find which one maximizes $S^2(\boldsymbol{\theta})$. Then a selection of points along the direction connecting this vertex to the centroid of the remaining vertices is examined to find a point which adequately reduces $S^2(\boldsymbol{\theta})$. This point is $\boldsymbol{\theta}^{(r+1)}$, which is the starting point for the next iteration.

Since calculations are not carried over from iteration to iteration, round-off errors do not propagate and multiply. Although time consuming, the simplex method has the reputation of being very reliable. For more details see Griffiths and Hill (1985).

C.2.4 Weighting

Nonlinear least squares is sometimes weighted and for much the same reasons that linear least squares is on occasion weighted. (For a fuller treatment of this subject, see Carroll and Ruppert, 1988.) In fact, as mentioned in Section 6.4, p. 118, we frequently perform weighted linear least squares using nonlinear methods when the weights depend on the parameters being estimated. In weighted NLS, instead of (C.2), we minimize

$$S^2(\boldsymbol{\theta}) = \sum_{i=1}^{n} w_i(y_i - f(\boldsymbol{x}_i, \boldsymbol{\theta}))^2 = \sum_{i=1}^{n} (\sqrt{w_i} y_i - \sqrt{w_i} f(\boldsymbol{x}_i, \boldsymbol{\theta}))^2 \qquad \text{(C.15)}$$

where w_i's are weights. Some computer packages require that this objective function (C.15), sometimes called a loss function, be explicitly described. Others accept a description of the weights w_i.

Now assume that f is linear in $\boldsymbol{\theta}$, i.e., $f(\boldsymbol{x}_i, \boldsymbol{\theta}) = \boldsymbol{x}_i'\boldsymbol{\theta}$. Further, let $w_i = w_i(\boldsymbol{\theta})$'s be functions of $\boldsymbol{\theta}$ and $W(\boldsymbol{\theta}) = \operatorname{diag}(w_1(\boldsymbol{\theta}), \ldots, w_n(\boldsymbol{\theta}))$. Now the objective function becomes

$$S^2(\boldsymbol{\theta}) = (\boldsymbol{y} - X\boldsymbol{\theta})' W(\boldsymbol{\theta})(\boldsymbol{y} - X\boldsymbol{\theta}).$$

The Gauss-Newton method and the methods of Section C.2 can easily be applied to this objective function. Computationally, all we need do is run a weighted NLS program with f set as $\boldsymbol{x}_i'\boldsymbol{\theta}$. This is what we did in Section 6.4.

We now show that the Gauss-Newton procedure applied to a weighted linear least squares model yields the steps of the iteratively reweighted least squares procedure described in Section 6.4. In this case, at the $(r+1)$th step of the Gauss-Newton procedure $\boldsymbol{e}^{(r)}$ becomes $\boldsymbol{e}^{(r)} = \boldsymbol{y} - X\boldsymbol{\theta}^{(r)}$ and, since $V^{(r)} = X$ for all iterations, we now use

$$\boldsymbol{\theta}^{(r+1)} - \boldsymbol{\theta}^{(r)} = [X'W^{(r)}X]^{-1}X'W^{(r)}\boldsymbol{e}^{(r)}$$

for our iterations, where $W^{(r)} = W(\boldsymbol{\theta}^{(r)})$. Therefore,

$$\begin{aligned}\boldsymbol{\theta}^{(r+1)} &= \boldsymbol{\theta}^{(r)} + [X'W^{(r)}X]^{-1}X'W^{(r)}\boldsymbol{e}^{(r)} \\ &= \boldsymbol{\theta}^{(r)} + [X'W^{(r)}X]^{-1}X'W^{(r)}[\boldsymbol{y} - X\boldsymbol{\theta}^{(r)}] \\ &= \boldsymbol{\theta}^{(r)} + [X'W^{(r)}X]^{-1}X'W^{(r)}\boldsymbol{y} - [X'W^{(r)}X]^{-1}X'W^{(r)}X\boldsymbol{\theta}^{(r)} \\ &= [X'W^{(r)}X]^{-1}X'W^{(r)}\boldsymbol{y}.\end{aligned}$$

In carrying out such iteratively reweighted linear least squares using the Gauss-Newton procedure, step halving is unnecessary and in fact, should be disabled if it is the default option.

C.3 Pitfalls

As Wilkinson (1987, p. NONLIN–33) puts it, "Nonlinear estimation is an art. One-parameter estimation is minimalist, multiparameter is rococo. There are numerous booby traps (dependencies, discontinuities, local minima, etc.) which can ruin your day." With simple f's and few parameters, point estimation of $\boldsymbol{\theta}$ can be quite straightforward; but when f is 'terribly nonlinear' with many parameters, the successful completion of a least squares exercise is often cause for celebration.

With continuously differentiable $S^2(\boldsymbol{\theta})$'s, the main problem is one of several local minima. This does not always occur. In some cases it is possible to show theoretically that there is only one local minimum which, therefore, is global. This is particularly true when f is linear and the weights are the cause of nonlinearity of $S^2(\boldsymbol{\theta})$. When the form of f has been well studied in the literature, chances are that the question of uniqueness of estimates has already been examined. When theoretical information is not available and there are relatively few parameters, values of $S^2(\boldsymbol{\theta})$ for different $\boldsymbol{\theta}$'s can be computed and these would provide us with an understanding of the behavior of $S^2(\boldsymbol{\theta})$. Alternatively, one could run an NLS procedure for several different and widely separated starting values to see if the respective iterations converge to different local minima. If they do, the $\boldsymbol{\theta}$ yielding the smallest minimum would be selected (after perhaps trying out some more initial values). Another suggestion (Freund and Littell, 1986) is to approximate f with a suitable polynomial which can then be fitted by linear least squares. If the residual sum of squares for the polynomial regression is close to the value of $S^2(\boldsymbol{\theta})$ for the $\boldsymbol{\theta}$ found by the nonlinear method, then the chances are that the desired global minimum of $S^2(\boldsymbol{\theta})$ has been found.

Unless one is very sure that the right estimate has been found and is secure about the quality of the estimates, it is essential to end an NLS exercise with an examination of plots of residuals against independent variables and predicted values.

Even apart from the issue of multiple minima, the selection of initial estimates deserves attention. When parameters appear in the exponent of large numbers within f, bad starting estimates can lead to underflows and overflows; i.e., to numbers too large or too small for the computer to handle.

A number of methods are available for choosing initial values if intuition is not adequate. When an OLS cognate is available, it can be run to obtain initial parameter estimates. For example, if f is linear and we are using NLS because weights depend on parameters, we could use OLS (setting all weights equal to 1) to obtain initial estimates. If by ignoring the error term, it is possible to transform the variables to make $y = f(\boldsymbol{x}, \boldsymbol{\theta})$ linear, we could apply OLS to this transformed model.

Often an examination of the behavior of f as different independent variables go to zero or infinity can give us an idea of what initial values to choose. Draper and Smith (1981, p. 474) describe a generalization of this approach in which they suggest solving for $\boldsymbol{\theta}$, the k equations $\boldsymbol{y_i} = f(\boldsymbol{x}_i, \boldsymbol{\theta})$, where the k sets of independent variable values have been chosen from among all $\boldsymbol{x}_i$'s to be reasonably widely separated, and then using this solution as the initial choice.

Example C.2

A reasonable model for population densities y_i in an urban area is $y_i = \theta_1 + \theta_2 d_i^{-\theta_3} + \epsilon_i$, where d_i is the distance from city center and $\theta_3 > 0$. When d_i is large, $y_i \approx \theta_1$, the density in the urban fringe and rural areas. Ignoring the subscripts and the error terms, the model can be written as

$$\log[y_i - \theta_1] = \log[\theta_2] - \theta_3 \log[d].$$

Therefore, if we know θ_1 we can plot $\log[y_i - \theta_1]$ against $\log[d]$. The intercept will give us $\log[\theta_2]$ and the slope θ_3. These estimates should be good enough as initial values. ■

Even with reasonable choices of initial estimates, NLS methods sometimes give trouble. Improvements in $S^2(\boldsymbol{\theta})$ from iteration to iteration could become small, or, in the case of the Gauss-Newton procedure, step halving might have to be done too often. Overflows or underflows can still occur. In such cases, switching methods often helps and fortunately most serious statistical packages offer a choice of methods.

Up to now we have concerned ourselves solely with the problem of getting estimates at all, not with their quality. In general, NLS estimates are at least as susceptible to the ill effects of outliers, incorrect specification, and multicollinearity as OLS estimates. While sometimes OLS methods can be adapted for identification and alleviation of these problems, such methods are not in general use and are not available in the popular statistical packages.

C.4 Bias, Confidence Regions and Measures of Fit

Assume that all went well with one of the procedures described in Sections C.1 and C.2 and the iterations converged at step u so that $\boldsymbol{\theta}^{(u)}$ is the point $\hat{\boldsymbol{\theta}}$ at which the global minimum of $S^2(\boldsymbol{\theta})$ occurs. Actually, $\boldsymbol{\theta}^{(u)}$ will only be a close approximation to such a point but let us pretend that it is the minimizing point. When f is nonlinear, this final estimate $\hat{\boldsymbol{\theta}}$ will nearly always be a biased estimate of $\boldsymbol{\theta}^*$ in the model (C.1), i.e., in general $\mathrm{E}[\hat{\boldsymbol{\theta}}] \neq \boldsymbol{\theta}^*$ (Box, 1971).

However, under some mild conditions, $\hat{\boldsymbol{\theta}}$ may be shown to be a consistent and asymptotically normal estimator of $\boldsymbol{\theta}^*$ in the model (C.1) — see Amemiya (1983) or Judge *et al.* (1985). Moreover, the asymptotic covariance matrix of $\boldsymbol{\theta}^{(u)}$ is $\sigma^2[(V^{(u)})'V^{(u)}]^{-1}$ and $s^2 = S^2(\hat{\boldsymbol{\theta}})/(n-k)$ is a consistent estimator of σ^2. These facts are used in testing linear hypotheses and in the construction of confidence intervals and regions with $V^{(u)}$ playing the role X did in Chapter 3. For example, if $s^2[(V^{(u)})'V^{(u)}]^{-1} = (a_{ij})$, then $\hat{\theta}_j/a_{jj}^{1/2}$ is treated as if it had the t distribution. As for the OLS case, computer packages provide these t values.

Several asymptotically equivalent alternative expressions for the covariance matrix are also available (e.g., estimates of the Hessian matrix can also be used to obtain an estimate for the covariance). For these alternatives and for tests of nonlinear hypotheses, see Judge *et al.* (1985, p. 209 *et seq.*)

From the discussion of Section C.1.1, we can see where the covariance matrix $\sigma^2[(V^{(u)})'V^{(u)}]^{-1}$ came from. In that section we had approximated the function f by its tangent plane $V^{(r)}(\boldsymbol{\theta}-\boldsymbol{\theta}^{(r)})$ and thereby obtained the estimate

$$\begin{aligned} \boldsymbol{t}^{(r)} &= [(V^{(r)})'V^{(r)}]^{-1}(V^{(r)})'\boldsymbol{e}^{(r)} \\ &\approx [(V^{(r)})'V^{(r)}]^{-1}(V^{(r)})'[V^{(r)}(\boldsymbol{\theta}-\boldsymbol{\theta}^{(r)})+\boldsymbol{\epsilon}] \\ &= (\boldsymbol{\theta}-\boldsymbol{\theta}^{(r)}) + [(V^{(r)})'V^{(r)}]^{-1}(V^{(r)})'\boldsymbol{\epsilon}, \end{aligned}$$

whence $\mathrm{cov}(\boldsymbol{t}^{(r)}) = \sigma^2[(V^{(r)})'V^{(r)}]^{-1}$.

All tests and confidence regions using asymptotic results are actually based on the tangent plane. The level curves of $S^* = (\boldsymbol{y} - V^{(u)}\boldsymbol{\theta})'(\boldsymbol{y} - V^{(u)}\boldsymbol{\theta})$, i.e., curves over which the values of S^* are the same, can easily be seen (from the form of S^*, which is quadratic in $\boldsymbol{\theta}$) to be ellipses around $\boldsymbol{\theta}^{(u)}$. But $S^2(\boldsymbol{\theta})$, which is approximated by S^*, will rarely have elliptic level curves. Since level curves bound confidence regions, the asymptotic confidence regions can in practice be quite different from the actual confidence

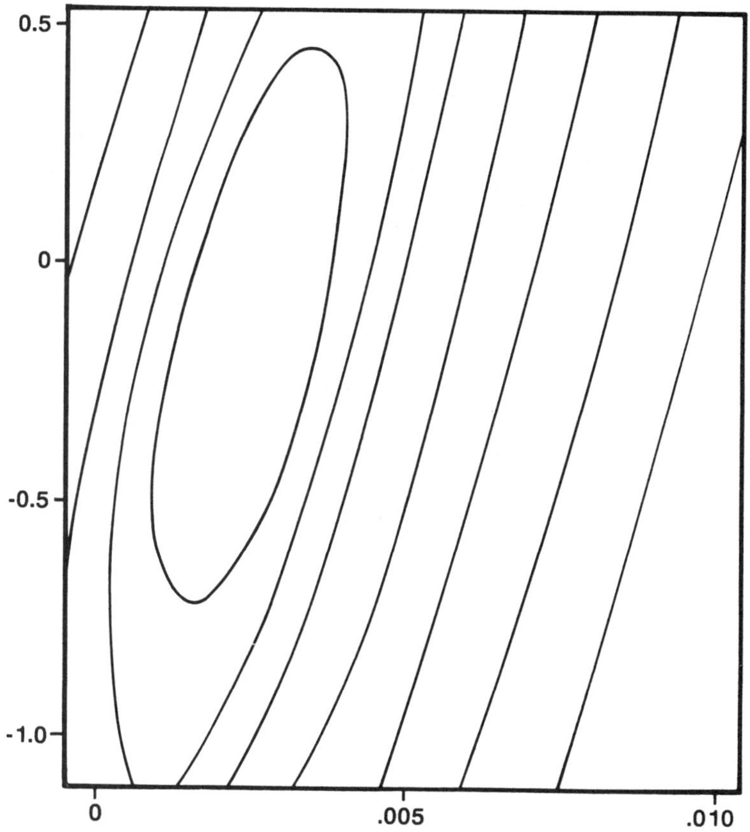

EXHIBIT C.1: Level Curves of Weighted Sum of Squares for Travel Mode Choice Data

regions. [Obviously, if two or less parameters are involved, we can obtain approximate level curves by computing $S^2(\boldsymbol{\theta})$ for several values of $\boldsymbol{\theta}$ near $\hat{\boldsymbol{\theta}}$, but we would not know the confidence levels associated with the regions the curves enclose.] The upshot of all this is that test and confidence regions associated with nonlinear regression should be interpreted with several grains of salt.

Most computer packages provide the value of $S^2(\hat{\boldsymbol{\theta}})$ and/or $S^2(\hat{\boldsymbol{\theta}})/(n-k)$ which can be used as a measure of fit. They will either give $R^2 = 1 - S^2(\hat{\boldsymbol{\theta}})/\sum_{i=1}^n y_i^2$ or provide the ingredients to compute it (either as part of the NLS program or some other portion of the statistical package). Notice that this is the no-intercept form of R^2, which would appear to be relevant here since frequently there are no intercepts in the functions we fit, and

Step	θ_0	θ_1	$S^2(\boldsymbol{\theta})$
0	0	0	1381.670051071950
1	-0.116583990	0.0020871670	129.720690631157
2	-0.123448664	0.0024688319	112.656625048396
	-0.124069413	0.0024941630	113.229792492524
	-0.123759038	0.0024814975	112.916241995827
	-0.123603851	0.0024751647	112.779692655420
	-0.123526257	0.0024719983	112.716473728977
	-0.123487460	0.0024704151	112.686128119070
	-0.123468062	0.0024696235	112.671271267681
	-0.123458363	0.0024692277	112.663921829192
	-0.123453513	0.0024690298	112.660266856604
	-0.123451088	0.0024689309	112.658444306955
	-0.123449876	0.0024688814	112.657534266290
	-0.123449270	0.0024688567	112.657079554497
	-0.123448967	0.0024688443	112.656852275735
	-0.123448815	0.0024688381	112.656738655638
	-0.123448739	0.0024688350	112.656681850410
	-0.123448702	0.0024688335	112.656653449002
	-0.123448683	0.0024688327	112.656639248599
	-0.123448673	0.0024688323	112.656632148472
	-0.123448668	0.0024688321	112.656628598428
	-0.123448666	0.0024688320	112.656626823411
	-0.123448665	0.0024688320	112.656625935903

EXHIBIT C.2: Iterations of the Gauss-Newton Procedure for Travel Mode Choice Data

even if there were, their role would be different. But many packages will also give the intercept form of R^2; i.e., $1 - S^2(\hat{\boldsymbol{\theta}})/\sum_{i=1}^{n}(y_i - \bar{y})^2$ or the means to compute it.

C.5 Examples

Several of the manuals that accompany statistical packages provide numerous examples of NLS. We shall augment these with two more.

Example C.3

As mentioned in Section 9.2.4, p. 186, we can also use NLS to estimate parameters of a logit model. Therefore, consider the data of Exercise 9.8, p. 214, and let us estimate θ_0 and θ_1 in the model

$$y_i = \frac{\exp(\theta_0 + \theta_1 x_{i1})}{1 + \exp(\theta_0 + \theta_1 x_{i1})} + \epsilon_i = 1 - [1 + \exp(-\theta_0 - \theta_1 x_{i1})]^{-1} + \epsilon_i, \quad \text{(C.16)}$$

Step	θ_0	θ_1	$S^2(\boldsymbol{\theta})$
0	0	0	1381.670051071950
1	-0.117900463	0.0020828501	129.936954803759
2	-0.123479203	0.0024685013	112.651795383351
	-0.124069491	0.0024941602	113.229723134302
	-0.124070493	0.0024941562	113.229666855365
	-0.124080477	0.0024941165	113.229105112580
	-0.124176787	0.0024937288	113.223589469723
	-0.124870685	0.0024906104	113.176435170649
	-0.125648332	0.0024800114	112.960160699828
	-0.124094688	0.0024708301	112.714054674204
	-0.123551655	0.0024687639	112.658718755054
	-0.123486576	0.0024685279	112.652495508745
	-0.123479941	0.0024685039	112.651865474667
	-0.123479277	0.0024685015	112.651802393271
	-0.123479210	0.0024685013	112.651796084351
	-0.123479204	0.0024685013	112.651795453451
	-0.123479203	0.0024685013	112.651795390361
	-0.123479203	0.0024685013	112.651795384052
	-0.123479203	0.0024685013	112.651795383421
	-0.123479203	0.0024685013	112.651795383358
	-0.123479203	0.0024685013	112.651795383352
	-0.123479203	0.0024685013	112.651795383351
3	-0.123479203	0.0024685013	112.651795383351

EXHIBIT C.3: Iterations of the Marquardt Procedure for Travel Mode Choice Data

Parameter	Estimate	Asymp. s.e.
θ_0	-.12345	.066272
θ_1	.00247	.000166

EXHIBIT C.4: Parameter Estimates and Standard Errors for Model Choice Data

where $x_{i1} = t_r - t_a$ is the difference in travel times between transit and car. Since y_i is a proportion of counts, we need to run a weighted nonlinear least squares procedure with weights w_i which are $n[\mathrm{E}(y_i)]^{-1}[1 - \mathrm{E}(y_i)]^{-1}$. Level curves of a weighted version of $S^2(\boldsymbol{\theta})$ (see (C.15)) for intervals of parameter values that we felt were reasonable are shown in Exhibit C.1. The curves indicate that there is possibly a single reasonable local minimum value. What we need to do now is find it.

Let us use a Gauss-Newton procedure for which it is usually necessary to specify the derivatives of the right side of (C.16) with respect to θ_0 and θ_1, and a set of initial values for the parameters. The derivatives are,

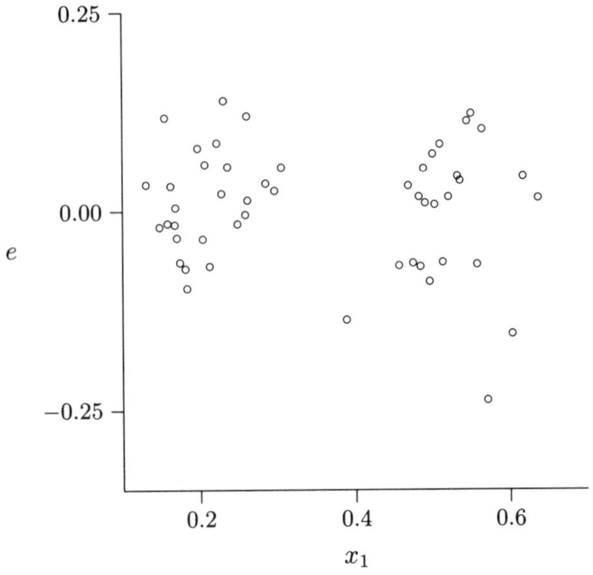

EXHIBIT C.5: Plot of Residuals Against Independent Variable Values for Travel Mode Choice Data

respectively,

$$\frac{\exp(-\theta_0 - \theta_1 x_{i1})}{[1+\exp(-\theta_0 - \theta_1 x_{i1})]^2} \quad \text{and} \quad \frac{x_{i1}\exp(-\theta_0 - \theta_1 x_{i1})}{[1+\exp(-\theta_0 - \theta_1 x_{i1})]^2}.$$

We tried a number of different sets of initial values, including those found by using OLS, i.e., by completing Exercise 9.8. Exhibit C.2 shows the values of the parameters and $S^2(\boldsymbol{\theta})$ for a set of iterations corresponding to the starting value $\boldsymbol{\theta} = (0, 0)$. The rows with no iteration number represent step halving. The iterations end with ten step halvings producing no reduction in $S^2(\boldsymbol{\theta})$.

For each of the starting values we tried, after three or fewer steps, the parameter values reached a point close to that at Step 2 of Exhibit C.2. This and Exhibit C.1 led us to conjecture that the iterations had indeed taken us to a point very near the minimum point.

Exhibit C.3 shows iterations for the Marquardt procedure applied to this problem. Unnumbered rows indicate increasing c by a factor of 10 (see Section C.1.4). These iterations did ultimately meet the convergence criterion. The differences in estimates for the two procedures as they stand are barely perceptible and therefore we felt quite sure that we had essentially found the desired value $\hat{\boldsymbol{\theta}}$ of $\boldsymbol{\theta}$. Our estimates along with their standard errors are given in Exhibit C.4. These estimates are, in fact, very close to those that would be obtained from Exercise 9.8. Exhibit C.5 shows a plot of residuals against x_{i1}'s. While it illustrates the possible existence of an outlier, it also shows that for the data used we have a good fit. There is nothing to indicate that the local optimum found does not provide good parameter estimates.

The program we used also gives values of

$$\sum_{i=1}^{n} w_i y_i^2 \text{ and } \sum_{i=1}^{n} w_i (y_i - \bar{y})^2,$$

which are 2701.00 and 694.51. Therefore, the no-intercept version of R^2 is .96 and the intercept version is .84. Since the weighting essentially divides each $[y_i - f(\boldsymbol{x}_i, \boldsymbol{\theta})]^2$ by the variance of y_i, $S^2(\boldsymbol{\theta})/[n-k]$ should be about one if the fit were perfect. Here, because several factors besides the very roughly adjusted travel times obviously influence choice of mode, the value of $S^2(\boldsymbol{\theta})/[n-k] = 2.3$ has to be considered good. The correlation matrix was found to be

$$\begin{pmatrix} 1.0000 & 0.6654 \\ 0.6654 & 1.0000 \end{pmatrix},$$

indicating a modest amount of multicollinearity, which is also apparent from Exhibit C.1. ■

Example C.4 (Continuation of Example 4.4, Page 92)
Suppose we felt somewhat uncomfortable estimating the break point x in the broken line regression example purely by eye (Example 4.4, p. 92). Then we could estimate x with nonlinear regression. Recall that the model (4.8) used there was of the form

$$y_i = \beta_0 + \beta_1 x_{i1} + \beta_2 \delta_i (x_{i1} - x) + \epsilon_i. \qquad \text{(C.17)}$$

Because of the presence of δ_i which depends on x, the right side of (C.17) is not only nonlinear, it is also not differentiable at $x = x_{i1}$. Therefore, it is preferable to use a derivative-free method like the simplex method (Section C.2.3) or the DUD method (Section C.1.3). In the latter case, since derivatives are numerically computed, the procedure often behaves like a derivative-free procedure.

Coding of dummy variables for a nonlinear least squares package is not a problem. $\delta_i(x_{i1} - x)$ can be written as $\max(x_{i1} - x, 0)$ if the package allows a 'max' function. Packages that do not have such a function often permit functions with a statement as argument. These functions return the value 1 when the statement is true and 0 when it is not.

Since from Example 4.4 we already have reasonable parameter estimates, we can use them as starting values. Exhibit C.6 gives parameter estimates for each iteration of a simplex procedure. The first line gives the starting values and the last line the final estimates. Exhibit C.7 gives values of final

Step	$S^2(\boldsymbol{\theta})$	β_0	β_1	β_2	x
0	4704.2	-2.000	9.000	-3.000	7.000
1	4607.3	-0.168	8.730	-4.472	11.22
2	4482.2	0.028	8.951	-3.033	7.969
3	4354.2	-0.336	9.065	-2.800	6.985
4	4334.7	-1.770	9.277	-2.635	6.732
5	4325.1	-2.835	9.487	-2.836	6.711
6	4322.2	-3.234	9.564	-3.044	6.726
7	4321.2	-3.320	9.580	-3.163	6.705
8	4320.9	-3.436	9.601	-3.154	6.689
9	4320.9	-3.483	9.609	-3.157	6.685
10	4320.8	-3.489	9.610	-3.144	6.683
11	4320.8	-3.498	9.612	-3.147	6.683
12	4320.8	-3.494	9.611	-3.148	6.683
13	4320.8	-3.487	9.610	-3.147	6.683
14	4320.8	-3.481	9.609	-3.146	6.683
15	4320.8	-3.486	9.610	-3.147	6.683
16	4320.8	-3.487	9.611	-3.148	6.683
17	4320.8	-3.487	9.611	-3.148	6.683
18	4320.8	-3.487	9.610	-3.147	6.683

EXHIBIT C.6: Estimates by Iteration: Simplex Method Applied to Life Expectancy Data

parameter estimates from a DUD procedure. Although the starting values were identical, it is obvious that the final estimates are not the same, except for the all-important x. Its value of 6.683 in both cases is close to the 7 we had chosen in Example 4.4.

Parameter	Estimate	Std. Error
β_0	-2.9528	4.358
β_1	9.5242	0.793
β_2	-3.0870	0.860
x	6.6833	1.097

EXHIBIT C.7: Estimates and Standard Errors: DUD Procedure

The main reason for the dissimilarity of estimates was the high level of multicollinearity, which is illustrated by the correlation matrix

$$\begin{pmatrix} 1.0000 & -0.9844 & -0.7486 & 0.7588 \\ -0.9844 & 1.0000 & 0.8537 & -0.8402 \\ -0.7486 & 0.8537 & 1.0000 & -0.9189 \\ 0.7588 & -0.8402 & -0.9189 & 1.0000 \end{pmatrix}$$

of the parameters at the final estimate of the DUD procedure. In fact, at some of the earlier iterations the correlation matrix was even worse!

Since f is not even differentiable at places, it is desirable to see if we have indeed found the desired estimate and not a less than optimal local minimum. One way to more or less convince ourselves of this was to make several runs with different starting values within a reasonable range. When the starting value of x was in the range 5 to 7.4 and the other parameters were in the range ± 25 per cent of their final values, convergence always occurred, the final value of x was always 6.683, and the final values of the other estimates were acceptably close to those given in Exhibits C.6 and C.7. But for initial values outside these ranges, iterations did on occasion converge to other local optima. Most were unrealistic; e.g., x would be negative or greater than 10. One exception was the local minimum where estimates of β_0, β_1, β_2 and x were -5.56, 9.03, -4.41 and 7.6. The value of $S^2(\boldsymbol{\theta})$ in this case was 4359 and in all cases was higher than for the estimates in Exhibits C.6 and C.7. Therefore, at least the estimate of x of 6.683 seems to be the one we were after. The reader is invited to repeat this exercise, plot residuals and comment on the quality of fit of the chosen model and its closest alternative in Exercise C.5 ∎

Problems

Exercise C.1: In Example C.3 we considered a model corresponding to Exercise 9.8, p. 214, Part 1. Using NLS, construct models corresponding to Parts 2 and 3. Are these much different from the models you estimated in Problem 9.8?

Exercise C.2: The data set given in Exhibit C.8 shows, by country of origin, numbers of successful (APR) and unsuccessful (DEN) applicants for political asylum to the Immigration and Naturalization Service of the United States during the period 1983–85. Also given are the independent variables E, which takes the value 1 if the country is in Europe or is populated mainly by people of European descent, and 0 otherwise, and H, which takes value 1 if the country is considered hostile to the U.S. and 0 if non-hostile. Use both weighted linear least squares and nonlinear least squares to construct logit models for the proportion of successful applicants. How different are the two models? Test the two hypotheses that the probability of gaining political asylum is affected (1) by being European and (2) by coming from a hostile country.

Exercise C.3: Let $y(t)$ be the U.S. population considered as a differentiable function of time t. Then from the reasoning given in Exercise 7.8, p. 147, it follows that we can write, approximately,

$$\frac{dy}{dt} = \beta_0 + \beta_1 y,$$

whence $\log[\beta_0 + \beta_1 y] = \beta_1 t + \alpha$ where α is a parameter. Therefore, solving for y and rearranging parameters, we get

$$y = \theta_0 + \theta_1 \exp[\theta_2 t].$$

Estimate the parameters θ_j using the data of Exercise 7.7 and ignoring serial correlation. Plot residuals and comment.
[**Hint on getting initial values:** Notice that when $t \to -\infty$, $y = \theta_0$, which leads to an initial value of 0 for θ_0. Now taking logs of both sides of $y = \theta_1 \exp[\theta_2 t]$ and using OLS, the other parameters can be estimated.]

Exercise C.4: Another model often proposed for population forecasting is the logistic model

$$y = \alpha \frac{\exp[\beta_0 + \beta_1 t]}{1 + \exp(\beta_0 + \beta_1 t)} = \alpha[1 + \exp(-\beta_0 - \beta_1 t)]^{-1}.$$

Because of the presence of α it is difficult to apply linear least squares as we described in Section 9.2.4, p. 186. Apply nonlinear least squares to estimate the parameters of this model using U.S population data (Exercise 7.7, p. 146). Here $y = \alpha$ when $t \to \infty$. Therefore, this value needs to be guessed. After that, the transformation of Section 9.2.4, p. 186, can

Country	APR	DEN	H	E
AFGHANISTAN	296	504	1	0
ALGERIA	0	5	0	0
ANGOLA	4	12	1	0
ARGENTINA	2	104	0	0
BAHAMAS	0	3	0	0
BANGLADESH	0	391	0	0
BELGIUM	0	4	0	1
BELIZE	0	3	0	0
BOLIVIA	0	16	0	0
BRAZIL	2	4	0	0
BULGARIA	20	27	1	1
BURMA	1	7	0	0
BURUNDI	0	2	0	0
CAMEROON	1	4	0	0
CAMBODIA	3	4	1	1
CHAD	0	1	0	0
CHILE	9	62	0	1
CHINA (PRC)	67	297	1	0
COLUMBIA	5	25	0	0
CONGO	0	1	0	0
COSTA RICA	1	28	0	0
CUBA	82	1257	1	0
CYPRUS	0	4	0	0
CZECHOSLAVKIA	77	110	1	1
DJIBOUTI	0	1	0	0
DOMINICAN REP.	1	3	0	0
ECUADOR	1	4	0	0
EGYPT	2	685	0	0
EL SALVADOR	473	18258	0	0
ETHIOPIA	559	1572	1	0
FIJI	0	1	0	0
FRANCE	1	10	0	1
GAMBIA	0	3	0	0
G.D.R.	14	20	1	1
GERMANY (FDR)	0	9	0	1
GHANA	28	80	0	0
GREECE	0	10	0	1
GRENADA	0	20	0	0
GUATEMALA	9	1252	0	0
GUINEA	1	4	0	0
GUYANA	4	13	0	0
HAITI	28	1117	0	0
HONDURAS	6	206	0	0
HUNGARY	115	268	1	1
INDIA	1	241	0	0
INDONESIA	2	16	0	0
IRAN	9556	7864	1	0
IRAQ	83	619	0	0
IRELAND	3	18	0	1
ISRAEL	1	63	0	1
ITALY	0	4	0	1
IVORY COAST	0	3	0	0
JAMAICA	0	3	0	0
JAPAN	1	2	0	0
JORDAN	2	119	0	0
KAMPUCHEA	4	6	1	0
KENYA	0	10	0	0
KOREA	5	13	1	0
KUWAIT	0	11	0	0
LAOS	5	25	1	0
LEBANON	20	1107	0	0
LIBERIA	15	200	0	0
LIBYA	70	55	1	0
LITHUANIA	1	1	1	1
MALAWI	1	1	0	0
MALAYSIA	0	3	0	0
MADAGASCAR	0	2	0	0
MALI	0	2	0	0
MEXICO	1	30	0	0
MOROCCO	0	12	0	0
MOZAMBIQUE	1	50	0	0
NAMIBIA	3	6	1	0
NEW ZEALAND	0	2	0	1
NEPAL	0	1	0	0
NICARAGUA	1520	12983	1	0
NIGER	0	1	0	0
NIGERIA	0	23	0	0
PAKISTAN	24	337	1	0
PANAMA	0	13	0	0
PERU	1	19	0	0
PHILIPPINES	68	216	0	0
POLAND	1433	3119	1	1
PORTUGAL	0	4	0	0
ROMANIA	297	364	1	1
RWANDA	0	3	0	0
SEYCHELLES	8	4	0	0
SIERRA LEONE	2	5	0	0
SINGAPORE	1	0	0	0
SOMALIA	56	353	0	0
SOUTH AFRICA	12	29	0	0
SPAIN	0	13	0	1
SRI LANKA	0	90	0	0
SUDAN	0	17	0	0
SURINAM	0	29	0	0
SWAZILAND	0	3	0	0
SYRIA	64	225	1	0
TAIWAN	2	9	0	0
TANZANIA	0	14	0	0
THAILAND	3	15	0	0
TURKEY	4	60	0	1
U.S.S.R.	89	78	1	1
UGANDA	69	176	0	0
U.K	0	7	0	1
HONG KONG	0	1	0	0
URAGUAY	2	7	0	0
VENEZUELA	0	14	0	0
VIETNAM	42	133	1	0
YEMEN ADEN	1	2	1	0
YEMEN SANAA	6	5	1	0
YUGOSLAVIA	27	201	1	1
ZAIRE	7	20	0	0
ZIMBABWE	1	13	0	0

EXHIBIT C.8: Data on Asylum Requests to the U.S. by Country of Origin of Applicant
SOURCE: Prof. Barbara Yarnold, Dept. of Political Science, Saginaw Valley State University, Saginaw, Michigan.

be used to obtain the other parameters by OLS. Try different values of α, including 400,000 and 600,000, and obtain parameter estimates. In each case plot residuals. Comment on the suitability of the logistic function as a method for describing U.S population growth.

Exercise C.5: Repeat Example C.4. Plot residuals against independent variable values and predicted values for the chosen parameter estimates and their closest rival(s). Is there a good reason to choose one set of parameter values over other(s)?

Exercise C.6: Use nonlinear least squares to fit a model of the form (9.1) to the data of Example C.4 but now with proper weights and with outliers deleted (as for Exercise 9.6).

Exercise C.7: If in Exercise C.3, the errors were assumed to be first order autoregressive, what would you do?

Exercise C.8: Using NLS, can you improve on the model you constructed in Exercise 9.15, p. 216?

Tables

z	0.00	0.01	0.02	0.03	0.04	0.05	0.06	0.07	0.08	0.09
0.0	.50000	.50399	.50798	.51197	.51595	.51994	.52392	.52790	.53188	.53586
0.1	.53983	.54380	.54776	.55172	.55567	.55962	.56356	.56749	.57142	.57535
0.2	.57926	.58317	.58706	.59095	.59483	.59871	.60257	.60642	.61026	.61409
0.3	.61791	.62172	.62552	.62930	.63307	.63683	.64058	.64431	.64803	.65173
0.4	.65542	.65910	.66276	.66640	.67003	.67364	.67724	.68082	.68439	.68793
0.5	.69146	.69497	.69847	.70194	.70540	.70884	.71226	.71566	.71904	.72240
0.6	.72575	.72907	.73237	.73565	.73891	.74215	.74537	.74857	.75175	.75490
0.7	.75804	.76115	.76424	.76730	.77035	.77337	.77637	.77935	.78230	.78524
0.8	.78814	.79103	.79389	.79673	.79955	.80234	.80511	.80785	.81057	.81327
0.9	.81594	.81859	.82121	.82381	.82639	.82894	.83147	.83398	.83646	.83891
1.0	.84134	.84375	.84614	.84850	.85083	.85314	.85543	.85769	.85993	.86214
1.1	.86433	.86650	.86864	.87076	.87286	.87493	.87698	.87900	.88100	.88298
1.2	.88493	.88686	.88877	.89065	.89251	.89435	.89617	.89796	.89973	.90147
1.3	.90320	.90490	.90658	.90824	.90988	.91149	.91309	.91466	.91621	.91774
1.4	.91924	.92073	.92220	.92364	.92507	.92647	.92786	.92922	.93056	.93189
1.5	.93319	.93448	.93574	.93699	.93822	.93943	.94062	.94179	.94295	.94408
1.6	.94520	.94630	.94738	.94845	.94950	.95053	.95154	.95254	.95352	.95449
1.7	.95543	.95637	.95728	.95818	.95907	.95994	.96080	.96164	.96246	.96327
1.8	.96407	.96485	.96562	.96638	.96712	.96784	.96856	.96926	.96995	.97062
1.9	.97128	.97193	.97257	.97320	.97381	.97441	.97500	.97558	.97615	.97670
2.0	.97725	.97778	.97831	.97882	.97932	.97982	.98030	.98077	.98124	.98169
2.1	.98214	.98257	.98300	.98341	.98382	.98422	.98461	.98500	.98537	.98574
2.2	.98610	.98645	.98679	.98713	.98745	.98778	.98809	.98840	.98870	.98899
2.3	.98928	.98956	.98983	.99010	.99036	.99061	.99086	.99111	.99134	.99158
2.4	.99180	.99202	.99224	.99245	.99266	.99286	.99305	.99324	.99343	.99361
2.5	.99379	.99396	.99413	.99430	.99446	.99461	.99477	.99492	.99506	.99520
2.6	.99534	.99547	.99560	.99573	.99585	.99598	.99609	.99621	.99632	.99643
2.7	.99653	.99664	.99674	.99683	.99693	.99702	.99711	.99720	.99728	.99736
2.8	.99744	.99752	.99760	.99767	.99774	.99781	.99788	.99795	.99801	.99807
2.9	.99813	.99819	.99825	.99831	.99836	.99841	.99846	.99851	.99856	.99861
3.0	.99865	.99869	.99874	.99878	.99882	.99886	.99889	.99893	.99897	.99900
3.1	.99903	.99906	.99910	.99913	.99916	.99918	.99921	.99924	.99926	.99929
3.2	.99931	.99934	.99936	.99938	.99940	.99942	.99944	.99946	.99948	.99950
3.3	.99952	.99953	.99955	.99957	.99958	.99960	.99961	.99962	.99964	.99965
3.4	.99966	.99968	.99969	.99970	.99971	.99972	.99973	.99974	.99975	.99976
3.5	.99977	.99978	.99978	.99979	.99980	.99981	.99981	.99982	.99983	.99983
3.6	.99984	.99985	.99985	.99986	.99986	.99987	.99987	.99988	.99988	.99989
3.7	.99989	.99990	.99990	.99990	.99991	.99991	.99992	.99992	.99992	.99992
3.8	.99993	.99993	.99993	.99994	.99994	.99994	.99994	.99995	.99995	.99995
3.9	.99995	.99995	.99996	.99996	.99996	.99996	.99996	.99996	.99997	.99997
4.0	.99997	.99997	.99997	.99997	.99997	.99997	.99998	.99998	.99998	.99998

EXAMPLE: For $z = 1.96 = 1.9 + 0.06$, the row marked '1.9' and the column marked '.06' yields $\Phi(1.96) = .97500$, where $\Phi(z)$ is the standard normal distribution function. Also note $\Phi(-z) = 1 - \Phi(z)$.

TABLE 1: Values of the Standard Normal Distribution Function.

ν	$\alpha = 0.2$	$\alpha = 0.125$	$\alpha = 0.05$	$\alpha = 0.025$	$\alpha = 0.005$	$\alpha = 0.0005$
1	1.3764	2.4142	6.3138	12.7062	63.6567	636.6192
2	1.0607	1.6036	2.9200	4.3027	9.9248	31.5991
3	0.9783	1.4228	2.3534	3.1824	5.8409	12.9240
4	0.9408	1.3445	2.1325	2.7774	4.6041	8.6103
5	0.9193	1.3011	2.0156	2.5714	4.0322	6.8688
6	0.9055	1.2735	1.9437	2.4477	3.7084	5.9588
7	0.8958	1.2544	1.8951	2.3653	3.5004	5.4079
8	0.8887	1.2404	1.8600	2.3066	3.3562	5.0420
9	0.8832	1.2298	1.8336	2.2628	3.2506	4.7815
10	0.8789	1.2214	1.8129	2.2287	3.1700	4.5874
11	0.8754	1.2146	1.7963	2.2016	3.1065	4.4375
12	0.8724	1.2089	1.7827	2.1794	3.0552	4.3183
13	0.8700	1.2042	1.7714	2.1609	3.0129	4.2213
14	0.8679	1.2002	1.7617	2.1453	2.9774	4.1409
15	0.8661	1.1968	1.7535	2.1320	2.9473	4.0732
16	0.8645	1.1938	1.7463	2.1204	2.9213	4.0154
17	0.8631	1.1911	1.7400	2.1103	2.8988	3.9655
18	0.8619	1.1888	1.7345	2.1014	2.8790	3.9220
19	0.8608	1.1867	1.7295	2.0935	2.8615	3.8838
20	0.8598	1.1849	1.7251	2.0865	2.8459	3.8499
21	0.8589	1.1832	1.7211	2.0801	2.8319	3.8196
22	0.8581	1.1816	1.7175	2.0744	2.8193	3.7925
23	0.8574	1.1802	1.7143	2.0692	2.8078	3.7680
24	0.8567	1.1790	1.7113	2.0644	2.7974	3.7457
25	0.8561	1.1778	1.7085	2.0600	2.7879	3.7255
26	0.8555	1.1767	1.7060	2.0560	2.7792	3.7069
27	0.8550	1.1757	1.7037	2.0523	2.7712	3.6899
28	0.8545	1.1748	1.7015	2.0489	2.7638	3.6742
29	0.8540	1.1740	1.6995	2.0457	2.7569	3.6597
30	0.8536	1.1732	1.6976	2.0428	2.7505	3.6463
31	0.8532	1.1724	1.6959	2.0400	2.7445	3.6338
32	0.8528	1.1717	1.6943	2.0374	2.7390	3.6221
33	0.8525	1.1711	1.6927	2.0350	2.7338	3.6112
34	0.8522	1.1704	1.6913	2.0327	2.7289	3.6010
35	0.8518	1.1699	1.6900	2.0306	2.7243	3.5914
36	0.8516	1.1693	1.6887	2.0286	2.7200	3.5824
37	0.8513	1.1688	1.6875	2.0267	2.7159	3.5740
38	0.8510	1.1683	1.6863	2.0249	2.7120	3.5660
39	0.8508	1.1678	1.6853	2.0232	2.7084	3.5584
40	0.8505	1.1674	1.6842	2.0215	2.7049	3.5513
42	0.8501	1.1666	1.6823	2.0185	2.6985	3.5380
44	0.8497	1.1658	1.6806	2.0158	2.6927	3.5261
46	0.8493	1.1652	1.6790	2.0134	2.6875	3.5152
48	0.8490	1.1645	1.6776	2.0111	2.6827	3.5053
50	0.8487	1.1640	1.6763	2.0090	2.6782	3.4963

NOTE: The table gives values of $t_{\alpha,\nu}$ such that $\mathrm{P}[t > t_{\alpha,\nu}] = \alpha$ for a variable t which has a Student's t distribution with ν degrees of freedom.

TABLE 2: Percentage Points for Student's t Distribution.

n	$\alpha = .01$	$\alpha = .05$	$\alpha = .1$	$\alpha = .3$	$\alpha = .5$	$\alpha = .7$	$\alpha = .9$	$\alpha = .95$	$\alpha = .99$
1	0.0002	0.0039	0.0157	0.1481	0.4545	1.0742	2.7067	3.8431	6.6370
2	0.0201	0.1026	0.2107	0.7133	1.3863	2.4079	4.6052	5.9915	9.2103
3	0.1148	0.3518	0.5844	1.4237	2.3660	3.6649	6.2514	7.8147	11.3449
4	0.2971	0.7107	1.0636	2.1947	3.3567	4.8784	7.7794	9.4877	13.2767
5	0.5543	1.1455	1.6103	2.9999	4.3515	6.0644	9.2364	11.0705	15.0863
6	0.8721	1.6354	2.2041	3.8276	5.3481	7.2311	10.6446	12.5916	16.8119
7	1.2390	2.1674	2.8331	4.6713	6.3458	8.3834	12.0170	14.0671	18.4753
8	1.6465	2.7326	3.4895	5.5274	7.3441	9.5245	13.3616	15.5073	20.0902
9	2.0879	3.3251	4.1682	6.3933	8.3428	10.6564	14.6837	16.9190	21.6660
10	2.5582	3.9403	4.8652	7.2672	9.3418	11.7807	15.9872	18.3070	23.2125
11	3.0535	4.5748	5.5778	8.1479	10.3410	12.8987	17.2750	19.6775	24.7284
12	3.5706	5.2260	6.3038	9.0343	11.3403	14.0111	18.5493	21.0285	26.2205
13	4.1069	5.8919	7.0415	9.9257	12.3398	15.1187	19.8131	22.3646	27.6919
14	4.6604	6.5706	7.7895	10.8215	13.3393	16.2221	21.0653	23.6874	29.1449
15	5.2293	7.2609	8.5468	11.7212	14.3389	17.3217	22.3084	24.9985	30.5817
16	5.8122	7.9616	9.3122	12.6244	15.3385	18.4179	23.5431	26.2990	32.0038
17	6.4078	8.6718	10.0852	13.5307	16.3382	19.5110	24.7703	27.5900	33.4126
18	7.0149	9.3905	10.8649	14.4399	17.3379	20.6014	25.9908	28.8722	34.8093
19	7.6327	10.1170	11.6509	15.3517	18.3377	21.6891	27.2049	30.1465	36.1950
20	8.2604	10.8508	12.4426	16.2659	19.3374	22.7745	28.4134	31.4135	37.5704
21	8.8972	11.5913	13.2396	17.1823	20.3372	23.8578	29.6165	32.6737	38.9364
22	9.5425	12.3380	14.0415	18.1007	21.3370	24.9361	30.8147	33.9276	40.2937
23	10.1957	13.0905	14.8480	19.0211	22.3369	26.0154	32.0084	35.1757	41.6428
24	10.8564	13.8484	15.6587	19.9432	23.3367	27.0930	33.1978	36.4183	42.9842
25	11.5240	14.6114	16.4734	20.8670	24.3366	28.1689	34.3831	37.6558	44.3186
26	12.1981	15.3792	17.2919	21.7924	25.3365	29.2432	35.5647	38.8885	45.6462
27	12.8785	16.1514	18.1139	22.7192	26.3363	30.3161	36.7428	40.1167	46.9675
28	13.5647	16.9279	18.9392	23.6475	27.3362	31.3877	37.9175	41.3406	48.2829
29	14.2565	17.7084	19.7677	24.5770	28.3361	32.4579	39.0891	42.5605	49.5926
30	14.9535	18.4927	20.5992	25.5078	29.3360	33.5269	40.2577	43.7765	50.8970
40	22.1643	26.5093	29.0505	34.8752	39.3353	44.1611	51.8069	55.7625	63.6960
50	29.7036	34.7615	37.6872	44.3170	49.3349	54.7186	63.1692	67.5092	76.1596
100	70.0600	77.9252	82.3559	92.1343	99.3341	106.900	118.501	124.348	135.814

TABLE 3: Percentage Points of the Chi-Square Distribution With n Degrees of Freedom.

n	$m=1$	$m=2$	$m=3$	$m=4$	$m=5$	$m=6$	$m=7$	$m=8$	$m=9$
1	3980.3	4842.8	5192.2	5377.1	5491.5	5570.0	5612.8	5652.6	5683.2
2	98.50	99.00	99.16	99.25	99.30	99.33	99.35	99.37	99.38
3	34.12	30.82	29.46	28.71	28.24	27.91	27.67	27.49	27.35
4	21.20	18.00	16.69	15.98	15.52	15.21	14.98	14.80	14.66
5	16.26	13.27	12.06	11.39	10.97	10.67	10.46	10.29	10.16
6	13.74	10.92	9.78	9.15	8.75	8.47	8.26	8.10	7.98
7	12.25	9.55	8.45	7.85	7.46	7.19	6.99	6.84	6.72
8	11.26	8.65	7.59	7.01	6.63	6.37	6.18	6.03	5.91
9	10.56	8.02	6.99	6.42	6.06	5.80	5.61	5.47	5.35
10	10.04	7.56	6.55	5.99	5.64	5.39	5.20	5.06	4.94
11	9.65	7.21	6.22	5.67	5.32	5.07	4.89	4.74	4.63
12	9.33	6.93	5.95	5.41	5.06	4.82	4.64	4.50	4.39
13	9.07	6.70	5.74	5.21	4.86	4.62	4.44	4.30	4.19
14	8.86	6.51	5.56	5.04	4.69	4.46	4.28	4.14	4.03
15	8.68	6.36	5.42	4.89	4.56	4.32	4.14	4.00	3.89
16	8.53	6.23	5.29	4.77	4.44	4.20	4.03	3.89	3.78
17	8.40	6.11	5.19	4.67	4.34	4.10	3.93	3.79	3.68
18	8.29	6.01	5.09	4.58	4.25	4.01	3.84	3.71	3.60
19	8.18	5.93	5.01	4.50	4.17	3.94	3.77	3.63	3.52
20	8.10	5.85	4.94	4.43	4.10	3.87	3.70	3.56	3.46
21	8.02	5.78	4.87	4.37	4.04	3.81	3.64	3.51	3.40
22	7.95	5.72	4.82	4.31	3.99	3.76	3.59	3.45	3.35
23	7.88	5.66	4.76	4.26	3.94	3.71	3.54	3.41	3.30
24	7.82	5.61	4.72	4.22	3.90	3.67	3.50	3.36	3.26
25	7.77	5.57	4.68	4.18	3.85	3.63	3.46	3.32	3.22
26	7.72	5.53	4.64	4.14	3.82	3.59	3.42	3.29	3.18
27	7.68	5.49	4.60	4.11	3.78	3.56	3.39	3.26	3.15
28	7.64	5.45	4.57	4.07	3.75	3.53	3.36	3.23	3.12
29	7.60	5.42	4.54	4.04	3.73	3.50	3.33	3.20	3.09
30	7.56	5.39	4.51	4.02	3.70	3.47	3.30	3.17	3.07
40	7.31	5.18	4.31	3.83	3.51	3.29	3.12	2.99	2.89
60	7.08	4.98	4.13	3.65	3.34	3.12	2.95	2.82	2.72
120	6.85	4.79	3.95	3.48	3.17	2.96	2.79	2.66	2.56
∞	6.66	4.63	3.80	3.34	3.03	2.82	2.66	2.53	2.42

TABLE 4: Upper 1 Per Cent Points of the F Distribution With m and n Degrees of Freedom: Continued on Next Page.

n	m=10	m=12	m=15	m=20	m=24	m=30	m=40	m=60	m=120	m = ∞
1	5706.4	5729.8	5765.3	5785.2	5777.2	5825.4	5745.6	5825.4	5825.4	5592.4
2	99.40	99.42	99.43	99.45	99.45	99.47	99.47	99.48	99.49	99.57
3	27.23	27.05	26.87	26.69	26.60	26.50	26.41	26.32	26.22	26.27
4	14.55	14.37	14.20	14.02	13.93	13.84	13.75	13.65	13.56	13.55
5	10.05	9.89	9.72	9.55	9.47	9.38	9.29	9.20	9.11	9.03
6	7.87	7.72	7.56	7.40	7.31	7.23	7.14	7.06	6.97	6.89
7	6.62	6.47	6.31	6.16	6.07	5.99	5.91	5.82	5.74	5.66
8	5.81	5.67	5.52	5.36	5.28	5.20	5.12	5.03	4.95	4.87
9	5.26	5.11	4.96	4.81	4.73	4.65	4.57	4.48	4.40	4.32
10	4.85	4.71	4.56	4.41	4.33	4.25	4.17	4.08	4.00	3.92
11	4.54	4.40	4.25	4.10	4.02	3.94	3.86	3.78	3.69	3.61
12	4.30	4.16	4.01	3.86	3.78	3.70	3.62	3.54	3.45	3.37
13	4.10	3.96	3.82	3.66	3.59	3.51	3.43	3.34	3.25	3.18
14	3.94	3.80	3.66	3.51	3.43	3.35	3.27	3.18	3.09	3.02
15	3.80	3.67	3.52	3.37	3.29	3.21	3.13	3.05	2.96	2.88
16	3.69	3.55	3.41	3.26	3.18	3.10	3.02	2.93	2.84	2.76
17	3.59	3.46	3.31	3.16	3.08	3.00	2.92	2.83	2.75	2.66
18	3.51	3.37	3.23	3.08	3.00	2.92	2.84	2.75	2.66	2.58
19	3.43	3.30	3.15	3.00	2.93	2.84	2.76	2.67	2.58	2.50
20	3.37	3.23	3.09	2.94	2.86	2.78	2.69	2.61	2.52	2.43
21	3.31	3.17	3.03	2.88	2.80	2.72	2.64	2.55	2.46	2.37
22	3.26	3.12	2.98	2.83	2.75	2.67	2.58	2.50	2.40	2.32
23	3.21	3.07	2.93	2.78	2.70	2.62	2.54	2.45	2.35	2.27
24	3.17	3.03	2.89	2.74	2.66	2.58	2.49	2.40	2.31	2.22
25	3.13	2.99	2.85	2.70	2.62	2.54	2.45	2.36	2.27	2.18
26	3.09	2.96	2.81	2.66	2.58	2.50	2.42	2.33	2.23	2.14
27	3.06	2.93	2.78	2.63	2.55	2.47	2.38	2.29	2.20	2.11
28	3.03	2.90	2.75	2.60	2.52	2.44	2.35	2.26	2.17	2.08
29	3.00	2.87	2.73	2.57	2.49	2.41	2.33	2.23	2.14	2.05
30	2.98	2.84	2.70	2.55	2.47	2.39	2.30	2.21	2.11	2.02
40	2.80	2.66	2.52	2.37	2.29	2.20	2.11	2.02	1.92	1.82
60	2.63	2.50	2.35	2.20	2.12	2.03	1.94	1.84	1.73	1.62
120	2.47	2.34	2.19	2.03	1.95	1.86	1.76	1.66	1.53	1.40
∞	2.34	2.20	2.06	1.90	1.81	1.72	1.61	1.50	1.35	1.16

TABLE 5: Upper 1 Per Cent Points of the F Distribution With m and n Degrees of Freedom: Continued From Last Page.

n	$m=1$	$m=2$	$m=3$	$m=4$	$m=5$	$m=6$	$m=7$	$m=8$	$m=9$
1	161.45	199.50	215.70	224.59	230.16	233.65	235.80	237.38	238.57
2	18.51	19.00	19.16	19.25	19.30	19.33	19.35	19.37	19.38
3	10.13	9.55	9.28	9.12	9.01	8.94	8.89	8.85	8.81
4	7.71	6.94	6.59	6.39	6.26	6.16	6.09	6.04	6.00
5	6.61	5.79	5.41	5.19	5.05	4.95	4.88	4.82	4.77
6	5.99	5.14	4.76	4.53	4.39	4.28	4.21	4.15	4.10
7	5.59	4.74	4.35	4.12	3.97	3.87	3.79	3.73	3.68
8	5.32	4.46	4.07	3.84	3.69	3.58	3.50	3.44	3.39
9	5.12	4.26	3.86	3.63	3.48	3.37	3.29	3.23	3.18
10	4.96	4.10	3.71	3.48	3.33	3.22	3.14	3.07	3.02
11	4.84	3.98	3.59	3.36	3.20	3.09	3.01	2.95	2.90
12	4.75	3.89	3.49	3.26	3.11	3.00	2.91	2.85	2.80
13	4.67	3.81	3.41	3.18	3.03	2.92	2.83	2.77	2.71
14	4.60	3.74	3.34	3.11	2.96	2.85	2.76	2.70	2.65
15	4.54	3.68	3.29	3.06	2.90	2.79	2.71	2.64	2.59
16	4.49	3.63	3.24	3.01	2.85	2.74	2.66	2.59	2.54
17	4.45	3.59	3.20	2.96	2.81	2.70	2.61	2.55	2.49
18	4.41	3.55	3.16	2.93	2.77	2.66	2.58	2.51	2.46
19	4.38	3.52	3.13	2.90	2.74	2.63	2.54	2.48	2.42
20	4.35	3.49	3.10	2.87	2.71	2.60	2.51	2.45	2.39
21	4.32	3.47	3.07	2.84	2.68	2.57	2.49	2.42	2.37
22	4.30	3.44	3.05	2.82	2.66	2.55	2.46	2.40	2.34
23	4.28	3.42	3.03	2.80	2.64	2.53	2.44	2.37	2.32
24	4.26	3.40	3.01	2.78	2.62	2.51	2.42	2.36	2.30
25	4.24	3.39	2.99	2.76	2.60	2.49	2.40	2.34	2.28
26	4.23	3.37	2.98	2.74	2.59	2.47	2.39	2.32	2.27
27	4.21	3.35	2.96	2.73	2.57	2.46	2.37	2.31	2.25
28	4.20	3.34	2.95	2.71	2.56	2.45	2.36	2.29	2.24
29	4.18	3.33	2.93	2.70	2.55	2.43	2.35	2.28	2.22
30	4.17	3.32	2.92	2.69	2.53	2.42	2.33	2.27	2.21
40	4.08	3.23	2.84	2.61	2.45	2.34	2.25	2.18	2.12
60	4.00	3.15	2.76	2.53	2.37	2.25	2.17	2.10	2.04
120	3.92	3.07	2.68	2.45	2.29	2.17	2.09	2.02	1.96
∞	3.85	3.00	2.61	2.38	2.22	2.11	2.02	1.95	1.89

TABLE 6: Upper 5 Per Cent Points of the F Distribution With m and n Degrees of Freedom: Continued on Next Page.

n	m=10	m=12	m=15	m=20	m=24	m=30	m=40	m=60	m=120	m = ∞
1	239.47	240.76	241.87	242.75	243.11	243.22	243.12	242.50	240.63	233.02
2	19.40	19.41	19.43	19.45	19.45	19.46	19.47	19.48	19.49	19.50
3	8.79	8.74	8.70	8.66	8.64	8.62	8.59	8.57	8.55	8.53
4	5.96	5.91	5.86	5.80	5.77	5.75	5.72	5.69	5.66	5.63
5	4.74	4.68	4.62	4.56	4.53	4.50	4.46	4.43	4.40	4.37
6	4.06	4.00	3.94	3.87	3.84	3.81	3.77	3.74	3.70	3.67
7	3.64	3.57	3.51	3.44	3.41	3.38	3.34	3.30	3.27	3.23
8	3.35	3.28	3.22	3.15	3.12	3.08	3.04	3.01	2.97	2.93
9	3.14	3.07	3.01	2.94	2.90	2.86	2.83	2.79	2.75	2.71
10	2.98	2.91	2.85	2.77	2.74	2.70	2.66	2.62	2.58	2.54
11	2.85	2.79	2.72	2.65	2.61	2.57	2.53	2.49	2.45	2.41
12	2.75	2.69	2.62	2.54	2.51	2.47	2.43	2.38	2.34	2.30
13	2.67	2.60	2.53	2.46	2.42	2.38	2.34	2.30	2.25	2.21
14	2.60	2.53	2.46	2.39	2.35	2.31	2.27	2.22	2.18	2.14
15	2.54	2.48	2.40	2.33	2.29	2.25	2.20	2.16	2.11	2.07
16	2.49	2.42	2.35	2.28	2.24	2.19	2.15	2.11	2.06	2.02
17	2.45	2.38	2.31	2.23	2.19	2.15	2.10	2.06	2.01	1.97
18	2.41	2.34	2.27	2.19	2.15	2.11	2.06	2.02	1.97	1.92
19	2.38	2.31	2.23	2.16	2.11	2.07	2.03	1.98	1.93	1.88
20	2.35	2.28	2.20	2.12	2.08	2.04	1.99	1.95	1.90	1.85
21	2.32	2.25	2.18	2.10	2.05	2.01	1.96	1.92	1.87	1.82
22	2.30	2.23	2.15	2.07	2.03	1.98	1.94	1.89	1.84	1.79
23	2.27	2.20	2.13	2.05	2.01	1.96	1.91	1.86	1.81	1.76
24	2.25	2.18	2.11	2.03	1.98	1.94	1.89	1.84	1.79	1.74
25	2.24	2.16	2.09	2.01	1.96	1.92	1.87	1.82	1.77	1.72
26	2.22	2.15	2.07	1.99	1.95	1.90	1.85	1.80	1.75	1.70
27	2.20	2.13	2.06	1.97	1.93	1.88	1.84	1.79	1.73	1.68
28	2.19	2.12	2.04	1.96	1.91	1.87	1.82	1.77	1.71	1.66
29	2.18	2.10	2.03	1.94	1.90	1.85	1.81	1.75	1.70	1.65
30	2.16	2.09	2.01	1.93	1.89	1.84	1.79	1.74	1.68	1.63
40	2.08	2.00	1.92	1.84	1.79	1.74	1.69	1.64	1.58	1.52
60	1.99	1.92	1.84	1.75	1.70	1.65	1.59	1.53	1.47	1.40
120	1.91	1.83	1.75	1.66	1.61	1.55	1.50	1.43	1.35	1.27
∞	1.83	1.75	1.67	1.57	1.52	1.46	1.39	1.32	1.22	1.00

TABLE 7: Upper 5 Per Cent Points of the F Distribution With m and n Degrees of Freedom: Continued From Last Page.

	$k=1$		$k=2$		$k=3$		$k=4$		$k=5$	
n	d_L	d_U	d_L	d_U	d_L	d_U	d_L	d_U	d_L	d_U
15	1.08	1.36	0.95	1.54	0.82	1.75	0.69	1.97	0.56	2.21
16	1.10	1.37	0.98	1.54	0.86	1.73	0.74	1.93	0.62	2.15
17	1.13	1.38	1.02	1.54	0.90	1.71	0.78	1.90	0.67	2.10
18	1.16	1.39	1.05	1.53	0.93	1.69	0.82	1.87	0.71	2.06
19	1.18	1.40	1.08	1.53	0.97	1.68	0.86	1.85	0.75	2.02
20	1.20	1.41	1.10	1.54	1.00	1.68	0.90	1.83	0.79	1.99
21	1.22	1.42	1.13	1.54	1.03	1.67	0.93	1.81	0.83	1.96
22	1.24	1.43	1.15	1.54	1.05	1.66	0.96	1.80	0.86	1.94
23	1.26	1.44	1.17	1.54	1.08	1.66	0.99	1.79	0.90	1.92
24	1.27	1.45	1.19	1.55	1.10	1.66	1.01	1.78	0.93	1.90
25	1.29	1.45	1.21	1.55	1.12	1.66	1.04	1.77	0.95	1.89
26	1.30	1.46	1.22	1.55	1.14	1.65	1.06	1.76	0.98	1.88
27	1.32	1.47	1.24	1.56	1.16	1.65	1.08	1.76	1.01	1.86
28	1.33	1.48	1.26	1.56	1.18	1.65	1.10	1.75	1.03	1.85
29	1.34	1.48	1.27	1.56	1.20	1.65	1.12	1.74	1.05	1.84
30	1.35	1.49	1.28	1.57	1.21	1.65	1.14	1.74	1.07	1.83
31	1.36	1.50	1.30	1.57	1.23	1.65	1.16	1.74	1.09	1.83
32	1.37	1.50	1.31	1.57	1.24	1.65	1.18	1.73	1.11	1.82
33	1.38	1.51	1.32	1.58	1.26	1.65	1.19	1.73	1.13	1.81
34	1.39	1.51	1.33	1.58	1.27	1.65	1.21	1.73	1.15	1.81
36	1.40	1.52	1.34	1.58	1.28	1.65	1.22	1.73	1.16	1.80
36	1.41	1.52	1.35	1.59	1.29	1.65	1.24	1.73	1.18	1.80
37	1.42	1.53	1.36	1.59	1.31	1.66	1.25	1.72	1.19	1.80
38	1.43	1.54	1.37	1.59	1.32	1.66	1.26	1.72	1.21	1.79
39	1.43	1.54	1.38	1.60	1.33	1.66	1.27	1.72	1.22	1.79
40	1.44	1.54	1.39	1.60	1.34	1.66	1.29	1.72	1.23	1.79
45	1.48	1.57	1.43	1.62	1.38	1.67	1.34	1.72	1.29	1.78
50	1.60	1.59	1.46	1.63	1.42	1.67	1.38	1.72	1.34	1.77
55	1.53	1.60	1.49	1.64	1.45	1.68	1.41	1.72	1.38	1.77
60	1.55	1.62	1.51	1.65	1.48	1.69	1.44	1.73	1.41	1.77
65	1.57	1.63	1.54	1.66	1.50	1.70	1.47	1.73	1.44	1.77
70	1.58	1.64	1.55	1.67	1.52	1.70	1.49	1.74	1.46	1.77
75	1.60	1.65	1.57	1.68	1.54	1.71	1.51	1.74	1.49	1.77
80	1.61	1.66	1.59	1.69	1.56	1.72	1.53	1.74	1.51	1.77
85	1.62	1.67	1.60	1.70	1.57	1.72	1.55	1.75	1.52	1.77
90	1.63	1.68	1.61	1.70	1.59	1.73	1.57	1.75	1.54	1.78
95	1.64	1.69	1.62	1.71	1.60	1.73	1.58	1.75	1.56	1.78
100	1.65	1.69	1.63	1.72	1.61	1.74	1.59	1.76	1.57	1.78

NOTE: For $d > 2$, compare $4 - d$ to the bounds given. k is the number of independent variables (excluding the constant term).

TABLE 8: Five Per Cent Point for the Durbin-Watson Statistic.
SOURCE: Durbin and Watson (1951), Reprinted from *Biometrika* with the permission of Biometrika Trustees.

References

Allen, D.M. (1971). *The Prediction Sum of Squares as a Criterion for Selecting Predictor Variables.* Technical Report #23, Department of Statistics, University of Kentucky.

Amemiya, T. (1980). Selection of Regressors. *International Economic Review* **21** 331–354.

Amemiya, T. (1983). Nonlinear Regression Models. In *Handbook of Econometrics.* Z. Griliches and M.D. Intriligator, eds. Amsterdam: North Holland.

Anderson, T.W. (1971). *The Statistical Analysis of Time Series.* New York: Wiley.

Andrews, D.F. (1971a). Significance Tests Based on Residuals. *Biometrika* **58** 139–148.

Andrews, D.F. (1971b). A Note on the Selection of Data Transformations. *Biometrika* **58** 249–254.

Andrews, D.F. and D. Pregibon (1978). Finding the Outliers That Matter. *Journal of the Royal Statistical Society, Series B* **40** 85–93.

Anscombe, F.J. (1948). The Transformation of Poisson, Binomial and Negative-Binomial Data. *Biometrika* **35** 246–254.

Anselin, L. (1988). *Spatial Econometrics: Methods and Models.* Dordrecht: Kluwer.

Apostol, T.M. (1957) *Mathematical Analysis.* Reading, MA: Addison-Wesley.

Atkinson, A.C. (1985). *Plots, Transformations, and Regression.* Oxford: Oxford University Press.

Atkinson, A.C. (1986). Diagnostic Tests for Transformations. *Technometrics* **28** 29–37.

Bartels, C.P.A. (1979). Operational Statistical Methods for Analyzing Spatial Data. In *Exploratory and Explanatory Statistical Analysis of Spatial Data.* C.P.A. Bartels and R.H. Ketellaper, eds. Boston: Martinus Nijhoff.

Bates, D.M. and D.G. Watts (1988). *Nonlinear Regression Analysis and Its Applications.* New York: Wiley.

Bellman, R.A. (1962). *Introduction to Matrix Analysis.* New York: McGraw-Hill.

Belsley, D.A. (1984). Demeaning Conditioning Diagnostics Through Centering. *American Statistician* **38** 158–168.

Belsley, D.A. (1986). Centering, the Constant, First-Differencing, and Assessing Conditioning. In *Model Reliability*. E. Kuh and D. Belsley, eds. Cambridge: MIT Press.

Belsley, D.A., E. Kuh and R.E. Welsch (1980). *Regression Diagnostics: Identifying Influential Data and Sources of Collinearity*. New York: Wiley.

Bhat, R., et al. (1989). Surfactant Therapy and Spontaneous Diuresis. *Journal of Pediatrics* **114** 443–447.

Bilodeau, M. and M.S. Srivastava (1988). Estimation of the MSE Matrix of the Stein Estimator. *The Canadian Journal of Statistics* **16** 153–159.

Blom, G. (1958). *Statistical Estimates and Transformed Beta Variates*. New York: Wiley.

Box, M.J. (1971). Bias in Nonlinear Estimation. *Journal of the Royal Statistical Society, B* **33** 171–201.

Box, G.E.P. and D.R. Cox (1964). An Analysis of Transformations, *Journal of the Royal Statistical Society, B* **26** 211–252.

Box, G.E.P. and G.M. Jenkins (1970). *Time Series Analysis, Forecasting and Control*. San Francisco: Holden Day.

Box, G.E.P. and P.W. Tidwell (1962). Transformations of the Independent Variable. *Technometrics* **4** 531–550.

Boyce, D.E., A. Farhi and R. Weischedel (1974). *Optimal Subset Selection: Multiple Regression, Interdependence and Optimal Network Algorithms*. Berlin: Springer-Verlag.

Burn, D.A. and T.A. Ryan (1983). A Diagnostic Test for Lack of Fit in Regression Models. In *American Statistical Association, Proceedings of the Statistical Computing Section*. 286–290.

Carroll, R.J. (1982). Two Examples of Transformations When There Are Possible Outliers. *Applied Statistics* **31** 149–152.

Carroll, R.J. and D. Ruppert (1985). Transformations in Regression: Robust Analysis. *Technometrics* **27** 1–12

Carroll, R.J. and D. Ruppert (1988). *Transformation and Weighting in Regression*. New York: Chapman and Hall.

Carter, R.A.L., M.S. Srivastava, V.K. Srivastava and A. Ullah (1990). Unbiased Estimation of the MSE Matrix of Stein-Rule Estimators, Confidence Ellipsoids and Hypothesis Testing. *Econometric Theory*. Forthcoming.

Chatterjee, S. and A.S. Price (1977). *Regression Analysis by Example*. New York: Wiley.

Chatterjee, S. and A.S. Hadi (1986). Influential Observations, High Leverage Points, and Outliers in Linear Regression, *Statistical Science* **1** 379–416.

Cliff, A.D. and J.K. Ord (1981). *Spatial Processes*. London: Pion.

Cook, D.G. and S.J. Pocock (1983). Multiple Regression in Geographic

Mortality Studies with Allowance for Spatially Correlated Errors. *Biometrics* **39** 361–371.

Cook, R.D. (1977). Detection of Influential Observations in Linear Regression. *Technometrics* **19** 15–18.

Cook, R.D. (1986). Assessment of Local Influence. *Journal of the Royal Statistical Society B* **48** 133–169.

Cook, R.D. and P.C. Wang (1983). Transformations and Influential Cases in Regression. *Technometrics* **25** 337–343.

Cook, R.D. and S. Weisberg (1980). Characterization of an Empirical Influence Function for Detecting Influential Cases in Regression. *Technometrics* **22** 495–508.

Cook, R.D. and S. Weisberg (1982). *Residuals and Influence in Regression.* New York: Chapman and Hall.

Cook, R.D. and S. Weisberg (1983). Diagnostic for Heteroscedasticity in Regression. *Biometrika* **70** 1–10.

Cox, D.R. (1970). *The Analysis of Binary Data.* London: Methuen.

Cragg, J.G. (1982). Estimation and Testing in Time-Series Regression Models with Heteroscedastic Disturbances. *Journal of Econometrics* **20** 135–157.

Daniel, C. and F.S. Wood (1971). *Fitting Equations to Data.* New York: Wiley.

Daniel, C. and F.S. Wood (1980). *Fitting Equations to Data.* Revised Edition. New York: Wiley.

Dacey, M.F. (1983). *Social Science Theories and Methods I: Models of Data* Evanston: Northwestern University.

Dixon, W.J. (1985) *BMDP Statistical Software,* Berkeley, University of California Press.

Doll, R. (1955). Etiology of Lung Cancer. *Advances in Cancer Research* **3**. Reprinted in *Smoking and Health.* (1964) Report of the Advisory Committee to the Surgeon General. Washington: U.S. Government Printing Office.

Dolby, J.L. (1963). A Quick Method for Choosing a Transformation. *Technometrics* **5** 317–325.

Draper, N.R. and R.C. Van Nostrand (1979). Ridge Regression and James Stein Estimators: Review and Comments. *Technometrics* **21** 451–466.

Draper, N.R. and H. Smith (1966). *Applied Regression Analysis.* New York: Wiley.

Draper, N.R. and H. Smith (1981). *Applied Regression Analysis.* Second Edition. New York: Wiley.

Durbin, J. and G.S. Watson (1951), Testing for Serial Correlation in Least Squares Regression, II. *Biometrika* **38** 159–178.

Efron, B. (1979). Bootstrap Methods, Another Look at the Jacknife. *Annals of Statistics* **7** 1–26.

Ellsworth, J. (1948). The Relationship of Population Density to Residential Propinquity as a Factor in Marriage Selection. *American Sociological*

Review **13** 444–448.

Eubank, R.L. (1984) Approximate Regression Models and Splines. *Communications in Statistical Theory and Methods A* **13** 433–484.

Ezekiel, M. and F.A. Fox (1959). *Methods of Correlation and Regression Analysis.* New York: Wiley.

Fletcher, R. and M.J.D. Powell (1963). A Rapidly Convergent Descent Method for Minimization. *The Computer Journal* **6** 163–168.

Forsythe, A.B. (1979). *BMDP-79, Biomedical Computer Programs P-series.* W.J. Dixon and M.B. Brown, eds. Berkeley: University of California Press.

Froehlich, B.R. (1973). Some Estimators for a Random Coefficient Regression Model. *Journal of the American Statistical Association* **68** 329–334.

Freund, J.R. and R.C. Littell (1986). *SAS System for Regression.* Cary, NC: SAS Institute.

Freund, J.R., R.C. Littell and P.C. Spector (1986). *SAS System for Linear Models.* Cary, NC: SAS Institute.

Freedman, D. (1981). Bootstrapping Regression Models. *Annals of Statistics* **9** 1218–28.

Furnival, G.M. and R.W.M. Wilson (1974). Regression by Leaps and Bounds. *Technometrics* **16** 499–511.

Geary, R.C. (1954). The Contiguity Ratio and Statistical Mapping. *The Incorporated Statistician* **5** 115–145.

Gnedenko, B.W. and A.N. Kolmogorov (1954). *Limit Distributions for Sums of Independent Variables* (Translated from Russian). Reading, MA: Addison-Wesley.

Griffith, D.A. (1988). *Advanced Spatial Statistics.* Dordrecht: Kluwer.

Griffith, D.A. (1989). *Spatial Regression Analysis on the PC.* Michigan Document Services (2790 Briarcliff, Ann Arbor, MI 48105).

Griffiths, P. and I.D. Hill (1985). *Applied Statistics Algorithms.* Chichester: Ellis and Horwood.

Hadi, A.S. (1985). K-clustering and the Detection of Influential Subsets (Letter to the Editor). *Technometrics* **27** 323–325.

Haggett, P., A.D. Cliff and A. Frey (1977). *Locational Models in Human Geography.* London: Arnold.

Hald, A. (1960). *Statistical Theory with Engineering Applications.* New York: Wiley.

Haining, R. (1987). Trend Surface Models with Regional and Local Scales of Variation with an Application to Areal Survey Data. *Technometrics* **29** 461–469.

Hsieh, D.A. (1983). A Heteroskedasticity-Consistent Covariance Matrix Estimator for Time Series Regressions. *Journal of Econometrics* **22** 281–290.

Hocking, R.R. (1976). The Analysis and Selection of Variables in Linear Regression. *Biometrics* **32** 1–51.

Hoerl, A.E. and R.W. Kennard (1970a). Ridge Regression: Biased Estimation for Nonorthogonal Problems. *Technometrics* **12** 55-67.

Hoerl, A.E. and R.W. Kennard (1970b). Ridge Regression: Applications to Nonorthogonal Problems *Technometrics* **12** 69-82.

Hsu, P.L. (1938). On the Best Unbiased Quadratic Estimate of Variance. *Statistical Research Memoirs* **2** 91–104.

Huber, M.J. (1957). Effect of Temporary Bridge on Parkway Performance. *Highway Research Board Bulletin* **167** 63–74.

Illinois Department of Transportation (1972). Illinois Motor Vehicle Accident Facts. Springfield, IL.

James, W. and C. Stein (1961). Estimation with Quadratic Loss. *Proceedings of the Fourth Berkeley Symposium on Mathematical Statistics and Probability.* 361-379. Berkeley: University of California Press.

John, J.A. and N.R. Draper (1980). An Alternative Family of Transformations. *Applied Statistics* **29** 190–197.

Judge, G.G., W.E. Griffiths, R.C. Hill, H. Lutkepohl and T-C. Lee (1985). *The Theory and Practice of Econometrics.* Second Edition. New York: Wiley.

Judge, G.G. and T. Takayama (1966). Inequality Restrictions in Regression Analysis. *Journal of the American Statistical Association* **61** 166–181.

Koerts, J. and A.P.J. Abrahamse (1969). *On the Theory and Application of the General Linear Model.* Rotterdam: Rotterdam University Press.

Krämer, W. and C. Donninger (1987). Spatial Autocorrelation Among Errors and the Relative Efficiency of OLS in the Linear Regression Model. *Journal of the American Statistical Association* **82** 577–87.

KRIHS (1985). *Study of Road User Charges.* Seoul: Korea Research Institute for Human Settlements.

Lawrence, A.J. (1988). Regression Transformation Diagnostics Using Local Influence. *Journal of the American Statistical Association* **83** 1067–1072.

L'Esperance, W.L., D. Chall and D. Taylor (1976). An Algorithm for Determining the Distribution Function of the Durbin-Watson Statistic. *Econometrica* **44** 1325–1346.

Leinhardt, S. and S.S. Wasserman (1979). Teaching Regression: An Exploratory Approach. *The American Statistician* **33** 196–203.

Levenberg, K. (1944). A Method for the Solution of Certain Nonlinear Problems in Least Squares *Quarterly of Applied Mathematics* **2** 164–168.

Longley, J.W. (1967). An Appraisal of Least Squares Programs for the Electronic Computer from the Point of View of the User. *Journal of the American Statistical Association* **62** 819–841.

Madansky, A. (1988).*Prescriptions for Working Statisticians.* New York: Springer-Verlag.

Mallows, C.L. (1973). Some Comments on C_p. *Technometrics* **15** 661–676.

Mallows, C.L. (1985). *Augmented Partial Residuals.* Unpublished Manuscript.

Mardia, K.V. and R.J. Marshall (1984). Maximum Likelihood Estimation of Models for Residual Covariance in Spatial Regression. *Biometrika* **71** 135–46.

Marquardt, D.W. (1963). An Algorithm for Least Squares Estimation of Nonlinear Parameters. *Journal of the Society of Industrial and Applied Mathematics* **2** 431–441.

McCullagh, P. and J.A. Nelder (1983). *Generalized Linear Models.* London: Chapman and Hall.

McNemar, Q. (1969). *Psychological Statistics.* New York: Wiley.

Milliken, G.A. and F.A. Graybill (1970). Extensions of the General Linear Hypothesis Model. *Journal of the American Statistical Association* **65** 797–807.

MINITAB (1988). *MINITAB Reference Manual.* State College, PA: MINITAB.

Moore, J. (1975). *Total Biochemical Oxygen Demand of Animal Manures.* Ph.D. Thesis. University of Minnesota, Dept. of Agricultural Engineering.

Moran, P.A.P. (1950). Notes on a Continuous Stochastic Phenomena. *Biometrika* **37** 17–23.

Mosteller, F. and J.W. Tukey (1977). *Data Analysis and Regression.* Reading, MA: Addison-Wesley.

Myers, R.H. (1986). *Classical and Modern Regression with Applications.* Boston, MA: Duxbury.

Neter, J.W., W. Wasserman, and M.H. Kutner (1985). *Applied Linear Statistical Models.* Homewood, IL: Irwin.

OECD (1982). *Car Ownership and Use.* Paris: Organization for Economic Co-operation and Development.

Ord, J.K. (1975). Estimation Methods for Models of Spatial Interaction. *Journal of the American Statistical Association* **70** 120-26.

Plackett, R.L. (1960). *Principles of Regression Analysis.* Oxford: Oxford University Press.

Plackett, R.L. (1972). The Discovery of the Method of Least Squares. *Biometrika* **59** 239–251.

Prais, S.J. and C.B. Winsten (1954). *Trend Estimators and Serial Correlation.* Cowles Commission Discussion Paper #383. Chicago.

Potthoff, R.F. and S.N. Roy (1964). A Generalized Multivariate Analysis of Variance Model Useful Especially for Growth Curve Models. *Biometrika* **51** 313–326.

Ralston, M.L. and R.I. Jennrich (1979). DUD, a Derivative-Free Algorithm for Nonlinear Least Squares. *Technometrics* **21** 7–14.

Rao, C.R. (1952). Some Theorems on Minimum Variance Estimation. *Sankhya* **12** 27–42.

Rao, C.R. (1970). Estimation of Heteroscedastic Variances in Linear Regression. *Journal of the American Statistical Association* **65** 161–172.

Rao, C.R. (1972). Estimation of Variance and Covariance Components in Linear Models. *Journal of the American Statistical Association* **67** 112–115.

Rao, C.R. (1973). *Linear Statistical Inference and Its Applications* New York: Wiley.

Rao, C.R. and S.K. Mitra (1971). *Generalized Inverse of Matrices and Its Applications.* New York: Wiley.

Riegel, P.S. (1981). Athletic Records and Human Endurance. *American Scientist* **69** 285–290.

Ripley, B.D. (1981). *Spatial Statistics.* New York: Wiley.

SAS (1982a). *SAS/ETS User's Guide.* Cary, NC: SAS Institute.,

SAS (1982b). *SAS User's Guide: Statistics, 1982 Edition.* Cary, NC: SAS Institute.

SAS (1985a). *SAS User's Guide: Basics, Version 5 Edition.* Cary, NC: SAS Institute.

SAS (1985b). *SAS User's Guide: Statistics, Version 5 Edition.* Cary, NC: SAS Institute.

SAS (1985c). *SAS/IML User's Guide, Version 5 Edition.* Cary, NC: SAS Institute.

SAS (1986). *SUGI Supplementary Library User's Guide, Version 5 Edition.* Cary, NC: SAS Institute.

SAS (1988). *SAS/STAT Users Guide, Release 6.03 Edition.* Cary, NC: SAS Institute.

Seber, G.A.F. (1977). *Linear Regression Analysis.* New York: Wiley.

Seber, G.A.F. and C.J. Wild (1989). *Nonlinear Regression.* New York: Wiley.

Sen, A. (1976). Large Sample-Size Distribution of Statistics Used in Testing for Spatial Correlation. *Geographical Analysis* **9** 175-184.

Sen, A. (1990). On the Distribution of Moran and Geary statistics of Spatial Autocorrelation. In *Spatial Statistics: Past, Present, and Future.* D. Griffith, ed. Ann Arbor, MI: IMaGe.

Sen, A. and C. Johnson. (1977). On the Form of a Bus Service Frequency Supply Function *Transportation Research* **11** 63–65.

Shapiro, S.S. and R.S. Francia (1972). An Approximate Analysis of Variance Test for Normality. *Journal of the American Statistical Association* **67** 21

Shapiro, S.S and M.B. Wilk (1965). An Analysis of Variance Test for Normality (Complete Samples). *Biometrika* **52** 591–611.

Smith, P.L. (1979). Splines as a Useful and Convenient Statistical Tool. *The American Statistician* **33** 57–62.

Spitzer, J.J. (1982). A Fast and Efficient Algorithm for the Estimation of Parameters in Models with the Box-and-Cox Transformation. *Journal of the American Statistical Association* **1982** 760–766.

SPSS (1986) *SPSS^X User's Guide.* Second Edition. Chicago: SPSS, Inc.

Srivastava, M.S. (1971). On Fixed Width Confidence Bounds for Regression Parameters. *Annals of Mathematical Statistics* **42** 1403–1411.

Srivastava, M.S. (1985). *Bootstrapping Durbin-Watson Statistics.* Technical Report, Department of Statistics, University of Toronto.

Srivastava, M.S. (1987). Asymptotic Distribution of Durbin-Watson Statistics. *Economics Letters* **24** 157–160.

Srivastava, M.S. and E.M. Carter (1983). *An Introduction to Applied Multivariate Statistics.* Amsterdam: North Holland.

Srivastava, M.S. and C.G. Khatri (1979). *An Introduction to Multivariate Statistics.* Amsterdam: North-Holland.

Srivastava, M.S. and B. Singh (1989). Bootstrapping in Multiplicative Models. *Journal of Econometrics* **42** 287–297.

Srivastava, M.S. and Y.K. Yau (1988). *Tail Probability Approximations of a General Statistic.* Technical Report #88-38, Center for Multivariate Analysis, University of Pittsburgh.

Srivastava, M.S. and F.K.L. Ng (1988). *Comparison of the Estimates of Intraclass Correlation in the Presence of Covariables.* Technical Report #15, Department of Statistics, University of Toronto.

Stead, A.G. (1987). A Macro For Estimating the Box-Cox Power Transformation. *Proceedings of the Twelfth Annual SAS Users Group International Conference* 942–946.

Stein, C. (1956). Inadmissibility of the Usual Estimator for the Mean of a Multivariate Normal Distribution. *Proceedings of the Third Berkeley Symposium on Mathematical Statistics and Probability.* Berkeley: University of California Press. 197–206.

Stevens, S.S. (1956). The Direct Estimation of Sensory Magnitudes. *American Journal of Psychology* **69** 1–25.

Stewart, G.W. (1987). Collinearity and Least Squares Regression. *Statistical Science* **2** 68–100 (including commentaries).

Stigler, S.M. (1975). Napoleonic Statistics: The Work of Laplace. *Biometrika* **62** 503–517.

Stigler, S.M. (1981). Gauss and the Invention of Least Squares. *The Annals of Statistics* **9** 465–474.

Stigler, S.M. (1984). Boscovich, Simpson and a 1760 Manuscript Note on Fitting a Linear Relation. *Biometrika* **71** 615–620.

Theil, H. (1971). *Principles of Econometrics.* New York: Wiley.

Thistead, R. (1978). *Multicollinearity, Information and Ridge Regression.* Technical Report #58, Department of Statistics, University of Chicago.

Tiku, M.L. (1975). *Selected Tables in Mathematical Statistics, Vol 2.* Institute of Mathematical Statistics, American Mathematical Society, Providence, RI.

Tufte, E.R. (1974). *Data Analysis for Politics and Policy.* Englewood Cliffs, N.J.: Prentice-Hall.

Tukey, J.W. (1949). One Degree of Freedom for Non-Additivity. *Biometrics* **5** 232–242.

Vecchia, A.V. (1988). Estimation and Model Identification for Continuous Spatial Processes. *Journal of the Royal Statistical Society B* **50** 297–312.

Warnes, J.J. and B.D. Ripley (1987). Problems with Likelihood Estimation of Covariance Functions of Spatial Gaussian Processes. *Biometrika* **74** 640–42.

Weisberg, S. (1980). *Applied Linear Regression.* New York: Wiley.

White, H. (1980). A Heteroskedasticity-Consistent Covariance Matrix Estimator and a Dircct Test for Heteroskedasticity. *Econometrica* **48** 817–838.

White, H. (1982). Instrumental Variables Regression with Independent Observations. *Econometrica* **50** 483–499.

White, K.J. (1978). A General Computer Program for Econometric Models — SHAZAM. *Econometrica* **46** 151–159.

Wilkinson, L. (1979). Tests of Significance in Stepwise Regression. *Psychological Bulletin* **86** 168–174.

Wilkinson, L. (1987). *SYSTAT: The System for Statistics.* Evanston: SYSTAT, Inc.

Wood, F.S. (1973). The Use of Individual Effects and Residuals in Fitting Equations to Data. *Technometrics* **15** 677–695.

Woodley, W.L., J. Simpson, R. Biondino and J. Berkeley (1977). Rainfall Results 1970–75: Florida Area Cumulus Experiment. *Science* **195** 735–742.

Index

$\mathbf{o}$, 270

$\mathbf{1}$, 270

Added variable plots, 243
Adjusted R^2, 40
Anscombe's corrections, 185
AR(r), 140
AR(1), 140
ARMA(r,m), 140
Asymptotic equivalence
  of GLS and EGLS, 134
Attribute, 83
Autoregressive process, 140
Averaged chi-square variables,
    293

b_j, 8
  components of variance of,
    224
  distribution of, 17
  effect of deleting variables
    on, 234, 235
  mean of, 15
  proportions of variance of,
    224
  sign of, 33, 219
  variance of, 15, 253
   effect of multicollinearity
    on, 219
$\boldsymbol{b}$, 30, 36
  consistency of, 36
  correlation matrix of, 37
  covariance of, 36
   effect of multicollinearity
    on, 221
  singular case, 30
$\boldsymbol{b}_{(0)}$, 42
$\boldsymbol{b}_{EGLS}$, 138
$\boldsymbol{b}_{GLS}$, 133
$\boldsymbol{b}(i)$, 156, 158
  derivation of, 173
Backward elimination procedure,
    240
Best linear unbiased estimator, 41
Best subset regression, 238
β_j, 6, 28
  confidence interval for, 17, 71
  estimate of variance of, 16
  standard error of, 16
  unbiased estimate of, 15
$\boldsymbol{\beta}$, 28
  confidence region for, 72
  least squares estimate of, 30
  ML estimate of, 62, 208
  tests involving, see Test
  unbiased estimator of, 35,
    138
$\boldsymbol{\beta}_{(0)}$, 42
Betas, 34
Bias, 233, 253, 284, 288
Bias ($\boldsymbol{t}$), 288
Bias ($\hat{\boldsymbol{y}}_p$), 237
Bias $[t(\boldsymbol{x})]$, 284
Binomial distribution, 116
BLUE, 41, 43, 133
BMDP, 239
Bonferroni inequality, 73, 158
Bootstrapping, 101, 107
Box-Cox transformations, 101,
    116, 204, 207
Box-Tidwell transformations, 209

Break point, 182
Broken curve regression, 182
Broken line regression, 89, 182, 313

C_p, 235, 238, 240
C.I., see Confidence interval
Case, 6, 29
Cauchy-Schwartz inequality, 270
Centered and scaled model, 44, 219, 253, 256
Centered model, 9, 42, 65, 156
Central limit theorem, 101, 108
Chebyshev's inequality, 47, 101, 284
Chi-square distribution, 115
    central, 290
    noncentral, 290
    table, 321
Cobb-Douglas model, 55, 183, 213
Cochran's theorem, 292
Collinearity index, 223
Comparison of regression equations, 67
Component plus residual plots, 200
Components of variance, 224
Condition number, 223
Confidence interval
    for β_j, 17, 71
    for expectation of predicted value, 71
    for future observation, 71
    nonlinear least squares, 308
Confidence intervals, 71
    joint, 73
    simultaneous, 73
Confidence regions, 72
    nonlinear least squares, 308
Consistent estimator, 47
Constant term, 29
Constrained least squares, 44
    and principal component regression, 255
    and ridge regression, 258
Contingency table, 92
Convergence in probability, 47
Cook's distance, 161
Correlation, 285
    partial, 241
    sample, 285
Correlation matrix
    of independent variables, 44
Counted variable, 93, 111, 116, 124, 186
Covariance, 285
Covariance matrix, 286
Covariance ratio, 160
$\text{cov}(\boldsymbol{b})$, 36
    consistent estimate of, 37
    effect of deleting variables on, 234, 236
    estimate of
        under heteroscedasticity, 115
    unbiased estimator of, 37

$\boldsymbol{d}$, 44
Data point, 6, 29
$\boldsymbol{\delta}$, 44
Δ, 164
Dependent variable, 1, 7
Design matrix, 29
Design point, 6, 29
$\det(A)$, 272
DFBETA_i, 158
DFBETA_{ij}, 159
DFBETAS_{ij}, 159
DFFIT_i, 159
DFFITS_i, 159
$\text{diag}\,(a_1, \ldots, a_n)$, 269
Dichotomous variable, 84
Dummy variable, see Indicator variable
Durbin-Watson test, 142
    table, 326

e_i, 3, 9, 30
$e_i^\star$, 156

distribution of, 157
$\boldsymbol{e}$, 30
Eigenvalues, 254, 276
and multicollinearity, 223
and near singularity, 219
Elasticity, 183
ϵ_i, 5, 6, 28
$\boldsymbol{\epsilon}$, 28, 298
Error covariance matrix
non-diagonal, 132
Error variance, 13
Errors, 6
Estimable, 41
Estimated generalized least squares, 134

F distribution, 62, 292
noncentral, 65, 293
tables, 322–325
$F_{m,n-k-1,\alpha}$, 62
F to remove, 240
Factors, 85
Forward selection procedure, 241
Future observation, 19, 158
confidence interval for, 71
distribution of, 72

G-M conditions, see Gauss-Markov conditions
γ_0, 42
$\hat{\gamma}_0$, 42
Gauss-Markov conditions, 11, 17, 35, 100, 111, 119, 132, 154, 218
Gauss-Markov theorem, 41, 121
generalization of, 41
General linear hypothesis, 61
test for, 111
using bootstrapping, 108
test statistic, 62, 63
asymptotic distribution, 106, 108
distribution of, 64
uniformly most powerful, 64
Generalized least squares, 121, 133
estimated, 134
residuals, 134
Generalized linear models, 181
GLS, see Generalized least squares
Growth curve model, 138

h_{ii}, 35
H, 31
Hat matrix, 107
Heteroscedasticity, 13, 111
detecting, 111
formal tests, 114
graphical methods, 112
with unknown variances, 135
Homoscedasticity, 111
Hypothesis
general linear, 61

iid, 62
Independent variables, 1, 7
correlation matrix of, 223
effects of deletion, 234
selection of, 233
Indicator variable, 33, 83, 193
as dependent variable, 92
Influential point, 13, 107, 155
Intercept, 6, 29
Iteratively reweighted least squares, 123
using NLS, 305

Jacobian of transformation, 207, 293

Knot, 182
Kolmogorov's test, 105

$\mathcal{L}$, 62
Lag, 140
Least squares

estimated generalized, 134
generalized, 121, 133
geometric interpretation of, 281
history of, 9
iteratively reweighted, 123
nonlinear, 123
ordinary, 119
weighted, 119
Least squares estimate, 7, 8, 30
centered and scaled model, 44
centered model, 42
constrained, see Constrained least squares
effect of multicollinearity on, 220
of $\boldsymbol{\beta}$, 158
variance of, 132, 253
Levels, 85
Leverage, 107, 155, 159
Likelihood function, 62
Likelihood ratio, 62, 63, 206
Linear estimators, 41
Linear independence, 270
Linear Least Squares Curve Fitting Program, 102, 239
Linear manifold, 280
Linearity, 7
Linearly dependent, 218
nearly, 218
Logistic distribution, 186
Logit model, 92, 186, 226
Logit transformation, 188
LS, see Least squares

M, 31
MA(m), 140
Mallows' C_p, 234
Markov's inequality, 47
Matrices
equality of, 267
orthogonal, 270
sum and product of, 267
Matrix, 267
characteristic roots of, 276
characteristic vectors of, 276
computation of inverse of, 273
determinant of, 272
diagonal, 269
differentiation, 30, 46
dimension of, 267
eigenvalues of, 276
generalized inverse of, 278
idempotent, 277
identity, 268
inverse of, 273
notation for, 267
null, 268
orthogonal, 270
partitioned, 271
positive definite, 279
positive semi-definite, 279
projection, 280
rank of, 270
singular, 270
trace of, 271
transpose of, 268
Maximum likelihood, 62, 188
Mean square error, 238, 261, 284
reducing, 261
Mean square error matrix, 259, 288
Measure of fit, 13, 39
MINITAB, 137, 192, 193, 197, 211
MINQUE, 132, 135
ML, see Maximum likelihood
Modulus transformations, 208
Moving average process, 140
MSE, 288
MSE($\hat{\boldsymbol{y}}_p$), 238
MSE($\hat{\boldsymbol{\beta}}$), 261
Multicollinearity, 218
detection of, 222
effect of deleting variables on, 233

reducing, 253
Multiple correlation, 295
Multiplicative models, 182
need for, 189
with additive errors, 184
with multiplicative errors, 183

Nested errors, 136
NLS, see Nonlinear least squares
Nonlinear least squares, 123, 188, 209, 298
algorithms, 299
choosing initial values, 307
bias, 308
confidence intervals and regions, 308
DUD procedure, 301
Gauss-Newton algorithm, 299
step halving, 300
Marquardt procedure, 302
measures of fit, 309
model, 298
quasi-Newton algorithms, 303
simplex method, 305
steepest descent method, 303
tests of hypotheses, 308
weighted, 305
Normal distribution
multivariate, 288
standard, 288
table, 319
Normal plots, 103
Normal probability paper, 103
Normality of b_j's, 101
Normality of observations, 17, 42, 62, 71–73, 100
checking for, 101
formal tests, 105
graphical methods, 101
transformations to achieve, 101

Observations, 3, 6
Observed value, 3
OLS, see Ordinary least squares
Ω, 132
unbiased estimator of, 138
Ordinal variable, 85
Ordinary least squares, 119
Outlier, 13, 155

p-value, 66
Partial correlation, 241
Partial F statistic, 240
Partial plots, 243
Partial regression leverage plots, 243
Partial regression plots, 243
Plot
added variable, 243
component plus residual, 200
normal, 103, 158
of absolute values of residuals, 114
of logarithms of absolute values of residuals, 114
of residuals, 4, 112, 157, 189, 235
of squares of residuals, 114
of standardized residuals, 112
of Studentized residuals, 112, 157
partial, 243
partial regression, 243
partial regression leverage, 158, 243
partial residuals, 200
probability, 101
rankit, see Rankit plot
Poisson distribution, 111
Polychotomous variable, 84
Polynomial regression, 181
Power transformations, 198, 204
Predicted value, 3, 9, 31, 36, 255
confidence interval, 71
distribution of, 71

effect of deleting variables on, 222, 234, 237
effect of multicollinearity on, 221
Predicteds, 4
Predictions, 18
Predictor variable, 1
PRESS residuals, 161
PRESS statistic, 239
Principal component regression, 253
bias of estimates, 255
variance of estimates, 255
Probit model, 92
Probit transformation, 188
Proportion of counts, 93, 116, 186
Proportions of variance, 224

Quadratic forms, 31, 279
of normal variables, 290
independence of, 292
Qualitative variable, 33, 85

R^2, 13, 39, 111
as multiple correlation, 40
when intercept missing, 14, 40
$\mathcal{R}$, 44
R_p^2, 235
R_a^2, 40
R^2
adjusted, 40
Random variables, 6, 284
correlated, 285
Random vector, 286
covariance matrix of, 286
Rankit plot, 102
of standardized residuals, 102
of studentized residuals, 102
Regression
history of, 1, 9
Regression model, 5, 28
multiple, 6, 29
simple, 5
Regression parameters, 6
Regressor, 1
Reparameterization, 86
Resampling, 107
Residual sum of squares, see Sum of squares, residual
Residuals, 3, 9, 30, 31, 155
generalized least squares, 134
non-zero expectations of, 235
PRESS, 161
ranks of absolute values of, 114
standardized, 157
Studentized, 156, 159
weighted regression, 120, 122
Response variable, 1
Ridge regression, 253, 256
bias of estimates, 258
physical interpretations, 257
variance of estimates, 258
Ridge trace, 259
RSS, 37
RSS_p, 238
RSTUDENT, 156
indicator variable interpretation, 157, 174

s^2, 16, 37, 40, 111
consistency of, 47
effect of deleting variables on, 233, 234
effect of deleting variables on expectation of, 236
$s_{(i)}^2$, 156
proof of alternative expression, 173
s_p^2, 235
Sample correlation coefficient, 285
Sample mean, 285
Sample multiple correlation, 47, 295
Sample standard deviation, 285
Sample variance, 285
SAS, 115, 223, 241, 242

PROC AUTOREG, 142
PROC IML, 211
PROC LOGIST, 187
PROC MATRIX, 137, 211, 260
PROC NLIN, 123, 124, 210
PROC REG, 72, 115, 171
PROC RIDGEREG, 260
PROC RSQUARE, 239
PROC STEPWISE, 240, 244
PROC UNIVARIATE, 102
Scaled model, 44, 219, 253
s.e.(b_j), 66
Search over all possible subsets, 238, 239
Serial correlation, 132, 140
test for, 142
Shapiro-Wilk test, 105
Shrinkage estimator, 253, 261
Shrinkage parameter, 257
σ^2, 13
consistent estimate of, 37, 134, 157
ML estimate of, 62, 63, 208
unbiased estimate of, 16, 157
for WLS, 120
intercept absent, 37
Significant, 66
Singularity, 218
near, 218
Slope, 6
Spatial autoregressive model, 144
Spatial correlation, 132, 143, 195
test for, 143
Spatial moving average model, 145
Spearman correlation, 114
Splines, 182
SPSS-X, 242
Stagewise procedure, 243
Standardized coefficients, 34
Standardized residuals, 102, 112
Statistic, 284
Stepwise procedure, 239, 242
Stepwise procedures, 240
Student's t distribution, 17
Studentized residuals, 102, 112
Sum of squares
corrected total, 43, 65
due to $\beta_1, \ldots, \beta_k$, 43
due to $\boldsymbol{\beta}_{(0)}$, 65
due to intercept, 43
due to regression, 43
error, 65
model, 43
predicted, 38
residual, 37, 38, 43, 65
total, 38
weighted, 119
SYSTAT
NONLIN, 123, 210

$t_{m,\alpha/2}$, 17
t distribution, 292
noncentral, 293
table, 320
t-value, 66
Test, 100
for $\beta_j = 0$, 17, 66
for $\boldsymbol{\beta}_{(0)} = 0$, 65
for nonlinearity, 189, 197
near neighbor approach, 195
Theil's approach, 194
Tukey's approach, 192
use of additional terms, 190, 192
use of repeated measurements, 192
nonlinear least squares, 308
of general linear hypothesis, 62
$\boldsymbol{\theta}$, 298
$\boldsymbol{\theta}^*$, 298
TMSE, 288
TMSE$(\hat{\boldsymbol{y}}_p)$, 238
TMSE$(\hat{\boldsymbol{\beta}})$, 261
TOL$_j$, 223
Tolerance, 223

Total mean square error, 238, 259, 261
Total variance, 287
tr(A), 271
Transform-both-sides method, 101
Transformation
  Jacobian of, 207, 293
Transformations, 180
  bias-reducing corrections for, 185
  both sides, 101
  Box-Cox, see Box-Cox transformations, 116
  Box-Tidwell, 209
  choosing, 197
    analytical methods, 204, 209
    Box-Cox method, see Box-Cox transformations
    graphical methods, 197, 200
  deciding on need for, 188
    graphical methods, 189
    near neighbor approach, 194, 195
    Theil's approach, 194
    use of additional terms, 190, 192
    use of indicator variables, 193
    use of repeated measurements, 192
  effect on error variance, 180
  power, 198
    dependent and independent variables, 209
  to achieve normality, 101
  variance stabilizing, 115
Translog model, 213
Two-sample testing, 83

Uniformly minimum variance, 38
Uniformly most powerful, 64

Variable selection, 233
  in weighted regression, 126
Variable selection procedures, 238
  backward elimination, 240
  best subset regression, 238
  forward selection procedures, 241
  search over all possible subsets, 238, 239
  stagewise procedure, 243
  stepwise procedure, 239
  stepwise procedures, 240, 242
Variance components, 224
Variance inflation factor, 223
Variance proportions, 224
Vector, 269
  length of, 270
  notation, 269
  null, 270
  random, see Random vectors
Vector space, 280
  orthogonal complement of, 280
Vectors
  linear combination of, 280
  orthogonal, 270
VIF_j, 223

Weighted least squares, 119
Weights, 118
  suggested, 122
  when variance is a step function, 126
WLS, see Weighted least squares

x_j, 6
x_{ij}, 6, 28
$\boldsymbol{x}_i$, 36, 298
$\boldsymbol{x}_{[j]}$, 218
X, 29, 298
$X_{(i)}$, 156

y, 6
y_i, 3, 6, 28
$\hat{y}_i$, 3, 9
$\boldsymbol{y}$, 28, 298
$\hat{\boldsymbol{y}}$, 31, 36
  covariance matrix of, 36
$\boldsymbol{y}(i)$, 156

Z, 42
$Z_{(s)}$, 44

Author Index

Abrahamse, A.P.J., 142, 143
Allen, D.M., 239
Amemiya, T., 239, 308
Anderson, T.W., 140
Andrews, D.F., 161, 192, 206
Anscombe, F.J., 185
Anselin, L., 145
Atkinson, A.C., 206

Bartels, C.P.A., 143
Bates, D.M., 301
Belsley, D.A., 155, 160, 171, 219, 224, 243
Bhat, R., 149
Bilodeau, M., 261
Blom, G., 101
Box, G.E.P., 140, 204, 209, 308
Boyce, D.E., 240, 242
Burn, D.A., 192

Carroll, R.J., 101, 114, 126, 206, 305
Carter, E.M., 96
Carter, R.A.L., 261
Chatterjee, S., 161, 171, 243
Cliff, A.D., 144
Cook, D.G., 145
Cook, R.D., 112, 114, 157, 161, 171, 178, 206, 243
Cox, D.R., 188, 204

Dacey, M.F., 15, 23–25, 29, 130
Daniel, C., 102, 154, 194, 195, 200, 239
Dolby, J.L., 200
Doll, R., 177
Donninger, C., 144
Draper, N.R., 208, 239, 260, 307

Efron, B., 107
Ellsworth, J., 25
Eubank, R.L., 182
Ezekiel, M., 112

Fletcher, R., 304
Forsythe, A.B., 241
Fox, F.A., 112
Francia, R.S., 105
Freedman, D., 107
Freund, J.R., 245, 306
Froehlich, B.R., 136
Furnival, G.M., 239

Geary, R.C., 143
Gnedenko, B.W., 108
Graybill, F.A., 192
Griffith, D. A., 145
Griffiths, P., 305

Hadi, A.S., 161, 171, 243
Haggett, P., 143
Haining, R., 145
Hald, A., 121
Hill, I.D., 305
Hocking, R.R., 239, 260
Hoerl, A.E., 256, 259
Hsu, P.L., 38
Huber, M.J., 2

James, W., 261
Jenkins, G.M., 140
Jennrich R.I., 301

John, J.A., 208
Johnson, C., 95
Judge, G.G., 44, 114, 127, 134–136, 140, 239, 257, 260, 308

Kennard, R.W., 256, 259
Khatri, C.G., 276, 277, 279, 289, 291, 292, 295
Koerts, J., 142, 143
Kolmogorov, A.N., 108
Krämer, W., 144
Kuh, E., 155, 160, 171, 219, 224, 243

L'Esperance, W.L., 142
Lawrence, A.J., 161
Leinhardt, S., 90
Levenberg, K., 302
Littell, R.C., 245, 306
Longley, J.W., 232, 274

Madansky, A., 102, 105, 114
Mallows, C.L., 201, 234, 237, 238
Mardia, K.V., 145
Marquardt, D.W., 302
Marshall, R.J., 145
McCullagh, P., 181
McNemar, Q., 130
Milliken, G.A., 192
Mitra, S.K., 279
Moore, J., 79
Moran, P.A.P., 143
Mosteller, F., 198, 208, 243

Nelder,J.A., 181
Ng, F.K.L., 137

Ord, J.K., 144

Plackett, R.L., 9
Pocock, S.J., 145
Potthoff, R.F., 139
Powell, M.J.D., 304
Prais, S.J., 141
Pregibon, D., 161

Ralston, M.L., 301
Rao, C.R., 38, 42, 47, 132, 185, 270, 279, 280
Riegel, P.S., 70
Ripley, B.D., 143, 144
Roy, S.N., 139
Ruppert, D., 101, 114, 126, 206, 305
Ryan, T.A., 192

Seber, G.A.F., 182, 239, 274
Sen, A., 95, 143, 144
Shapiro, S.S., 105
Singh, B., 184
Smith, H., 239, 307
Smith, P.L., 182
Spitzer, J.J., 210
Srivastava, M.S., 96, 106, 137, 142, 184, 261, 276, 277, 279, 289, 291, 292, 295
Stead, A.G., 211
Stein, C., 261
Stevens, S.S., 23, 199, 206
Stewart, G.W., 219, 223, 224
Stigler, S.M., 1, 9

Takayama, T., 44
Theil, H., 134
Thistead, R., 259
Tidwell, P.W., 209
Tiku, M.L., 65
Tufte, E.R., 177
Tukey, J.W., 192, 198, 208, 243

Van Nostrand, R.C., 260
Vecchia, A.V., 145

Wang, P.C., 206
Warnes, J.J., 144
Wasserman, S.S., 90
Watts, D.G., 301
Weisberg, S., 112, 114, 157, 161, 171, 178, 243

Welsch, R.E., 155, 160, 171, 219, 224, 243
White, H., 114, 115, 142
Wild, C.J., 182
Wilk, M.B., 105
Wilkinson, L., 123, 210, 234, 306
Wilson, R.W.M., 239
Winsten, C.B., 141
Wood, F.S., 102, 154, 194, 195, 200, 239
Woodley, W.L., 178

Yau Y.K., 142

Springer Texts in Statistics *(continued from page ii)*

Santner and Duffy: The Statistical Analysis of Discrete Data
Saville and Wood: Statistical Methods: The Geometric Approach
Sen and Srivastava: Regression Analysis: Theory, Methods, and Applications
Whittle: Probability via Expectation, Third Edition
Zacks: Introduction to Reliability Analysis: Probability Models and Statistical Methods